# The Science of Naples

# The Science of Naples

*Making knowledge in Italy's pre-eminent city, 1500–1800*

Edited by Lorenza Gianfrancesco and
Neil Tarrant

UCLPRESS

First published in 2024 by
UCL Press
University College London
Gower Street
London WC1E 6BT

Available to download free: www.uclpress.co.uk



A CIP catalogue record for this book is available from The British Library.



ISBN: 978-1-80008-675-3 (Hbk.)
ISBN: 978-1-80008-674-6 (Pbk.)
ISBN: 978-1-80008-673-9 (PDF)
ISBN: 978-1-80008-676-0 (epub)
DOI: https://doi.org/10.14324/111.9781800086739

# Contents

# List of figures

# List of contributors

**Daniel Canaris** is an ARC DECRA Research Fellow and Lecturer in Italian Studies at the University of Sydney.

**Daniela Caracciolo** is a University Researcher at the University of Salento.

**James Clifton** is the Director of the Sarah Campbell Blaffer Foundation, Houston, Texas.

**Silvana D'Alessio** is Associate Professor in Early Modern History at the University of Salerno.

**Lorenza Gianfrancesco** is a Senior Lecturer in Early Modern History at the University of Chichester.

**Frank A. J. L. James** is Professor of History of Science in Science and Technology Studies at UCL.

**Alfonso Paolella** is an independent scholar.

**Gennaro Rispoli** is Director of the Museum of Sanitary Arts, Naples.

**Neil Tarrant** is Honorary Research Fellow in the School of History at the University of Leeds.

**Felix Waldmann** is a Fellow of Corpus Christi College, University of Cambridge.

# Acknowledgements

This volume originated in a workshop held at UCL in 2018. We would like to thank the UCL Institute for Advanced Studies, The British Italian Society and Cassalini Libri for their generous support that enabled us to hold this event. We would also like to thank Andrew Campbell for his role in putting on the workshop and his contributions to planning this volume.

The editors and contributors to this volume would like to acknowledge the following:

Chapter 1 (Daniela Caracciolo): This chapter is translated from the original by Neil Tarrant and Lorenza Gianfranesco.

Chapter 3 (Alfonso Paolella): This chapter is translated from the original by Neil Tarrant and Lorenza Gianfrancesco.

Chapter 4 (Lorenza Gianfrancesco): The author would like to take this opportunity to thank Ariel Hessayon and Neil Tarrant for their constructive criticism. She also wishes to gratefully acknowledge receipt of the 'Research and Innovation Funds for Research Facilitation' grant from the University of Chichester, which has enabled the author to conduct research on Vesuvius at several Italian archives and libraries. She also wishes to express her gratitude to the Bibliothèque Méjanes, Aix-en-Provence and to Mr Thomas Duvergé for generously providing photographs of Ms 0212 (1030), which constitute one of the key sources analysed in this study.

Chapter 5 (Silvana D'Alessio): This chapter is translated from the original by Lorenza Gianfrancesco, Neil Tarrant and Andrew Campbell.

Chapter 7 (Daniel Canaris): The author would like to thank Prof. Thierry Meynard and A/Prof. Pan Dawei, as well as the co-editors Lorenza Gianfrancesco and Neil Tarrant, for their suggestions on this paper.

Chapter 8 (Frank A. J. L. James): The author would like to acknowledge the gracious permission of Her Majesty Queen Elizabeth II to study items in the Royal Collection, Windsor Castle. He would also like to thank the following archives for affording access to their collections: British Library, The National Archives, Westminster School, Royal Society of London, Natural History Museum, National Library of Scotland, The Sir John Soane Museum, The Royal Institution, National Library of Ireland, Bodleian Library, Huntington Library, Victoria University of Toronto, Lincolnshire Archives, UCL and Yale University. He is also grateful to the editors for many valuable comments which have improved this chapter.

# Introduction
# The science of early modern Naples: a missing city?

Although Naples is now acknowledged as a leading centre of learning in the period 1500–1800, English language studies of its scientific culture remain fragmented. Indeed, the city of Naples has a small presence in the literature on early modern science. Although Anglophone historians have long assigned Italians an essential role in the development of science in the period leading up to the so-called Scientific Revolution of the seventeenth century, when discussing the developments that took place after the Galileo Affair in the 1630s their attention often transfers to the establishment of experimental and mathematical forms of natural philosophy in centres such as England and France. This point is especially clear in general surveys of early modern science.[1] While Anglophone historians of science do consider Italian sources in the seventeenth century, one of their primary concerns is whether the Catholic Church impeded the contribution of Italian scholars to scientific activity.[2] When they consider science per se, their studies often focus on individuals and erudite circles such as the Cimento academy – whose practices appeared to resemble those that one might find in the Royal Society or the Royal Academy – or the Italian reception of northern scientific ideas such as Newtonianism. Examples of intellectual activity that do not directly conform to this narrative of scientific development have often been overlooked.[3]

The general tendency to overlook Italy in English language accounts of early modern science is more pronounced in the case of Naples. Anglophone scholarship on late medieval and early modern Italy has been profoundly shaped by the seminal work of Jacob Burckhardt (1818–97), *The Civilisation of the Renaissance of Italy* (1861). His influential account has encouraged historians to lavish attention upon the cities of central and northern Italy such as Rome, Florence, Venice, Padua, Mantua and Ferrara, celebrating their role in the great outpouring of political, philosophical, scientific, artistic and musical culture known today as the Renaissance. While Burckhardt's influence has partly discouraged subsequent scholars from engaging with the history of late medieval and early modern Italy Naples, there are also other practical difficulties that have hampered modern scholars' efforts to reconsider the city's significance in these periods. Perhaps most notably, research has been hindered by the damage inflicted on the State Archive of Naples by Allied bombing during the Second World War. Although these devastating events did not result in the total destruction of the archival holdings, they nevertheless obstructed the recovery and reconstruction of aspects of the city's history. Despite these practical challenges, scholars have begun to pay increasing attention to the city of Naples and to reintegrate it into accounts of early modern histories of science. These important historiographical innovations notwithstanding, scholarly attention in the Anglophone tradition tends to focus more on Naples's political, social and artistic relevance, rather than on its significance as a centre of scientific learning.[4]

Recent scholarship – predominantly, although not exclusively, produced by Italian historians – has begun to construct a new picture of the city and its significance in and beyond the Italian peninsula. The city's scientific milieu has featured in several academic works published in the last decades, although its scientific culture is rarely treated as a subject in its own right. Existing monographs and some edited collections have either dealt with specific aspects of the Neapolitan debate on science or have used Neapolitan examples as case studies to illustrate developments that took place within a broader Italian and European context.[5]

Naples has also featured in works on European men of science such as Athanasius Kircher (1602–1680), Johann Faber (1574–1629), Johannes Kepler (1571–1630) and Robert Boyle (1627–1691), to name but a few.[6] In some cases the city has featured in research that considers the 'spaces' of science, namely research laboratories, museums, private collections and cabinets of curiosities that became famous throughout Europe.[7] Further studies and recent research in the field of digital humanities have analysed the role of Neapolitan academies in fostering a multidisciplinary scientific debate on disciplines such as alchemy, astrology, astronomy, cosmogony, medicine and volcanology that played a central role in the exchange of knowledge and the development of a European Republic of Letters.[8] Moreover, contemporary scholarship has considered Vesuvius and the Bay of Naples as a subject of scientific investigation.[9]

## Centres of Neapolitan science

By the beginning of the sixteenth century, the Kingdom of Naples – a large territory that extended from areas north of Naples to Sicily – was under Spanish control. As the capital of this large territory, Naples hosted a Spanish viceregal court between 1503 and 1707 (there was another Spanish viceroy at Palermo). The city was a major metropolitan centre. With a population that reached over 400,000 inhabitants in the 1630s, Naples ranked as Italy's largest and Europe's most densely populated city.[10] As one of the busiest Mediterranean ports, the city formed a key part of the Spanish monarchy's sprawling overseas empire, while acting as a point of intersection and exchange between east, west, north and south.[11] Indeed, the city's complex and multicultural social fabric fostered intense trading, cultural exchange and intellectual mobility. Moreover, during this period Naples cemented its position as a European centre of learning. It is well known that the city attracted foreign visitors. Even before the advent of the Grand Tour, major scholars such as Marin Mersenne (1588–1648), Nicolas-Claude Fabri de Peiresc (1580–1637) and Athanasius Kircher (1602–1680), to name but a few, all visited, studied and worked in Naples.[12]

These visitors would have encountered a number of local institutions engaged in forms of scientific research. Perhaps most obvious among them was the University of Naples. The universities were once considered marginal to histories of the development of early modern science, but this picture has been extensively revised.[13] The city's university was founded in 1224 by Emperor Frederick II, who forbade his subjects from studying elsewhere. This stipulation was reiterated by subsequent rulers, though in practice it was rarely observed. From its foundation the University of Naples offered many subjects, although it concentrated on law and medicine. Yet it was often neglected during the Angevin period and became subject to frequent closure in the fifteenth century. The start of Spanish viceregal rule in 1503, however, ushered a new period of relative stability. The student population grew steadily through the course of the sixteenth century, as did scholarly mobility. Important physicians from the Kingdom of Naples such as Donato Antonio Altomare (1506–1562) and Giovanni Filippo Ingrassia (1510–1580), to name but two, received their medical education at different universities: Naples, Padua and Rome.[14]

With some justification, Paul Grendler has observed that the University of Naples was 'a second-tier' institution throughout the Renaissance.[15] Although by the seventeenth century the university could boast teachers of the calibre of Marco Aurelio Severino (1580–1656), professor of anatomy and surgery, and Nicola Cirillo (1671–1735), professor of *medicina pratica*, neither its arts provision nor its medical school provided a serious rival to such institutions as Padua, Bologna or Rome.[16] As Maria Conforti has shown, two major factors led to the marginalisation of the city's university as a leading institution for the teaching of medical disciplines.[17] First, the decrease of public funding, which prioritised the faculty of law. Second, medical teaching and practical research were mostly conducted privately. The result of this was twofold. On the one hand, private teaching became a lucrative business for professors of medicine; on the other, spaces such as hospitals and the shops of apothecaries and barber-surgeons facilitated 'a freer debate, as well as opportunities to conduct active experimentation, especially in the field of anatomy'.[18]

Nonetheless, Conforti's examination of several manuscripts concerned with lecture notes taken by students of medicine at the city's university also highlights that during lectures professors engaged with recent developments in medical science. From these manuscripts we learn, for instance, of Nicola Cirillo's engagement with René Descartes' (1596–1650) theories on memory during his lectures on the human brain.[19] In this context, it is worth briefly highlighting the effort to reform some areas of teaching and research within the city's university. In the second half of the eighteenth century, for instance, new chairs in arithmetic, algebra and chemistry were introduced, not to mention the relocation of the faculty of medicine to the Hospital of the Incurables, which was equipped with a large anatomical theatre.[20]

Alongside the university, Naples hosted other scholarly centres that flourished in both secular and ecclesiastical contexts. These venues included museums and sites of scientific experimentation that functioned as hubs for networking and the dissemination of knowledge. These venues also hosted small museums, libraries and scientific instruments. Contradicting the received narrative that the Church suppressed scientific activity, some of the most vibrant venues for pharmaceutical research developed within religious houses. Laboratories based in these places became gathering spaces (like academies) where friars met with philosophers, literati and physicians. Participating in a wider debate on science and medicine, members of religious orders set up libraries and engaged in various forms of scientific experimentation. Some Neapolitan religious houses also hosted apothecary shops that catered for the medical needs of the city's population, especially the poor. Apothecary shops were located in various areas of the city centre and often situated near major hospitals.

In the early seventeenth century Neapolitan convents hosted about 20 apothecary shops, which sold legally approved medical remedies.[21]

In this lively microcosm, the most important centre for the production of medicines was located in the Dominican convent of Santa Caterina a Formello. The convent hosted an apothecary shop, a library – which contained texts by Giambattista Della Porta (1535–1615), Ferrante Imperato (*c.*1525–*c.*1615) and Galileo Galilei (1564–1642) – and a large space dedicated to empirical investigation and equipped with alembics, retorts, vases and furnaces. During the period in which the Dominican friar Donato d'Eremita (?–1630) was resident in Santa Caterina a Formello, the convent's research site became a gathering point for scientists and intellectuals. D'Eremita arrived in Naples in 1609 from Florence, where he had served Grand Duke Cosimo II (1590–1621) as a spagyrist. The friar's networks extended beyond Naples; his collaboration with Lincean academicians such as Johann Faber (1574–1629) and Fabio Colonna (1567–1640) is well known.[22]

In early modern Naples the development of scientific knowledge was also linked to trans-continental trade. The importation of objects and species from the New World fostered research on natural specimens and botanical species that featured in museums and research centres. For instance, the American *Passiflora* (passionflower) was used as a mild anaesthetic and employed for the production of medicine. Likewise, conventual scientific cabinets contained a variety of objects, with some including oriental bezoar stones, 'American seashells' and stuffed animals such as 'serpents from Brazil and crocodiles'.[23] Knowledge of the New World was also transmitted by texts. In 1644, for instance, Marco Aurelio Severino translated Antonio Colmenero de Ledesma's treatise on the medicinal properties of chocolate from Spanish into Latin.[24]

A figure who rose to prominence at the turn of the seventeenth century was Ferrante Imperato. A major collector, researcher and an expert in the apothecary's art, Imperato established in Naples a museum of science and natural philosophy that acquired European fame. As shown in some of the essays contained in this volume, the importance of this museum extended beyond its rich collection. Indeed, its structure – as Imperato discussed in his *Historia Naturale* – can surely be considered as one of the earliest examples of a taxonomical arrangement of 'the natural world'. Possibly referring to specimens kept in Neapolitan museums such as Imperato's, the French physician and surgeon Ambroise Paré observed that: 'You may see Elephant's teeth of a monstruous and stupendous bigness, at Venice, Rome, Naples and Paris'.[25] Classifications of nature featured in encyclopaedic collections too. The work of Maurizio

Di Gregorio (?–1652) provides an important example. A Dominican theologian from Sicily who moved to Naples and became a member of the Oziosi academy, Di Gregorio compiled *Encyclopedia*: a large manuscript that features sections concerned with the classification of plants; marine creatures; animals; birds; stones; medicinal substances and their properties; not to mention botanical and animal species from the New World.[26]

A similar encyclopaedic approach informed other anatomical and medical studies produced in Naples. One of the key figures working within this field was Marco Aurelio Severino. As Oreste Trabucco has shown, Severino's medical research and practice embraced two traditions. On the one hand, he had established himself within a Calabrian scholarly tradition that included figures such as Telesio, Campanella, Mario Schipano (1581? –?) and the Paracelsian physician Giovanni Battista Capucci (*c.*1600–*c.*1680); on the other, he was connected to a circle of German physicians active in Padua and the European school of anatomy that revolved around William Harvey (1578–1657). Harvey in fact visited Severino in Naples and corresponded with him.[27] In addition to teaching at the University of Naples, Severino operated within the city's extensive medical infrastructure, which included several hospitals catering for both locals and foreigners.[28] Among such institutions was the Hospital of the Incurables, which offered assistance to terminally ill patients. The city hosted a similar medical facility for women too.[29]

Some of Naples's hospitals and medical centres became renowned venues for anatomical experimentation.[30] Severino practised dissection at the hospital and established an extensive network that linked major medical figures in Europe.[31] He regarded Campanella as one of his teachers and embraced an empirical and 'practical' approach, intended – as recent scholarship has shown – to reform surgery by shedding the constraints of conventional medical theory.[32] Indeed, Severino's innovations in anatomy and physiology were sparked as much by his university training as by his practice of dissection. As Leonardo Di Capua (1617–1695) stated in his *Ragionamento*, Severino's human dissections successfully challenged Galen and other ancient scholars, thereby exposing the errors 'that their Greek, Arab and Latin followers had committed'.[33] Severino's cutting-edge research, surgical practices and growing reputation may have been a motivating factor for the denunciations brought against him to the Holy Office. Accused of grave religious irreverence by fellow surgeons and his own servant – allegations included poor mass attendance and failure to 'prescribe confession' to the terminally ill patients who he would routinely visit – Severino was tried, imprisoned and forced to leave Naples until 1641.[34]

In a milieu that, as discussed in this volume, increasingly questioned traditional medical views, the body remained, as David Gentilcore put it, 'a battleground for differing interpretations of disease: natural, divine and diabolic'.[35] Indeed, in Naples at that time dissections were occasionally performed on bodies of figures who had died in an odour of sanctity. As a side note, it is noteworthy that practices concerned with human embalming were also performed in the city. It is well known that Antonio Santorelli (1583–1653) – a physician from Nola and professor of medicine in Naples – published a text on this topic.[36] Physicians often worked in the service of ecclesiastical authorities and in 1608 Severino himself took part in the autopsy performed on the blessed Theatine Andrea Avellino (1520–1608). This endeavour cemented his fame in the city, although by this time Severino's reputation had already extended beyond Naples.[37]

In 1632 he published in Naples *De recondita abscessuum natura*: an illustrated treatise of surgical anatomy which contains analysis and classification of certain tumours that he had diagnosed both in male and female bodies.[38] So successful was Severino's text that in 1643 a new edition appeared in Frankfurt.[39] Likewise, his *Zootomia democritaea* was published in Nuremberg in 1645.[40] A monumental work of comparative anatomy – which included a final section on 'the correct method to teach dissection' – Severino's text analysed different species, with chapters detailing (and illustrating) some organs and their function. Interestingly, Severino located his work within the broader context of contemporary scientific research. Here he praised the empirical approach adopted in recent studies of the heavens and the stars which 'demonstrated motions' and sought to legitimise his own research methodology by stating that scholars should not hesitate to investigate 'the parts of the animals' that can be touched and handled.[41]

As a leading scholarly figure, Severino also promoted works of anatomy that appeared in Naples at that time. One example is the book by Jean Germain (dates unknown): a French Minim friar who was resident in the Neapolitan convent of Santa Maria della Stella in the 1620s. In 1625 Germain published in Naples a richly illustrated book of comparative anatomy.[42] Complementing locally published literature on animal anatomy (which included Pirro Antonio Ferraro's successful treatise on horses),[43] Germain's text was published under the auspices of some local academies, as attested by several dedicatory verses by members of the Erculei academy of Naples, including a brief eulogy by Severino himself. As a side note, friar Germain translated a medical treatise by Andrea Lorenzo (physician to King Henry IV of France) from French to Italian, which appeared in Naples in 1626 under the auspices of the Oziosi and Infuriati academies of Naples.[44]

The broader Neapolitan medical community, which embraced different categories of professionals and practitioners, also made important contributions to the city's intellectual life. Barber-surgeons, for instance, were active in the city and often collaborated with physicians and surgeons. As recently shown, this collaboration also extended to providing medical expertise to determine causes of death in legal cases.[45] Some barber-surgeons acquired fame and worked in the service of aristocrats and viceroys.[46] Figures such as Cinzio D'Amato (dates unknown) and Tiberio Malfi (dates unknown) were well-known professionals who authored successful manuals for barber-surgeons. These texts acted as professional guides for practising bloodletting; provided recipes to prepare embalming fluids for use on human cadavers; contained anatomical illustrations and even images of barbers performing human dissections. The dedicatory verses contained in D'Amato's and Malfi's texts also testify to the support that they received from acclaimed physicians such as Severino and Muzio Capalbo (dates unknown), not to mention scholars such as Giambattista Basile (1583–1632) and Giambattista Bergazzano (*c.*1576–1640s).[47]

As some of the sources presented in this introduction indicate, scientific literature was often published under the auspices of academies. Indeed, these institutions were central in positioning the city as a scholarly hub. Within the realm of science, early modern Neapolitan academies – whether public or private – provided an opportunity for debate, networking and knowledge transfer. Contemporary historiography has looked at early modern Neapolitan academies with a growing interest that has showcased their importance not only as centres of learned sociability but also as institutions that fostered research and experimentation.[48]

In the early modern period some Neapolitan academies such as the Oziosi and the Incauti were formally constituted under the auspices of the authorities or the city's elite. These and other circles – more than 50 were active in Naples in the seventeenth century – usually met in public spaces such as religious houses, churches or public palaces. Other circles followed a less formalised set of rules and held their gatherings in private venues – usually the residence of their founders. The Segreti academy, founded around 1560 by Giovan Vincenzo (dates unknown) and Giambattista Della Porta (1535–1615), is a case in point.[49] Similarly, a century later, the Investiganti academy initially gathered in the home of Tommaso Cornelio (1614–1684) and later in that of Marquis Andrea Concublet (1648–1675). This network of academies fostered debates within disciplines that included cosmology, optics, medicine, alchemy, botany, mineralogy, metallurgy and earth sciences.

The Neapolitan house of the Della Porta brothers acquired fame both as an intellectual hub and as a venue where the Segreti academy gathered. An area in Della Porta's house was equipped with scientific instruments such as 'a globe, bronze mathematical instruments, an astrolabe, stones of different colours, a perspective of mirrors and several small flasks and jars of various medicines'.[50] This chamber was also adorned with 'a portrait of Federico Cesi and a portrait of Galileo Galilei'.[51] There scholars would present the results of their research for the scrutiny of fellow academicians. Indeed, as Giambattista remarked in an Italian edition of *Magia Naturale* that appeared in Naples in 1611, 'in my house there has never been a lack of curious men, whose task was to investigate and experiment with the things studied'.[52] Scholars who frequented or visited Della Porta's house included Nicolas Claude Fabri de Peiresc, who was in Naples in 1601, Paolo Sarpi (1552–1623), Nicolò Antonio Stigliola (1546–1623), Giulio Cortese (1530s–1598), Tommaso Campanella, Antonio Mizaldo (1510–78), Ferrante Imperato and Federico Cesi (1585–1630).[53]

Although both Della Porta brothers were renowned intellectuals, it was Giambattista who acquired European fame. A polymath and prolific writer, he travelled extensively and networked with scholars and academicians such as Federico Cesi, the president of the Lincei academy of Rome, with whom he formed a long-lasting friendship and engaged in scholarly collaboration. Della Porta published texts on distillation, medicine, botany, optics, astrology, meteorology, pneumatics, hydraulics and physiognomy, to name but some of his interests. His most famous work, *Magia Naturale*, became a bestseller and was translated into several languages, including an English edition in 1658.[54]

Della Porta's fervent intellectual activity also generated controversy. As an expert in the field of optics, he devised a combination of concave and convex lenses allegedly used by Galileo Galilei for his telescope. Della Porta's reaction to the lack of recognition for his research resonated widely, and did so for a long time. In a speech delivered at the Incauti academy of Naples in the 1630s, more than 15 years after Della Porta's death, Filocalo Caputo (1582–1644) stated that:

> Challenging those who believe that Galileo Galilei is the author of such invention [i.e. the telescope], I will always call upon *Magia* [*Naturale*] of our Porta [...] testimony will also come from the learned Johannes Kepler [...] who showed Galileo the same lens that Giovanni Battista [Della Porta] had discovered. Testimony will come from me too, for when I spoke with Galileo and told him what I had seen in a copy of a letter directed to him from Kepler, he replied that he did not deny that [...] our Neapolitan [...] philosopher designed it, but he constructed the instrument.[55]

Della Porta's research methodology also contributed to shaping a Neapolitan scholarly tradition that positioned empiricism at the core of scientific research. Working at a time when the study of nature and cosmology was subject to scrutiny and censorship, figures such as Paolo Antonio Foscarini (1565–1616), Nicolò Antonio Stigliola, Francesco Fontana (*c.*1580–90–1665) and Filippo Finella (1584–1650s) all belonged to the scientific milieu that emerged after Della Porta's death. Indeed, Stigliola's and Fontana's studies on optics, the telescope and astronomy, not to mention Foscarini's support for Heliocentrism on religious grounds, all show the importance of Naples as a prime scholarly centre during Galileo's lifetime.[56]

The influence of Neapolitan science on European scholarly debates is also illustrated by the activities of the Investiganti academy of Naples during the second half of the seventeenth century. As Maurizio Torrini has shown, the Investiganti academy embraced an empirical approach that was in line with wider European practices. When Philip Skippon (1641–1691) and John Ray (1627–1705), both future members of the Royal Society, travelled to Naples in the 1660s they attended the academy's gathering – during which, Skippon wrote,

> the Academici Investigantes [...] discoursed about several things, and brought in the experiment of water ascending in glass *tubuli*, or small pipes; which they reasoned upon. After that, Leonardus à Capua discoursed about heat and cold; then Lucas Anion[ius] Portius seated himself in a chair, at the upper end of the room, and read a discourse on the same subject.[57]

As the Investigante Luca Antonio Porzio (1637–1715) asserted, experimentation was key to understanding nature. For only by these means could the limitations of knowledge derived from human senses be superseded.[58] Investiganti academicians combined studies in such fields as physics, earth sciences and astronomy with an interest in medicine that challenged the authority of Galenism and instead promoted chemistry and the importance of 'chemical medicaments'.[59]

As shown in this volume, the Neapolitan discourse on medicine and pharmacology was partly informed by an outbreak of plague that had hit Naples in 1656. This epidemic, which decimated the city's population, sparked a medical debate that, as Silvana D'Alessio has recently shown, led some physicians to acknowledge the validity of 'chymical remedies'.[60] By contesting the authority of received scholarly models, Neapolitan intellectuals developed the belief that knowledge was not static but instead ever evolving. It was in this context that the physician and Investigante Leonardo Di Capua (1617–1695) proclaimed 'the uncertainty' of medicine.[61]

The activities promoted within some of the city's academies also demonstrate how Neapolitan intellectuals kept abreast of wider scientific developments. The ideas of figures such as Galileo, Descartes, Jan Baptist Van Helmont (1580–1630), Pierre Gassendi (1592–1655), William Harvey and Isaac Newton (1642–1726/27) were discussed and disseminated within Naples, allowing at least some of the city's scholars to participate in a European scientific debate on mechanical philosophy.

As already mentioned, Nicola Cirillo is a case in point. A follower of Descartes and professor of medicine at the University of Naples in the early eighteenth century, Cirillo left a corpus of medical and scientific work, set up a large botanical garden and conducted extensive observations of natural phenomena.[62] In the words of his student and biographer Francesco Serao, Cirillo studied 'geometry, mechanics and those branches of mathematics which are necessary to a good philosophical enquiry'. As Serao noted, Cirillo supported his studies by assembling a large library, including ground-breaking books that 'were disseminated in Europe'.[63] Cirillo's wide scientific interests embraced meteorological observations and accounts of the eruptions of Mount Vesuvius, not to mention his studies on the use of cold water to cure fever.[64] He discussed such topics in correspondence with fellows of the Royal Society James Jurin (1684–1750) and Cromwell Mortimer (*c.*1698–1752). Cirillo himself became a fellow of the Royal Society in 1718. Fourteen years later Cirillo and Celestino Galiani (1681–1753) were among the founding members of the Neapolitan Academy of Sciences (Accademia delle Scienze). As Elvira Chiosi put it, this academy became a platform for the Neapolitan dissemination of 'the ideas of Newton and the empiricism of Locke'.[65] Galiani himself became a corresponding member of The Royal Society in 1735. Other academics operating at the University of Naples also became fellows of the Royal Society around this time.[66]

During the early eighteenth century the Kingdom of Naples underwent a series of political changes. However, the city of Naples still retained its scholarly importance. Following the end of Spanish rule and a brief period of Austrian power, the kingdom was conquered by Charles of Bourbon in 1734. He established an independent monarchy with a royal court in Naples. Charles also sought to implement a series of reforms. While not always successful, the monarchy's reformist efforts were informed and celebrated by the city's cohort of Enlightened thinkers. The work of scholars such as Giambattista Vico (1668–1744), Pietro Giannone (1676–1748), Antonio Genovesi (1713–1669) and Gaetano Filangieri (1753–1788), to name but some, undoubtedly reflected the continued vibrancy of Neapolitan intellectual life and the city's connections to the intellectual centres of Europe.[67] As Girolamo Imbruglia has shown, these connections proudly featured in periodicals that were published in Naples. In that sense, the foundation of the *Giornale Enciclopedico di Napoli* (*Encyclopaedic Journal of Naples*) functioned as a platform intended to keep its readers informed about the technological and scientific achievements within the kingdom and beyond.[68]

As Antonio Borrelli put it, despite the modest presence of scientific instruments in eighteenth-century Naples the local scholarly milieu was informed about a broader scientific development. For instance, the innovations introduced by Benjamin Franklin (1706–1790) fostered experimentation in the field of electricity.[69] Similarly, Neapolitan scholars produced texts on physiology and public health that aligned with the latest European research in the field of immunology.[70] The health emergencies caused in the Kingdom of Naples by diseases such as typhus and smallpox generated a debate in favour of public vaccination that positioned Naples as a leading European centre for its immunisation campaign against smallpox.[71]

The existence of these transnational scholarly networks also helped to facilitate a European discourse on disciplines such as volcanology and earth sciences. Following the catastrophic eruption that occurred between December 1631 and early January 1632, Vesuvius became the subject of increasing artistic and scholarly interest. Such interest became a sensational case that resonated throughout Europe when the excavations of the ancient cities of Pompeii and Herculaneum officially began in 1738. Archaeologists unearthed the urban structure of these cities; historians were overwhelmed by the discovery of countless treasures, such as papyrus scrolls, objects and inscriptions, that the eruption of Vesuvius had buried deep in the ground. Within the realm of science, these excavations re-shaped knowledge of disciplines such as archaeology, stratigraphy and, of course, volcanology.

The international interest generated by these excavations and the establishment in Naples of a Royal academy all contributed to the status of the city (and of the Bay of Naples more broadly) as a leading European site for scientific investigation. Visitors to Naples, such as William Hamilton (1730–1803) further enhanced scholarly and artistic interest in the city's potential contributions to the study of volcanology and antiquity. In a letter sent from Naples in October 1770 and addressed to the secretary of the Royal Society, Matthew Maty (1718–1776), Hamilton discussed 'the nature of the soil of Naples', remarking that

> The Kingdom of the Two Sicilies offers certainly the fairest field for observations of this kind, of any in the whole world; here are volcanoes existing in full force, some on their decline, and others totally extinct.[72]

Indeed, Vesuvius and the Phlegraean Fields offered scientists an opportunity to study the geomorphology of the Bay of Naples: an 'open air' laboratory in which to investigate chemistry and the earth sciences more broadly.[73]

The case of volcanology demonstrates how Neapolitan scientific research in the eighteenth and nineteenth centuries developed on parallel tracks. On the one hand, there existed a local scholarly community that operated within private venues and the city's university, as well as academic circles such as the Accademia Reale delle Scienze e Belle Lettere (Royal Academy of Science and Letters), founded under the auspices of King Ferdinand of Bourbon in 1780. On the other, a parallel milieu that had emerged within intellectual circles such as Hamilton's enabled some foreigners to study the artistic, scientific, historical and literary significance of the Bay of Naples. Scholars also observed and conducted experiments on sites such as Solfatara in Pozzuoli and, of course, the slopes of Vesuvius – all of which continue to offer endless opportunities for scientific investigation.[74]

## The science of Naples

It is hoped that the brief overview offered in this introduction and the essays contained in this volume will contribute to positioning Naples as an important centre of scientific learning. The aim here is to address the relative marginalisation of Neapolitan science in the existing Anglophone literature by showcasing a series of works produced by historians working both in Italy and the Anglosphere. This volume does not seek to offer a comprehensive view of the history of Neapolitan science, but rather a series of snapshots that illustrate the breadth of the city's scientific culture. In addition, this volume seeks to eschew teleological approaches to the study of the city of Naples while paying close attention to its contextual specificity as a site for the production of scientific knowledge.[75] This broader approach dovetails with modern historiography that has sought to reject present-centred narratives of the development of modern science. Instead our focus is on local sites for the production of scientific knowledge and the interpretation of its contemporary uses, meanings and strategies for legitimation.[76]

Accordingly, the essays in this volume are divided into four parts. The first, entitled 'Neapolitan science and its institutions', considers where science was practised within Naples and how these spaces influenced the knowledge produced therein. It opens with Daniela Caracciolo's discussion of the role played by Ferrante Imperato's private museum in the organisation of natural knowledge. She highlights the importance of early modern Neapolitan interest in collecting material from the natural world and reflects on the means by which these practices were established. Providing a detailed account of both Imperato's museum and the written works produced by him and his son (these works themselves being based upon the contents of Imperato's collection), Caracciolo also discusses the complex relationship between the curation and display of physical objects and the production of textual encyclopaedias. Moving the focus to public institutions, Felix Waldman's essay analyses the role of political science within the University of Naples. It examines the history of the teaching and study of politics in the context of Celestino Galiani's curricular reform. Waldman's essay also offers an explanation of why political science was not among the sciences studied at this institution.

The volume's second part, entitled 'Environment, Disaster and Disease in the Bay of Naples', analyses the multidisciplinary scholarly debate that emerged from the consideration of the urban and natural environment of the Bay of Naples. Alfonso Paolella offers an account of one of Giambattista Della Porta's lesser-known works, *De aëris transmutionibus*. In this work Della Porta sought to explain, among other phenomena, the causes of earthquakes. As Paolella shows, his ideas were informed in part by the experience of living through the seismic events that occurred in the Bay of Naples in 1583. Della Porta drew on these observations to engage with the explanations for earthquakes offered not only by his peers, but also by ancient authorities. In her essay, Lorenza Gianfrancesco analyses the environmental impact of, and the human response to, the catastrophic eruption of Mount Vesuvius that occurred between December 1631 and January 1632. Gianfrancesco also demonstrates how this disaster sparked a scientific debate on the earth sciences that would soon acquire a European dimension. Finally, Silvana D'Alessio discusses in her essay how the outbreak of plague that struck Naples in 1656 influenced physicians' understanding of the disease. Focusing upon the work of two physicians, Geronimo Gatta and Carlo Morexano, her essay analyses how their observations on the course of the plague through their city led them to reject miasmatic theories of disease causation and instead to develop alternative explanations, partly rooted in Santorio Santorio's concept of contagion.

The third section of the book, entitled 'Naples and the Early Modern World', considers the various networks that connected the city to learned circles beyond the Italian peninsula. The case studies presented in this section offer a new perspective on what scholars resident in other European states knew about, and believed that they could learn from, Neapolitan science – and, in turn, how Neapolitan intellectuals could utilise knowledge produced in foreign scholarly contexts. In his essay James Clifton discusses Johann Daniel Major's knowledge of Neapolitan cabinets of curiosities. A professor of medicine at the Christian Albrecht University in Kiel in the 1670s, Major produced a series of studies on *Kunst und Naturalien-Kammern* (cabinets of art and naturalia). Despite Major's enduring interest in institutions such as Imperato's museum, Clifton suggests that by the later seventeenth century the connections between Naples and Germany were relatively weak.

On the other hand, Daniel Canaris discusses Naples's extra-European connections in his study of Leonardo Di Capua's understanding of Chinese medicine presented in his work *Parere* (1681). Resisting a tendency common among his contemporaries to assimilate Chinese medicine to a Galenic framework, Di Capua's study of Chinese medicine was intended instead to support his critique of this body of medical knowledge. In the final chapter of this section, Frank James continues the story of Naples' connections with the wider world into the nineteenth century by discussing the involvement of British men of science, including Humphry Davy (1778–1829), in efforts to unroll papyri discovered during the archaeological excavations at Herculaneum. He reveals a story of courtly diplomacy between Naples and London that involved the sharing of scientific and technical knowledge. Echoing Clifton's conclusions, James depicts Davy operating in a changed world, one in which the ties that had once bound together an international republic of letters, and the values of its members themselves, were growing weaker.

In the volume's final essay Gennaro Rispoli describes the history of one of Naples' foremost medical institutions: the Hospital of the Incurables ('Ospedale degli Incurabili'). He offers an account of its gradual evolution, the multiple ways in which the hospital has been utilised and the creation of its rich art collections. In addition to showcasing the contribution made by the hospital's physicians, surgeons and patrons, Rispoli – himself a trained surgeon and Director of the *Museo delle Arti Sanitarie e di Storia della Medicina* in Naples – reflects on the significance of the institution's history in contemporary Neapolitan life and the enduring capacity of art to heal.

Collectively, these essays reinforce the volume's central argument that the importance of Naples as a centre of scientific and medical study has been unduly neglected in Anglophone scholarship. In sum, Naples was – in several respects – a leading Italian site of learning. As the case studies analysed in this volume demonstrate, prominent scholars such as Ferrante Imperato, Giambattista Della Porta, Marco Aurelio Severino or Nicola Cirillo – to name but a few – did not operate in isolation. Della Porta's collaboration with Federico Cesi (who hoped to open in Naples a branch of the Lincean academy of Rome), Donato D' Eremita's Florentine and Roman contacts, Marco Aurelio Severino's European prominence or the Neapolitan contribution to the debate that emerged within the Royal Society in relation to disciplines such as volcanology or climatology combine to provide strong evidence that the science of Naples made the city pre-eminent within the Italian peninsula and beyond throughout the early modern period.

## Notes

1 This approach was set out in the foundational literature on the Scientific Revolution. See for example Boas Hall, *The Scientific Renaissance*; Hall, *The Scientific Revolution*.

2 There has been increased interest in the ecclesiastical censorship of science over the last 20 years in response to the opening of the archives of the Roman Inquisition and Index. This interest has sought to revise older accounts of a necessary antagonism between post-Tridentine Catholicism and science. An excellent introduction to this topic is Baldini and Spruit, *Catholic Church and Modern Science*. Recent works include Donato ed., *Medicine and the Inquisition*; Marcus, *Forbidden Knowledge*; Vermeir and Regier (eds), *Boundaries, Extents and Circulations*; Tarrant, *Defining Nature's Limits*.

3 Segre, *In the Wake of Galileo*; Findlen, 'Controlling the experiment'; Beretta, 'At the source of Western science'. On the reception of scientific knowledge see for example Ferrone, *Scienza*

*Natura Religione*; Casini, *Newton e la Coscienza Europea*; Brigaglia and Nastasi, 'Bologna e il Regno delle Due Sicilie'; Macuglia, 'Newtownianism'. An important development in historiography of early modern Italian science has been the study of women engaged in the sciences during the eighteenth century, although these women have often been shown to have been engaged in practising 'northern' science. See for example Findlen, 'Science as a career'; Findlen, 'Translating the new science'; Berti Logan, 'The desire to contribute'; Cavazza, "Between modesty and spectacle'; Bertucci, 'The in/visible woman'.

4 Important contributions include Musi, *La Rivolta di Masaniello*; de Miranda, *Una quiete operosa*; Robertson, *The Case for the Enlightenment*; Guarino, *Representing*; Boccadamo, *Napoli e l'Islam*; Marino, *Becoming Neapolitan*; Rak, *A dismisura d'uomo*; Villari, *Un sogno di libertà*; Calaresu and Hills (eds), *New Approaches*; Astarita (ed.), *A Companion*; Hills, *The Matter of Miracles*; Rak, *Napoli Civile*; Mondola, Riga and Allocca (eds), *Il Manso*; de Divitis (ed.), *Companion*; D'Alessio, *Masaniello*.

5 Findlen, *Possessing Nature*; Gentilcore, *Healers and Healing*; Rao, *Editoria e cultura*; Fusco, *Peste*; Mazzola (ed.), *Le scienze*; Musi, *La disciplin*; D'Alessio, 'On the Neapolitan plague'; Brevaglieri, *Natural Desiderio*; Fehrenbach and van Gastel (eds), *Nature and the Arts*.

6 Findlen (ed.), *Athanasius Kircher*; Fletcher, *A Study*; Gianfrancesco, 'From propaganda'.

7 Findlen, *Possessing Nature*; Di Gregorio, '*Le coselline*'; Bertucci, 'Architecture'; Bertucci, 'Designing'; Bonan and Occhi (eds), *Environment*.

8 Della Porta. *Coelestis Physiognomonia*; Piccari, *Giovan Battista Della Porta*; Clericuzio, *Elements*; Freedberg, *The Eye of the Lynx*; Gianfrancesco, 'Books'; Arecco, 'Pitagorismo'; Verardi, *La scienza*; Borrelli, 'Heat'. In the field of digital humanities, the 'Italian Academies database' (IAD) contains extensive material on early modern Neapolitan academies.

9 Nazzaro, 'L'eruzione'; Ricciardi, *Diario*; Nazzaro, *Il rischio Vesuvio*; Scarth, *Vesuvius*; Everson, 'The melting pot'; Cocco, *Watching Vesuvius*; Casapullo and Gianfrancesco (eds), *Napoli e il Gigante*; Nazzaro, 'Implicazioni'; Rodríguez Fernández, *El Vesubio*; Tortora, Cassano and Cocco (eds), *L'Europa moderna*; Viceconte, Schiano and Cecere (eds), *Heroes*.

10 Fusco, *Peste*; Gianfrancesco, 'Books', 251.

11 Calaresu and Hills, 'Introduction', 3–4, Musi, 'The Kingdom of Naples'; Caputo, Gianfrancesco and Palmieri (eds), *Tales*, introduction.

12 Gianfrancesco, 'From propaganda to science'; Miller, *Peiresc's Mediterranean World*; Findlen (ed.), *Athanasius Kircher*; Fletcher, *A Study*. Interestingly, in 1638 Friar Salvatore Imbroll (possibly a pseudonym) published in Naples a book explaining Kircher's invention of a scientific device that in Fletcher's words 'could answer questions in mathematics, geometry, medicine and calculations of the calendar', *A Study*, 36. The book in question is Imbroll, *Specula*.

13 Gascoigne, 'A reappraisal'; Feingold and Navarro-Brotons, *Universities*.

14 Musi, *La disciplina*, 18–20.

15 Grendler, *Universities*, 45; the following is based on 41–5.

16 The teaching of *Medicina pratica* was divided into three branches: dietetic, pharmaceutical and chirurgical. See Conforti, 'Medicina', 32–3.

17 Conforti, 'Medicina', 23–39.

18 Conforti, 'Medicina', 25–6.

19 Conforti, 'Medicina', 34.

20 Galasso, 'Scienze', 213–15.

21 See, for instance, Caracciolo, *Napoli Sacra*; Chichierchia and Papa, *Storia*; Gentilcore, *Healers*.

22 L. Gianfrancesco, 'Books', 257–60. On cultural encounters between Rome and Spanish Naples see Brevaglieri, *Natural Desiderio*, 31–70.

23 d'Onofri, *Istruzione*, 9; 14.

24 Colmenero De Ledesma, *Chocolata Inda*. An English translation of this text appeared in London in 1652. See Wadsworth, *Chocolate*. See Marini-Bettòlo, *La collaborazione*.

25 Paré, *The Works*, 689.

26 Lombardi, *Enciclopedia*; Di Gregorio, '*Le coselline*'.

27 See Trabucco, '"L'invenzione della tradizione"', 184.

28 Capaccio, *Il Forastiero*, 912–16; Caracciolo, *Napoli Sacra*; Musi, *La disciplina*, 9–66; Gianfrancesco, 'Books, gold, and elixir', 255–6; Novi Chavarria, *Accogliere e curare*, 117–26 in particular.

29 Capaccio, *Il Forastiero*, 913–14. More recent publications include Agostino Falanga, *The Venerable*; Borrelli, '*Le origini*'.

30 Silvia De Renzi has highlighted that in the 1670s some sources considered Naples, Rome, Padua, Leiden and Amsterdam to be cities 'with easy access to dissected bodies, a resource for investigating the causes of diseases'. See De Renzi, 'Seats and series', 108.
31 For an overview of Severino, his library and the milieu within which he operated see Conforti, 'Surgery'.
32 Conforti, 'Surgery', 293. For an interesting study of aspects of the reception of Campanella's *Medicinalium Libri* see Conforti, 'Tracce campanelliane'.
33 Di Capua, *Parere*, 68.
34 The documentation detailing the accusations against Severino is in Amabile, *Il Santo Officio*, 66–8.
35 Gentilcore, 'Contesting illness', 119.
36 Santorelli, *Postpraxis medica*. On 23–5 Santorelli discusses the case of the Theatine Andrea Castaldo Pescara, who died in 1629 and was considered to have died in an odour of sanctity. See Gotor, 'La devozione'. On the relationship between Harvey and Severino see, for instance, Schmitt and Webster, 'Harvey'.
37 See Trabucco, 'Marco Aurelio Severino'.
38 Severino, *De recondita*.
39 Severino, *De recondita*.
40 Severino, *Zootomia Democritea*; Bertoloni Meli, 'Of snails and horsetails'.
41 Severino, *Zootomia Democritea*, 6.
42 Germain, *Breve*.
43 Ferraro, *Cavallo*. This text contains an illustrated section detailing the anatomy of horses, as well as a section on equine veterinary treatments by Giovanni Battista Ferraro (Pirro Antonio's father).
44 Germain, *Discorsi*.
45 See Carnevale 'Visum', 257–8.
46 D'Amato, *Prattica Nuova*, 7.
47 D'Amato, *Prattica Nuova*; Malfi, *Il Barbiere*.
48 Giustiniani, 'Breve'; Minieri Riccio, 'Cenno storico'; Maylender, *Storia*; Musi, '"Non pigra quies"'; de Miranda, *Una quiete operosa*; Gianfrancesco, 'From propaganda to science'; Gianfrancesco, 'Accademie'; Gianfrancesco, 'Dal palazzo'. For a list of early modern Neapolitan academies see also IAD.
49 References to the Segreti academy are in Della Porta, *Della Magia Naturale*; Della Porta, *Della Chirofisonomia*, 3–6. References to the *Investiganti* include Luca Antonio Porzio, *Del Sorgimento*; Borrelli, *De Motionibus*; Bulifon, *Lettere Memorabili*, 281. More recent studies include Torrini, 'L'Accademia'; Eamon, *Science and the Secrets of Nature. Books of secrets in medieval and early modern culture* (Princeton: Princeton University Press, 1994), 199, 270; Robertson, *The Case for the Enlightenment*, 94–147.
50 See Fulco, 'Per il "Museo"'.
51 See Fulco, 'Per il "Museo"', 290.
52 '*Ne ha mancato mai nella mia casa Academia di curiosi huomini, il cui compito era'investigar, e far esperienza delle cose investigate*', Della Porta, *Della Magia Naturale*, Preface to the reader.
53 Imperiale, *Musaeum*, 122–4. A reference to Antonio Mizaldo and Giambattista Della Porta's interest in the occult and the 'secrets of nature' is in Garzoni, *La Piazza Universale*, 184; Gianfrancesco, 'From propaganda to science', 15.
54 Porta, *Natural Magick*.
55 '*Dica pur chi vuole che il Galileo Galilei di sì raro trovato sia stato l'autore, ch'io sempre chiamerò in testimonio gli fioritissimi commentarij della magia del nostro Porta … testimonio ne sarà il dottissimo Giovanni Keplero … che lo stesso occhiale mostrò al Galileo, che Gio. Battista haveva ritrovato; testimonio ne sono io che discorrendo co' l'istesso Galileo, e narrandoli ciò che visto havevo nella copia di una lettera del Keplero diretta a lui, rispose che no' negava essere stato il nostro napoletano l'Architetto, ma esso il fabro, il nostro filosofo fece il disegno, ma esso costruì lo strumento.*' Caputo, *Oratione*, 18–19. See also *Le Opere di Galileo Galilei*, Vol. 11; Gianfrancesco, 'From propaganda to science', 15–16.
56 See, for instance, Foscarini, *Lettera*; Finella, *Phisonomia*; Finella, *De Planetaria*; Stigliola, *Il Telescopio*; Fontana, *Novae Coelestium*.
57 Skippon, *An Account*, vol. 6, 607. A reference to this passage is in Torrini, 'L'accademia', 852; Testa, *Italian Academies*, 164–5.

58 Porzio, *Del sorgimento*, vol. 2, 344. See also Torrini, *L'accademia*, 855. For an excellent overview of the medical world in early modern Naples, see Musi, *La disciplina*, 9–66.
59 Di Capua, *Parere*, 507, 555–9.
60 D'Alessio, 'On the Neapolitan plague of 1656', 199–204. On the plague in Naples see also De Renzi, *Napoli*; Fusco, *Peste*; Fusco, *La Grande epidemia*. For recent studies on the Neapolitan epidemics in 1656 and 1764 see the contributions of D'Alessio, Fusco and Palmieri contained in Caputo, Gianfrancesco and Palmieri (eds), *Tales of Two Cities*, 123–56; 227–46.
61 Zurlini, 'The uncertainty of medicine', 110–11.
62 del Gaizo, 'Contributo', 76–7.
63 Serao, *Introduzione*, in Cirillo, *Consulti Medici*, Sig. B4$^{1\text{-}v}$.
64 Cirillo, 'De Frigide', 142–51.
65 Chiosi, 'Intellectuals and academies', 122.
66 Domenico Cirillo (1739–1799), a scientist and professor of medical pathology at the University of Naples, became a Fellow of the Royal Society. He sent reports to London which were concerned with the observations that he had conducted in areas of the Kingdom of Naples. For instance, his letter to Sir William Watson (1715–1787) on the medical benefits of the manna tree (*Fraxinus Ornus*) and some curative methods against the bite of the tarantula were published in the *Philosophical Transactions* in 1770. See Cirillo, 'A Letter'.
67 Imbruglia, *Naples*; Robertson, *The Case for Enlightenment.*
68 Imbruglia, 'Enlightenment', 85.
69 Borrelli, 'Istituzioni', 141–3.
70 Borrelli, 'Dall'innesto'; Palmieri, 'Two Tales'.
71 Borrelli, 'Dall'innesto', 85.
72 Hamilton, 'Remarks', 3.
73 Guerra, 'If you don't'.
74 To trace the history of the Royal Academy of Naples see Beltrani, '*La Reale Accademia*', memoria n. 5; Chiosi, '"Humanitates" e scienze'; Chiosi, 'Intellectuals and Academies', 118–35; Scarth, *Vesuvius*, 173–244. An overview of the scientific milieu that emerged in eighteenth-century Naples is in Rao, 'Politica'; Borrelli, 'Medicina'; Borelli, 'Istituzioni'. For a recent contribution on the role of the geomorphological structure of the Bay of Naples in the history of early modern Neapolitan medicine see Conforti, *Medicina sotto il volcano.*
75 Calaresu and Hills, 'Introduction', 2–3; Marino, 'Constructing', especially 12–15.
76 On the problem of teleology in history of science see Cunningham and Williams, 'Decentering'. For an introduction to the approaches that have informed modern Anglophone history of science see Golinski, *Making Natural Knowledge*; see too Shapin, *Never Pure*. Recent studies have also considered the varieties of scientific and medical knowledge that were used in daily life in early modern Italy rather than charting the successive 'advances' that led to modern science. See for example Cavallo and Storey, *Healthy Living*; Strocchia, *Forgotten Healers.*

# Bibliography

## Websites

'The Italian Academies database' (IAD) (https://www.bl.uk/catalogues/ItalianAcademies/default.aspx).

## Primary sources

*Le Opere di Galileo Galilei*, Edizione Nazionale, 20 vols. Florence: Barbera, 1809–1909.

Baldini, Ugo and Leen Spruit. *Catholic Church and Modern Science: Documents from the archives of the Roman congregations of the Holy Office and the Index Volume 1: Sixteenth-century documents.* Rome: Libreria Editrice Vaticana, 2009.

Borrelli, Alfonso Giovanni. *De Motionibus Naturalibus A Gravitate Pendentibus Liber Io: Alphonsi Borrelli in Academia Pisana Matheseos Professoris*. Regio Iulio: 1670.

Bulifon, Antonio. *Lettere Memorabili, Istoriche, Politiche, ed Erudite*. Naples: 1698.

Capaccio, Giulio Cesare. *Il Forastiero*. Naples: 1634.
Caracciolo, Engenio. *Napoli Sacra*. Naples: 1623.
Caputo, Filocalo. *Oratione*. Naples: 1632.
Niccolo Cirillo, *Consulti Medici*, Vol. I. Naples: 1738.
Cirillo, Nicola. 'De Frigide in Febribus Usu', *Philosophical Transactions* 36 (1729): 142–51.
Cirillo, Domenico. 'A letter to Dr. William Watson, F.R.S. giving some account of the manna tree, and of the tarantula', *Philosophical Transactions* 60 (1770): 233–8.
De Ledesma, Antonio Colmenero. *Chocolata Inda. Opusculum de qualitate & naturâ Chocolatae [...] Hispanico antehac idiomate editum: nunc verò curante Marco Aurelio Severino [...] in Latinum translatum*. Nuremberg: 1644.
Porta, John Baptista. *Natural Magick*. London: 1658.
D'Amato, Cinzio. *Prattica Nuova, et utilissima: Di tutto quello ch'al diligente Barbiero s'appartiene*. Venice: 1669.
Della Porta, Giambattista. *Coelestis Physiognomonia*, edited by Alfonso Paolella. Naples: ESI, 1996.
Della Porta, Giambattista. *Della Magia Naturale*. Naples: 1611.
Della Porta, Giambattista. *Della Chirofisonomia Overo Di quella Parte della humana Fisonomia, che si appartiene alla Mano. Libri Due del Signor Gio: Battista della Porta Napolitano Tradotti da un Manoscritto Latino Dal Signor Pompeo Sarnelli*. Naples: 1677.
Di Capua, Leonardo. *Parere del Signor Lionardo Di Capua divisato in otto ragionamenti, ne' quali partitamente narrandosi l'origine, e'l progresso della medicina, chiaramente l'incertezza della medesima si fa manifesta*. Naples: 1681.
Di Gregorio, Maurizio. *'Le coselline di un ometto curioso': L' idea per fare le gallerie universali di tutte le cose del mondo, naturali, artificiali e miste*, edited by Daniela Caracciolo. Galatina Bari: Congedo, 2008.
Ferraro, Pirro Antonio. *Cavallo Frenato*. Naples: 1602.
Finella, Filippo. *Phisonomia Naturale di Filippo Finella*. Naples: 1625.
Finella, Filippo. *De Planetaria Naturali Phisonomia*. Naples: 1649.
Fontana, Francesco. *Novae Coelestium Terrestriumq[ue] Rerum Observationes. Et Fortasse Hactenus Non Vulgatæ. à Francisco Fontana, specillis à se inventis, et ad summam perfectionem perductis, editæ*. Naples: 1646.
Foscarini, Paolo Antonio. *Lettera* [...] *Sopra l'Opinione de' Pittagorici e del Copernico. Della Mobilità della terra e Stabilità del Sole*. Naples: 1615.
Imbroll, Salvatore. *Specula Militensis Encyclica*. Naples: 1638.
Imperiale, Giovanni. *Musaeum Historicum Et Physicum*. Venice: 1640.
Tommaso Garzoni, *La Piazza Universale Di Tutte Le Professioni Del Mondo*. Seravalle Di Venetia: 1605.
Germain, Jean. *Breve e sustatiale trattato intorno alle figure anathomiche delli più principali animali terrestri, aquatili, et volatili, con la simpatia er convenienza che hanno, ó in parte, ó in tutto, con il corpo humano* [...]. Naples: 1625.
Germain, Jean. *Discorsi della conservatione della vista, delle malattie melanconiche, delli cataratti, e della vecchiaia. Composti in lingua Francese dal Sig. Andrea Lorenzo, Medico Fisico del Christ.* [mo] *Herrico IIII. Rè di Francia*. Tradotti in lingua Italiana, e commentati da Fr. Gio. Germano Francese Medico Chirurgo. Naples: 1626.
Hamilton, William. 'Remarks upon the nature of the soil of Naples, and its neighbourhood; in a letter from the Honourable William Hamilton, His Majesty's Envoy Extraordinary at Naples, to Mathew Maty, M.D. Sec. R. S.', *Philosophical Transactions* 61 (1771): 3.
Malfi, Tiberio. *Il Barbiere di Tiberio Malfi* [...] *Barbiere, consule dell'arte in Napoli*. Naples: 1626.
d'Onofri, Pietro. *Istruzione al forestiere e al dilettante intorno a quanto di antico, e di raro si contiene nel Museo del Real Convento di S. Caterina a Formello de' PP. Domenicani Lombardi in questa Città di Napoli*. Napoli: 1791.
Paré, Ambroise. *The works of that famous chirurgion Ambrose Parey*. Translated out of Lattine and compared with the French by Tho[mas] Johnson. London: 1649.
Porzio, Luca Antonio. *Del Sorgimento De' Licori Nelle Fistole Aperte d'ambidue gli estremi, et intorno à molti corpi, che tocchino la loro superficie. Discorso Di Luc'Antonio Porzio Accademico Investigante*. Venice: 1667.
Porzio, Luca Antonio. *Opera Omnia*, Vol. 2. Naples: Felice Carlo Mosca, 1736.

Santorelli, Antonio. *Postpraxis medica, seu de medicando defuncto liber*. Naples: 1629.
Severino, Marco Aurelio. *De recondita abscessuum natura libri VIII*. Naples: Ottavio Beltrano, 1632.
Severino, Marco Aurelio. *De recondita abscessuum natura libri VIII*. Frankfurt: Johannem Moenum, 1643.
Severino, Marco Aurelio. *Zootomia Democritea. Idest, anatome generalis totius animantium opificii, libris quinque distincta, quorum seriem sequens facies delineabat*. Nuremberg: 1645.
Skippon, Philip. 'An account of a journey made through part of the Low Countries, Germany, Italy and France'. In *A Collection of Voyages and Travels*. London: 1746.
Stigliola, Nicola Antonio. *Il Telescopio over Ispecillo Celeste*. Naples: 1627.
Wadsworth, James. *Chocolate: or an Indian Drinke*. London: 1652.

## Secondary sources

Abulafia, David. 'The South'. In *Italy in the Age of the Renaissance*, edited by John M. Najemy, 208–25. Oxford: University of Oxford Press, 2004.
Amabile, Luigi. *Il Santo Officio della Inquisizione in Napoli*. Soveria Mannelli: Rubbettino Editore, 1987.
Arecco, D. 'Pitagorismo e tradizione esoterica dall' Europa a Napoli (1516–1783)', *Atrium* 17 (2015): 68–84.
Astarita, Tommaso (ed.) *A Companion to Early Modern Naples*. Leiden; Boston: Brill, 2013.
Beltrani, Giovanni. 'La Reale Accademia di scienze e belle lettere fondata in Napoli nel 1778', *Atti dell'Accademia Pontaniana* 30 (1900).
Beretta, Marco. 'At the source of Western science: The organisation of experimentalism at the Accademia del Cimento (1657–1667)', *Notes and Records of the Royal Society of London* 54 (2000): 131–51.
Bertoloni Meli, Domenico. 'Of snails and horsetails: Anatomical empiricism in the early modern period', *Early Science and Medicine* 18 (2013): 435–52.
Bertucci, Paola. 'The in/visible woman: Mariangela Ardinghelli and the circulation of knowledge between Paris and Naples in the Eighteenth Century', *Isis* 104 (2013): 226–49.
Bertucci, Paola. 'The architecture of knowledge: Science, collecting and display in eighteenth-century Naples'. In *New Approaches to Naples c.1500–c.1800: The power of place*, edited by Melissa Calaresu and Helen Hills, 149–74. Abingdon and New York: Routledge, 2013.
Bertucci, Paola. 'Designing the house of knowledge in eighteenth-century Naples: The Ephemeral Museum of Ferdinando Spinelli, Prince of Tarsia'. In *Cabinets of Experimental Philosophy in Eighteenth-Century Europe*, edited by Jim Bennett and Sofia Tallas, 119–36. Leiden: Brill, 2013.
Boas Hall, Marie. *The Scientific Renaissance; 1450–1630*. New York: Dover Publications, 1994.
Boccadamo, Giuliana. *Napoli e l'Islam. Storie di Musulmani, Schiavi e Rinnegati in Età Moderna*. Naples: D'Auria Editore, 2010.
Bonan, Giacomo and Katia Occhi (eds). *Environment and Infrastructure*. Berlin and Boston: De Gruyter, 2023.
Borrelli, Antonio. 'Medicina e società a Napoli nel secondo Settecento', *Archivio storico per le province napoletane* 112 (1994): 123–77.
Borelli, Antonio. 'Istituzioni e attrezzature scientifiche a Napoli nell'età dei Lumi', *Archivio storico per le province napoletane* 114 (1996): 131–83.
Borrelli, Antonio. 'Dall'innesto del vaiolo alla vaccinazione jenneriana: il dibattito scientifico napoletano', *Nuncius* 12 (1997): 67–85.
Borrelli, Antonio. 'Le origini della scuola medica dell'Ospedale degli Incurabili di Napoli', *Archivio storico per le province napoletane* 143 (2000): 135–47.
Borrelli, Arianna. 'Heat and moving spirits in Telesio's and Della Porta's meteorological treatises'. In *Bernardino Telesio and the Natural Sciences in the Renaissance*, edited by Pietro Daniel Omodeo, 66–95. Leiden; Boston: Brill, 2019.
Brevaglieri, Sabina. *Natural Desiderio di Sapere: Roma Barocca tra Vecchi e Nuovi Mondi*. Rome: Viella, 2019.
Brigaglia, Aldo and Pietro Nastasi. 'Bologna e il Regno delle Due Sicilie: aspetti di un dialogo scientifico (1730–1760)'. In *Scienza e letteratura nella cultura italiana del Settecento*, edited by Walter Tega and Renzo Cremante, 145–78. Bologna: Il Mulino, 1984.
Burckhardt, Jacob. *The Civilisation of the Italian Renaissance in Italy*, translated by S. G. C Middlemore. London: Penguin, 1991.

Calaresu, Melissa and Helen Hills (eds). *New Approaches to Naples c.1500–c.1800: The power of place*. Abingdon and New York: Routledge, 2013.
Caputo, Vincenzo. Lorenza Gianfrancesco and Pasquale Palmieri (eds). *Tales of Two Cities. News, stories and media events in early modern Florence and Naples*. Rome: Viella, 2023.
Carnevale, Diego. 'Visum et repertum: Medical doctrine and criminal procedures in France and Naples (17th–18th centuries)'. In *The Body of Evidence. Corpses and Proofs in Early Modern European Medicine*, edited by Francesco Paolo De Ceglia, 245–69. Leiden-Boston: Brill, 2020.
Casapullo, Rosa and Lorenza Gianfrancesco (eds). *Napoli e il Gigante: Il Vesuvio tra immagine, scrittura e memoria*. Soveria Mannelli: Rubbettino, 2014.
Casini, Paolo. *Newton e la Coscienza Europea*. Bologna: Il Mulino, 1983.
Cavallo, Sandra and Tessa Storey. *Healthy Living in Late Renaissance Italy*. Oxford: Oxford University Press, 2013.
Cavazza, Marta. 'Between modesty and spectacle: Women and science in eighteenth-century Italy'. In *Italy's Eighteenth Century: Gender and culture in the age of the Grand Tour*, edited by Paula Findlen, Wendy Wassyng Roworth and Catherine Sama, 275–302. Stanford, CA: Stanford University Press, 2009.
Chichierchia, Luca and Simona Papa. *Storia della farmacia a Napoli. Dalla "Spetiaria"conventuale alle botteghe dell'Ottocento*. Milan: Electa, 1998.
Chiosi, Elvira. '"Humanitates" e scienze. La Reale Accademia napoletana di Ferdinando IV: Storia di un progetto', *Studi Storici* 30 (1989): 435–56.
Chiosi, Elvira. 'Intellectuals and academies'. In *Naples in the Eighteenth Century: The birth and death of a nation state*, edited by Girolamo Imbruglia, 118–34. Cambridge: Cambridge University Press, 2000.
Clericuzio, Antonio. *Elements, Principles and Corpuscles: A study of atomism and chemistry in the seventeenth century*. Dordrecht; London: Springer, 2000.
Cocco, Sean. *Watching Vesuvius: A history of science and culture in early modern Italy*. Chicago, IL: The University of Chicago Press, 2013.
Conforti, Maria. 'Surgery, medicine and natural philosophy in the library of Marco Aurelio Severino (1580–1656)', *Bruniana e Campanelliana* 10 (2004): 283–98.
Conforti, Maria. 'Medicina e università a Napoli tra Sei e Settecento: un progetto di ricerca', *Medicina nei secoli: Journal of History of Medicine and Medical Humanities* 17 (2005): 23–39.
Conforti, Maria. 'Tracce campanelliane nella medicina italiana del Seicento', *Bruniana & Campanelliana* 23 (2017): 587–91.
Conforti, Maria. *Medicina sotto il volcano. Corpi e salute a Napoli in età moderna*. Milan: Editrice Bibliogragica, 2021.
Cunningham, Andrew and Perry Williams, 'Decentering the "big picture": *The Origins of Modern Science* and the modern origins of Science', *The British Journal for the History of Science* 26 (1993): 407–32.
D'Alessio, Silvana. 'On the Neapolitan plague of 1656: Expedients and remedies'. In *Disaster Narratives in Early Modern Naples: Politics, communication and culture*, edited by Domenico Cecere, Chiara De Caprio, Lorenza Gianfrancesco and Pasquale Palmieri, 187–204. Rome: Viella, 2018.
D'Alessio, Silvana. *Masaniello: The life and afterlife of a Neapolitan revolutionary*. Amsterdam: Amsterdam University Press, 2023.
de Divitis, Bianca (ed.) *Companion to the Renaissance in Southern Italy 1350–1600*. Leiden; Boston: Brill, 2022.
del Gaizo, Modestino. 'Contributo allo studio delle fonti della storia della medicina', *Bollettino della Real Accademia Medico-Chirurgica di Napoli* (1891): 76–7.
de Miranda, Girolamo. *Una quiete operosa: Forme e Pratiche dell'Accademia Napoletana degli Oziosi 1611–1645*. Naples: Federiciana Editrice Universitaria, 2000.
De Renzi, Salvatore. *Napoli nell'anno 1656*. Naples: Domenico de Pascale, 1867.
De Renzi, Silvia. 'Seats and series. Dissecting diseases in the seventeenth century'. In *Pathology in Practice: Diseases and dissections in early modern Europe*, edited by Silvia De Renzi, Marco Bresadola and Maria Conforti, 96–115. London; New York: Routledge, 2018.
Donato, Maria Pia (ed.). *Medicine and the Inquisition in the Early Modern World*. Leiden; Boston: Brill, 2019.

Everson, Jane E. 'The melting pot of science and belief: Studying Vesuvius in seventeenth-century Naples', *Renaissance Studies* 26 (2012): 691–727.

Falanga, Agostino. *The venerable Maria Lorenza Longo: Foundress of the Hospital of the Incurables and of the Capuchin Poor Clares of Naples, 1463–1543.* New York; Mahwah, NJ: Paulist Press, 2009.

Ferrone, Vincenzo. *Scienza Natura Religione: Mondo Newtoniano e Cultura Italiana nel Primo Settecento*. Napoli: Jovene Editore, 1982.

Fehrenbach, Frank and Joris van Gastel (eds). *Nature and the Arts in Early Modern Naples.* Berlin; Boston: De Gruyter, 2020.

Findlen, Paula. 'Controlling the experiment: Rhetoric, court patronage and the experimental method of Francesco Redi', *History of Science* 31 (1993): 35–64.

Findlen, Paula. 'Science as a career in Enlightenment Italy: The strategies of Laura Bassi', *Isis* 84 (1993): 441–69.

Findlen, Paula. 'Translating the new science: Women and the circulation of knowledge in Enlightenment Italy', *Configurations* 3 (1995): 167–206.

Findlen, Paula. *Possessing Nature: Museums, collecting, and scientific culture in early modern Italy* (Berkeley, CA; London: University of California Press, 1996).

Findlen, Paula. *Athanasius Kircher: The last man who knew everything*. New York; London: Routledge, 2004.

Fletcher, John Edward. *A Study of the Life and Works of Athanasius Kircher, 'Germanus Incredibilis'*, edited by Elizabeth Fletcher. Leiden; Boston: Brill, 2011.

Fragnito, Gigliola. *La Bibbia al rogo: La censura ecclesiastica e i volgarizzamento della Scrittura (1471–1605)*. Bologna: Il Mulino, 1997.

Fragnito, Gigliola. *Proibito capire: la chiesa e il volgare nella prima età moderna.* Bologna: Il Mulino, 2005.

Fragnito, Gigliola. *Rinascimento perduto: La letteratura italiana sotto gli occhi dei censori (secoli XV–XVII)*. Bologna: Il Mulino, 2019.

Feingold, Mordechai and Victor Navarro-Brotons, *Universities and Science in the Early Modern Period*. Dordrecht: Springer, 2006.

Freedberg, David. *The Eye of the Lynx. Galileo, his friends, and the beginning of Natural History*. Chicago, IL: University of Chicago Press, 2002.

Fulco, Giorgio. 'Per il "Museo" dei Fratelli Della Porta'. In *La 'Meravigliosa' Passione. Studi sul Barocco tra Letteratura e Arte*, 251–325. Naples: Salerno Editrice, 2001.

Fusco, I. *Peste, demografia e fiscalità nel Regno di Napoli nel XVII secolo.* Milan: Franco Angeli, 2007.

Galasso, Giuseppe. 'Scienze, istituzioni, attrezzature scientifiche nella Napoli del Settecento'. In *L'età dei lumi. Studi storici sul Settecento europeo in onore di Franco Venturi*, 193–228. Naples: Jovene, 1985.

Gascoigne, John. 'A reappraisal of the role of the universities in the Scientific Revolution'. In *Reappraisals of the Scientific Revolution*, edited by David Linberg and Robert S. Westman, 207–60. Cambridge: Cambridge University Press, 1990.

Gentilcore, David. 'Contesting illness in early modern Naples: Miracolati, physicians and the congregation of rites', *Past & Present* 148 (1995): 117–48.

Gentilcore, David. *Healers and Healing in Early Modern Italy.* Manchester: Manchester University Press, 1998.

Gianfrancesco, Lorenza. 'Accademie, scienze e celebrazioni a Napoli nel primo Seicento', *Quaderni di Symbolon* 5 (2010): 175–209.

Gianfrancesco, Lorenza. 'From propaganda to science: Looking at the world of academies in early seventeenth-century Naples', *California Italian Studies* 3 (2012): 1–31.

Gianfrancesco, Lorenza. 'Books, gold, and elixir: Alchemy and religious orders in early modern Naples', *Ambix* 65 (2018): 250–74.

Gianfrancesco, Lorenza. 'Dal palazzo alla strada: l'accademia degli Oziosi tra cerimoniale e propaganda (1622–1637)'. In *Il Manso overo de l'amicitia. Vita e cultura di un gentiluomo nella Napoli spagnola (1567–1645)*, edited by Roberto Mondola, Pietro Giulio Riga and Chiara Allocca, 81–97. Naples: Pironti, 2021.

Giustiniani, Lorenzo. *Breve contezza delle Accademie istituite nel Regno di Napoli*. Naples, 1801.

Golinski, Jan. *Making Natural Knowledge: Constructivism and the history of science*. Cambridge: Cambridge University Press, 1998.

Gotor, Miguel. 'La devozione frustrata in onore del teatino Andrea Castanldo Pescara. I "Santificatori" napoletani e la manipolazione dei corpi carismatici nella prima età moderna', *Quaderni Storici* 1 (2021): 143–71.

Guarino, Gabriel. *Representing the King's Splendour: Communication and reception of symbolic forms of power in Viceregal Naples*. Manchester: Manchester University Press, 2010.
Guerra, Corinna. 'If you don't find a good laboratory find a good volcano: Mount Vesuvius as a natural chemical laboratory in eighteenth-century Italy', *Ambix* 62 (2015): 245–65.
Hall, A. Rupert. *The Scientific Revolution, 1500–1800: The formation of the modern scientific attitude*. Boston: Beacon Press, 1966.
Hills, Helen. *The Matter of Miracles: Neapolitan Baroque architecture and sanctity*. Manchester: Manchester University Press, 2016.
Imbruglia, Girolamo. *Naples in the Eighteenth Century: The birth and death of a nation state*. Cambridge: Cambridge University Press, 2000.
Imbruglia, Girolamo. 'Enlightenment in eighteenth-century Naples'. In *Naples in the Eighteenth Century: The birth and death of a nation state*, edited by Girolamo Imbruglia, 70–94. Cambridge: Cambridge University Press, 2000.
Lombardi, Carmela. *Enciclopedia e Letteratura*. Siena: Mediateca del Barocco, 1993.
Logan, Gabriella Berti, 'The desire to contribute: An eighteenth-century Italian woman of science', *The American Historical Review* 99 (1994): 785–812.
Macuglia, Daniele. 'Newtownianism and information control in Rome at the wake of the eighteenth century', *Annals of Science* 77 (2020): 108–26.
Marcus, Hannah. *Forbidden Knowledge: Medicine, science and censorship in early modern Italy*. Chicago; London: University of Chicago Press, 2020.
Marino, John A. *Becoming Neapolitan: Citizen culture in Baroque Naples*. Baltimore: Johns Hopkins University Press, 2011.
Marino, John A. 'Constructing the past of early modern Naples: Sources and historiography'. In *A Companion to Early Modern Naples*, edited by Tommaso Astarita, 11–34. Leiden; Boston: Brill, 2013.
Marini-Bettòlo, Giovanni Battista. *La collaborazione scientifica tra Italia e Spagna per la conoscenza delle risorse naturali del Nuovo Mondo*. Naples: Bibliopolis, 1993.
Maylender, Michele. *Storia delle Accademie d'Italia*. Bologna: Rava, 1926–1930.
Mazzola, R. (ed.). *Le scienze nel Regno di Napoli*. Rome: Aracne, 2009.
Miller, Peter N. *Peiresc's Mediterranean World.* Cambridge, MA; London; Harvard University Press, 2015.
Minieri Riccio, Carlo. 'Cenno storico delle accademie fiorite nella città di Napoli', *Archivio storico per le Province Napoletane* 3 (1879): 161–379.
Mondola, Roberto, Pietro Giulio Riga and Chiara Allocca (eds). *Il Manso overo de l'amicitia. Vita e cultura di un gentiluomo nella Napoli spagnola (1567–1645)*. Naples: Pironti, 2021.
Musi, Aurelio. *La Rivolta di Masaniello nella Scena Politica Barocca.* Naples: Guida, 1989.
Musi, Aurelio. '"Non pigra quies". Il linguaggio politico degli Accademici Oziosi e la Rivolta Napoletana del 1647–48'. In *I Linguaggi politici delle rivoluzioni in Europa*, edited by Eluggero Pii, 84–104. Florence: Leo S. Olschki, 1992.
Musi, Aurelio. 'The Kingdom of Naples in the Spanish imperial system'. In *Italy: Politics, Society and Religion 1500–1700*, edited by Thomas James Dandelet and John A. Marino, 73–97. Leiden; Boston: Brill, 2006.
Musi, Aurelio. *La disciplina del corpo: Le arti mediche e paramediche nel mezzogiorno moderno*. Rome: Guida, 2011.
Nazzaro, A. 'L'eruzione del 1631 ed il collasso del Vesuvio in base all'analisi delle fonti contemporanee', *Rendiconti della Società di Mineralogia e Petrologia* 43 (1989): 725–32.
Nazzaro, A. *Il rischio Vesuvio.* Naples: Guida, 2009.
Nazzaro, A. 'Implicazioni di una ermeneutica delle fonti vesuviane sull'eruzione del 1631: Forma del vulcano e risposta al rischio'. In *Napoli e il Gigante: Il Vesuvio tra immagine, scrittura e memoria*, edited by Rosa Casapullo and Lorenza Gianfrancesco, 137–78. Soveria Mannelli: Rubbettino, 2014.
Novi Chavarria, Elisa. *Accogliere e curare. Ospedali e culture delle nazioni nella Monarchia ispanica (secc. XVI–XVII)*. Rome: Viella, 2020.
Palmieri, Pasquale. 'Two tales of one city: Justice and famine in 1764 Naples'. In *Tales of Two Cities. News, stories and media events in early modern Florence and Naples*, edited by V. Caputo, L. Gianfrancesco and P. Palmieri, 227–44. Rome: Viella, 2023.
Piccari, P. *Giovan Battista Della Porta: Il filosofo, il retore, lo scienziato.* Milan: Franco Angeli, 2007.

Prosperi, Adriano. 'Se gli italiani oggi non sono lettori forti è colpa (anche) della censura di secoli fa.' *Tutto Libri*, 11 June 2022, 16.
Rak, Michele. *A dismisura d'uomo. Feste e spettacoli del Barocco tra Napoli e Roma* Palermo: Duepunti, 2012.
Rak, Michele. *Napoli Civile*. Lecce: Argo, 2021.
Rao, Anna Maria. 'Politica e scienza a Napoli fra Sette e Ottocento'. In *Atti del bicentenario del Real Museo Mineralogico 1801–2001*, edited by Maria Rosaria Ghiara and Carmela Petti, 16–35. Naples: Mediasitalia, 2001.
Rao, Anna Maria. *Editoria e cultura a Napoli nel diciottesimo secolo*. Naples: Liguori, 1998.
Ricciardi, G. P. *Diario del Monte Vesuvio: XVIII–XIX secolo*. Naples: Edizioni Scientifiche e Artistiche, 2009.
Robertson, John. *The Case for the Enlightenment: Scotland and Naples 1680–1760*. Cambridge: Cambridge University Press, 2005.
Rodríguez Fernández, Laura. *El Vesubio en llamas: Un texto napolitano en español sobre la erupción de 1631: Los incendios de la montaña de Soma (Nápoles, 1632)*. Naples: Pironti, 2014.
Scarth, Alwyn. *Vesuvius. A biography*. Princeton, NJ: Princeton University Press, 2009.
Schmitt, Charles B. and Charles Webster, 'Harvey and M. A. Severino: A neglected medical relationship', *Bulletin of the History of Medicine* 45 (1971): 49–75.
Segre, Michael. *In the Wake of Galileo*. New Brunswick, NJ: Rutgers University Press, 1991.
Shapin, Steven. *Never Pure: Historical studies of science as if it was produced by people with bodies, situated in time, space, culture and society, and struggling for credibility and authority*. Baltimore, MD: Johns Hopkins University Press, 2010.
Strocchia, Sharon T. *Forgotten Healers: Women and the pursuit of health in late Renaissance Italy*. Cambridge, MA: Harvard University Press, 2019.
Takayama, Hiroshi. 'Law and monarchy in the south'. In *Italy in the Central Middle Ages*, edited by David Abulafia, 58–81. Oxford: Oxford University Press, 2004.
Neil Tarrant, *Defining Nature's Limits: The Roman Inquisition and the boundaries of science* (Chicago; London: University of Chicago Press, 2022).
Testa, Simone. *Italian Academies and their Network, 1525–1700. From local to global*. Basingstoke; New York: Palgrave Macmillan, 2015.
Trabucco, Oreste. '"L'invenzione della tradizione" nella filosofia meridionale da Telesio agli Investiganti', *Bruniana & Campanelliana* 16 (2010): 177–87.
Trabucco, Oreste. 'Marco Aurelio Severino'. In *Dizionario Biografico degli Italiani*, Vol. 92, (2018), available on *Severino, Marco Aurelio in 'Dizionario Biografico' (treccani.it)*.
Torrini, Maurizio. 'L'Accademia degli Investiganti. Napoli 1663–1670', *Quaderni storici* 48 (1981): 845–83.
Tortora, A., D. Cassano and S. Cocco (eds). *L'Europa moderna e l'antico Vesuvio*. Battipaglia: Laveglia & Carlone, 2017.
Verardi, Donato. *La scienza e i segreti della natura a Napoli nel Rinascimento: La magia naturale di Giovan Battista Della Porta*. Florence: Firenze University Press, 2018.
Vermeir, Koen and Jonathan Regier (eds). *Boundaries, Extents and Circulations: space and spatiality in early modern natural philosophy, studies in history and philosophy of science*. Dordrecht: Springer, 2016.
Viceconte, Milena, Gennaro Schiano and Domenico Cecere (eds). *Heroes in Dark Times: Saints and officials tackling disaster (16th–17th centuries)*. Rome: Viella, 2023.
Villari, Rosario. *Un sogno di libertà: Napoli nel Declino di un Impero, 1585–1648*. Milan: Mondadori, 2012.
Zurlini, Fabiola. 'The uncertainty of medicine: Readings and reactions to Santorio between tradition and reformation (1615–1721)'. In *Santorio Santori and the Emergence of Quantified Medicine, 1614–1790*, edited by Jonathan Barry and Fabrizio Bigotti, 103–13. London: Palgrave Macmillan, 2022.

# Part I

# Neapolitan science and its institutions

# 1

# 'Attenta cura a mille bei secreti di natura': Ferrante Imperato's collection between scientific specimens and natural marvels

Daniela Caracciolo

> One treats of the whole world, and of all the things that are contained therein, I speak of God with all the intelligible things, of heaven, and all its parts, of fire and its nature; of the air and its regions, and of its other functions, of water with all the seas, lakes, rivers and springs, and other things that appertain to it, of the earth with all its lands, and its other forms, of all minerals, such as stones, half-minerals and metals, of all the plants, [such as] grasses as well as trees, of all the animals, whether sanguinous or non sanguinous, whether they dwell in the air, water or on the land, of Man with all the various parts of his body, and of the spirit, together with all of the natural operations of the body, and all the virtues, powers, motions, habits, knowledge and actions of the spirit, of all the sciences [...] as through reading [this text] one shall see.[1]

These words are taken from the *Tipocosmia* (1561), a work structured as a dialogue that takes place over the course of seven days. In this text the scholar and lexographer Alessandro Citolini (*c.*1500–*c.*1582) organised the sum of human knowledge according to a classificatory system based on an architectural model, which he considered capable of reflecting the order of God's creation.[2] His endeavour reflected a fashion for preparing compendia of history, geography and philosophy and collections of poetry, rhetoric and science, all of which were modelled on the genre of the 'encyclopedia', of the '[universal] systems' and of the 'universal sciences', and disseminated in various venues.[3] With his directory of all science, arts and crafts, Citolini was seeking to create a 'virtual world composed only of words',[4] a structure capable of containing the 'places of the memory'[5] from the 'intelligible world', comprising the words relating to the angels and blessed souls, to the 'sensible world'. The sensible world was divided into 'celestial' and 'elementary' regions, with the latter made up of the four elements, in either a pure or mixed state. It included all animate and inanimate beings, for instance, gems, stones, trees, birds, fish, animals, humans and sciences.

Citolini illustrated his ambition in the text when, on the seventh day of dialogue described in the *Tipocosmia*, Count Collaltino di Collato, proprietor of the Venetian villa in it was set, took his guests to a room where he showed them a globe of great size 'into which one may enter'.[6] Upon entering the globe, his stupefied visitors saw presented before them 'heaven and in the middle the earth'.[7] This device made the classificatory order available to all. In short, the Count shows them all of those 'things ordered in a manner more pleasing to the corporeal eye than to the mind's eye'.[8] Prior to their entrance through this contraption, the visitors went to the villa's studio where the Count 'opens a very large book and begins to show them this his new and artificial world'.[9] He would not have been able to explain his ideas completely to his guests without the indispensable aid of a book whose structure made it possible to arrive at complete comprehension of nature: the order of the world is visible and can be expressed through places and images.[10]

The topos of the book of nature was more than a mere metaphor.[11] It functioned not only as a literary device, but also constituted a genuine tool for the interpretation of reality, in line with the manner in which the physical world was comprehended and interpreted in the sixteenth century.[12] Reflecting on this interaction (nature/world = book) enables the modern scholar to understand better the epistemological disposition of early modern natural sciences and the operative methods used by their practitioners to read and decipher nature. The sixteenth century, as is well known, was an age of new and great discoveries in the fields of medicine, pharmacology, botany, mineralogy and zoology. It boasted a large array of scientists engaged in the hard work of reconciling classical science – received and reinvigorated by the editing, correction, translation and printing of the works produced by such ancient authorities as Pliny, Theophrastus, Dioscorides and Vitruvius[13] – and the early modern practice of investigating natural objects.[14]

The scientific culture of Naples from the end of the sixteenth century to the opening decades of the following century was similarly animated by an interaction between intellectual experience and ancient authority.[15] Despite some substantial differences in their individual approaches, the various protagonists engaged in this endeavour all shared a renewed interest in the investigation of nature. Giuseppe Galasso has effectively identified two poles between which this Neapolitan culture of investigating nature oscillated. The first was represented by a purely speculative approach, epitomised by the natural philosophy of Bernardino Telesio (1509–1588).[16] The second approach, more techno-scientifically oriented, was reflected in the work of Ferrante Imperato (1525–1615) and Giambattista Della Porta (1535–1615).[17]

## Ferrante Imperato and the secrets of nature

Much has been said about the work and the methods of the apothecary Ferrante Imperato.[18] To make knowledge about the natural world, he drew on the authority of the ancients, whose works he often translated with help,[19] adding personal comments derived from his empirical observations. He acknowledged that much still remained to be known. At the beginning of the first book of *Historia naturale* (Napoli, 1599), he observed that 'considering on the other hand the immensity of the subject, the grandeur of nature, that although attempts have been made to explain it, it nevertheless remains neither wholly or even mostly expressed, there is therefore left to scholars a wide array of new things [to investigate]'.[20] In this manner, Imperato signalled his intention to occupy himself with 'things that either remain unknown due to the antiquity of the authors, and the loss of facility in their language, or even those things which they have omitted, or in fact discussed imperfectly or obscurely'.[21]

Imperato supplemented the harvest of information that he had gathered from the ancient writers with direct empirical observations of the natural order and accurate representations of the natural objects found in his *Historia*. The work was in fact intended to make manifest Imperato's 'understanding of the secrets of nature',[22] that is, to disclose the occult qualities inherent in minerals, animals and plants. He was therefore working within the 'books of secrets' tradition, which used experiments to disclose the secret powers of nature. In this tradition, apothecaries and alchemists provided proofs of their bold preparations.[23] In a similar manner, Ferrante compiled an imposing natural encyclopedia that combined the authority of the classics with an empirical explanation of processes and instruments.

The approach outlined in Ferrante's encyclopedia showed that he possessed an experimental attitude equal to that of Della Porta, who in a passage from his *Magia naturalis* had declared:

> It is now necessary to be very aware of the nature of simples, that is, not to be a mere herbalist but instead a great investigator of plants. There is nothing more unbecoming to an artisan than not to understand properly the tools that he uses. So too it is necessary [for a magus] to have an exact understanding of metals, minerals, jewels and stones.[24]

For Della Porta, so too for Imperato, the investigation of empirical reality was to be achieved through the comprehension of the 'secrets of nature'.[25] This shared approach to the investigation of nature is attested in the well-known account of a trip to Naples sent by the young Federico Cesi (1585–1630) to his friend and companion Francesco Stelluti (1577–1652) in the spring of 1604. In this letter, Cesi recalled:

> I have conversed in this manner with Signore Battista Porta e il Signore Ferrante Imperato, who are both very good friends to me and the Lincei, and in truth they are miracles of nature, and one could say much more about them; I have learned a great deal from my discussions with them and I have, and *I will have most beautiful secrets*, and I have passed a good deal of my time in Naples with these two with great benefit.[26]

These sources demonstrate that Imperato was part of the philosophical, medical and naturalistic milieu of late sixteenth-century Naples. This community drew inspiration from the activities of the Bolognese scholar Leonardo Fioravanti (1517–1588), who in 1549 had arrived in the city where he contributed to the diffusion of Paracelsian ideas,[27] and from those of the Viterbese polymath Girolamo Ruscelli (1518–1566), who resided in the city from 1542 until 1552. While in Naples, Ruscelli was in the service of Alfonso D'Avalos (1502–1546) and acted as a sponsor of the Accademia dei Secreti.[28] Imperato also formed close friendships with Bartolomeo Maranta (1500–1571), a pupil of Luca Ghini (1480–1556), who was famous for developing theriac,[29] and Colantonio Stigliola (1546–1623), who was interested in botany, mathematics and chemistry, a friend of Maranta and later a follower of the ideas of Nicolaus Copernicus (1473–1543) and Giordano Bruno (1548–1600).[30] Imperato also became friends with the botanist Fabio Colonna (1567–1640). As is well known, Stigliola and Colonna were mentioned in the dedicatory letter of *Historia naturale*.[31]

## *Historia naturale*

*Historia naturale* was composed of 28 books, divided in turn into 622 chapters. The work was prefaced by two dedicatory letters. The first, dated 1 November 1599, was written by Ferrante Imperato's son Francesco and dedicated to Giovanni di Velasco – the Chief Constable of Castile, the governor of Milan and the son-in-law of the Spanish viceroy of Naples, Pedro Téllez-Girón, from 1582 to 1586. The dedication of Imperato's work was intended to earn Giovanni di Velasco's benevolence. As Imperato mentioned in *Historia*, Di Velasco visited his museum of natural objects. The second dedicatory letter, addressed to the 'Learned Reader' (*Studioso lettore*), was written by Ferrante Imperato himself. It is of great interest, providing valuable insight into his encyclopedic approach to making natural knowledge. It illustrated the arguments used in, and the ultimate purpose of, his book by stating that it was 'arranged according to the differences and order of things'.[32] Indeed, this letter provides a key to understanding his scientific project. It explained how Imperato arranged the content of the treatise according to two criteria: first, the partitioning of all things into four elements; second, that of the classification according to the three kingdoms of nature, orders that appear perfectly similar:

> Whereupon starting from Earth, and its differences and virtues, we pass thence to the element water, to the various conditions and impressions that it receives, and from water we pass to Air and the substances generated in it. The consideration of the operations of hot and cold follows from this, to which one can add the value of the furnaces, with which we judge the various operations of fire. From this follows the consideration of bodies from the first generation: I speak of salts and the nutrients of the soil and then of the considerations in Naples of the types of metals and stones, together with good methods to perfect them and at the end [of this book] the examination of plants as well as terrestrial and marine creatures.[33]

Imperato further elaborated his encyclopedic programme in Book 21, dedicated to 'Philosophical Medicine' (*Medicina philosophica*) – which, as Massimo Varro has observed, is useful for reconstructing his cosmological vision.[34] In this book, Imperato distinguished between what he termed the 'greater work' ('opera maggiore') or macrocosm and the 'lesser work' ('opera minore') or microcosm. He divided both into

16 levels, proceeding from the invisible to the visible: the 'universal matter contains in itself the four elements',[35] while the minor, that is the microcosm, contains 'the plant, animal and mineral [ether]'.[36]

The other 28 books of the treatise followed a similar sequence. They were arranged as follows: Books 1–5, The Earths, their Diverse uses and natures; Books 6–7, Waters, their differences and origins; Books 8–9, The Element of the Air, bodies that in air acquire solidity; Books 10–11, Effects of Fire and of Light. These were followed by Book 12, Generation of Fire, and the various operations of hot and cold. The remainder of the work considered the world of mixed things with their respective categories. Books 13–21 described Metals, Mineral and Stones. Book 27 explored Plants and Book 28 Animals. In this manner, Imperato described an image of the world, with all its divisions and distinctions.[37]

The encyclopedic system that the Ferrante Imperato used to conduct his research also manifested itself in a visual disposition that privileged observation and graphic representation. In the dedicatory letter, he remarked:

> We have, in addition to that aforementioned scholarship, added representations of things that have a particular appearance, which have not been published by others, so that it was possible for us to assist the reader's comprehension.[38]

Imperato, the faithful follower of nature, therefore elaborated a practice of writing based upon the close relations between knowledge, perceptions of natural reality and its visual representation. He fixed upon the printed page previously unknown aspects of natural reality, which he had derived from the observation of specimens and the elaboration of information gleaned from naturalists and explorers.[39] Expanding on this latter point, he noted

> I say this, that is to say, I confess, that our studies, and the things that we have discussed have made progress with the help of friends, who have gathered as supporters, helping me to procure a supply of things that have come to me from diverse parts of the world.[40]

The need to make visible a hidden nature necessitated the use of a hitherto unknown means to see it, write about it and represent it:[41] this could only be achieved if nature was observed through the mediation of the figurative arts.[42] The eye must grasp characteristics and analogies;

it must organise new knowledge within a coherent system to be bought back to unity. Meanwhile the acts of gathering artefacts, comparing and describing them are the fundamental means to produce knowledge.[43] Illustration, by virtue of the opportunity it affords to express precise and significant information, offers a considerable aid to those disciplines that derive some of their fundamental theoretical presuppositions from the direct observation of natural reality.[44] All who were daily confronted with the natural world were obliged to take account of illustration's importance, realising that written descriptions were no longer sufficient to be certain that they had correctly identified the object before them. In this manner, information derived from the texts of the ancient authors was brought into question.

*Historia naturale* contains 119 woodcuts, all produced by an unknown artist or artists. They include images of plants, animals and minerals, presumably all drawn from life, which were accompanied by explanatory captions. These illustrations were often incorporated into the text, but in some cases occupied an entire page.[45] Imperato thus founded his study of the 'Great Book of Nature' on the use of images, which he employed to illustrate and clarify the material presented in the body of the text.

The relationship between the natural world, the process of collection and the illustrated book thus seems clear: the detailed representation of an object, whether a plant, an animal, a crystal, a shell or a fossil, is an icon of the things that exist in the world reproduced in the space of a collection. Assuming an intermediate position between the object itself and that which can be discovered about it in written texts, the image constitutes an alternative means of documenting, verifying and knowing – and therefore of appropriating – the studied specimen.[46] This method was therefore based on the comparison of two levels of reality: the first was the textual reality and the second the physical reality of objects rendered as images. The naturalist described, ordered and explained objects, in an attempt to provide as much information as possible about each and every one. It is here that Imperato revealed his ambition: departing from the description of a single specimen to arrive at a complete inventory of natural knowledge.[47]

Imperato's writings are the fruit of a process of encyclopedic compilation, which was based on a process of accumulating knowledge. It proceeded by presenting a series of exemplars designed to guide the reader's comprehension. To take one example, Imperato noted that

> From the earth there are the gems, and more noble species of stone, and some soils are marked either by their colour or by their

clarity and smoothness, or their density, or by another of their properties.[48]

From the study of the generation of stones, minerals and metals, for example, there emerges quite a compact picture. Imperato examines the techniques of extracting and working metals and minerals, by studying their virtues, properties, appearances and external characteristics. His discussion unfolds principally through morphological observations.

Following the medical-practical sequence of investigation first propounded by Georgius Agricola (1494–1555) in his work *De re metallica*, Imperato was interested above all in the specimen's external form (its superficial appearance, colour, lustre and transparency), followed by its hardness, cleavage, weight and its uses in the fields of mineralogical research and therapeutics.[49] 'The Great Book of Nature', in which the rules and codes of tradition were reflected, could not be separated from the ceaseless observation of natural reality that the 'moderns' not only undertook but also considered necessary. Ferrante was, for this reason, part of a scientific culture that was constantly being recomposed, corrected and enriched. He provided a systematisation similar to that already devised by Michele Mercati (1541–1593), the Prefect of the Vatican Botanic Gardens, with whom Ferrante corresponded regularly and exchanged specimens.[50] We can briefly divide these specimens into the following categories of earths (clays), salts, allums, sulphides and sulphates, sulphur, bitumen, carbons, zinc, tuff, pumice, chalk, fossils, shells, stones contained in animals, marbles, crystals and ornamental gems.

During the alchemical process of rendering the natural world in prose, the wealth of information increased greatly. In order to incorporate the information received from tradition within a framework capable of assessing it anew, Imperato patiently gathered and displayed that information in a directory, which would function as a kind of 'guide'. *Historia naturale* was presented as a sequence of descriptions arranged in a manner that we might characterise as being based upon textual elaboration of the content: translated citations from ancient texts were followed by personal comments derived from experiments and direct observations, and then by specific insights to be utilised in the fields of medicine and pharmacy, as well as by artisans and practitioners such as farmers, metallurgists and builders. The principle underlying the arrangement of the text of the *Historia naturale* was a kind of textual *continuum* that produced, at times, a certain sense of an accumulation, even an excess of information. Imperato found himself working within a scholarly tradition constituted by a series of texts that presented diverse languages, codes and interpretations – derived from both the works of Greco-Latin antiquity and the zoological, botanical and mineralogical treatises of the sixteenth century.[51]

Imperato's attitudes to classical culture were ambivalent. On the one hand, he acknowledged the importance of ancient knowledge that had been rediscovered thanks to philological work of the Renaissance;[52] on the other, he recognised the numerous cracks that were appearing in the edifice of ancient scholarship, which had for centuries provided the basis for knowledge. His cultural foundations can thus be ascribed to the inheritance of Renaissance humanism: information was selected and collected from ancient texts, which were read and analysed with the intention of joining these monuments of the past to a need for novelty.[53] He nevertheless sought to replace a strenuous defence of received knowledge with a renewed culture, one capable of maintaining itself when confronted with empirical reality. The *Historia naturale* derived its essence from this imperative, and its form assumed an extraordinarily complex character. Information derived from the ancients was compared to newly acquired knowledge and corrected or revised accordingly.

Imperato's awareness of the vast scope of the material with which he was dealing induced him to avail himself of the advice and opinions of numerous others. In the treatise, he proposed a method based upon a form of active collaboration. In turn, this method would gradually allow those Neapolitan researchers who wished actively to participate in the discussions that distinguished the most advanced European cultural circles to adopt hitherto unknown methods of experimental research.[54] In seeking to legitimise his approach, Imperato did not hesitate to display his familiarity with the major naturalists of his age:[55]

> I say this, that is to say, I confess, that our studies and the things of which we treat, have made progress with the help of some friends, who are either assisting as agents who procure for me a supply of things from diverse parts of the world, or they have been companions and supporters in my labours: if there is something of value in our treatise, all of them must receive some of the praise.[56]

The text of *Historia naturale* attests to collaboration between scientists, who recognised the importance of sharing the results of empirical investigation as a tool for exchanging ideas, and a sign of their strong interest in understanding the issues involved in any speculation concerning the 'secrets of nature'.

*Historia naturale* is the product of what we might term a collective endeavour.[57] In the work's dedicatory letter Imperato's son Francesco[58] reiterated his father's declared intention to legitimate his actions in the sphere of a precise professional category, namely that of the pharmacist, when he noted:

> In our age, inclined more often to idleness than to curiosity, many of the noblest minds in several European cities have gathered in their writings and images of things drawn from life, all that it has been possible by human endeavour to illustrate such a noble profession.[59]

The fecund cultural atmosphere from which Imperato's writing emerged derived from a composite system of knowledge that directed itself towards the study of nature. This would converge only in part with the experimental method developed by Galileo Galilei (1564–1642).[60] When presenting his work, Imperato did not hesitate to associate it with the anti-Aristotelianism that characterised a large part of Neapolitan culture during the sixteenth century.[61] A breach had opened in the blind adhesion to the Aristotelian assumptions, because 'it remains to remind you of that which Aristotle and other *naïve* writers did not cease to acknowledge, that the human sciences grow from communication between one another'.[62]

## Imperato's Museum

In his enyclopedic work *Piazza di tutte le professioni del mondo* (Venezia, 1585), Tommaso Garzoni offered a systematic account of 150 different professions, each presented in the same manner. Every entry provided a history of relevant anecdotes concerning examples of exemplary figures within, as well as an account of the instruments used in – and other information relating to – the profession in question.[63] In his entry on apothecaries, Garzoni furnished a precise picture of how an apothecary's shop must have appeared in the second half of the sixteenth century. He noted

> The apothecaries today sell in their shops all the spices which come from the Levant to our lands, and for which they are sought out and suitably esteemed by all. These Apothecaries, that is aromatics, are called servants of the physicians, because they are those that

> draw together simples; they make syrups, unguents, decoctions, selections of compound drugs, *violebi*, *triscisi*, pills, drinks and other similar things relevant to medicine.[64]

Garzoni further added

> Above all, it is the apothecary's task to gather, dry, store and preserve plants, or other things and thus squeeze out the juice, place it in infusions, make decoctions or similar boiled products, to mix, to foam, to season, to buy medicines and similar things.[65]

He believed that the apothecaries shop was an important centre for scientific activity. Investigation, observation and laboratory experimentation took place in the pharmacy. It was a place of research and discovery, where it was possible to study the curative properties of plants, animals and herbs, all collected, displayed and suitably preserved.[66] Here, in a city such as Naples,

> many skilled/ and loyal apothecaries /[...]/ make infusions/ pills, alongside a thousand other things, all wondrous; / preparing here also comfits, syrupy conserves, / with which they purge fevers and the intestines.[67]

According to Fioravanti, Ferrante Imperato distinguished himself within this community of apothecaries as a:

> most well read and learned man [...] who has a most beautiful palazzo in the Santa Chiara area of Naples, where he maintains a most superb workshop of aromatic spices, and in this palazzo he has a studio with so many diverse things that it is a wonder of the world, where one can see so many herbs, stones, animal that they appear infinite.[68]

Was Ferrante's studio a *Wunderkammer* or an archetype of the modern natural history museum? Thanks to Enrica Stendardo's recovery of two manuscripts conserved in the Biblioteca Ambrosiana in Milan, it is possible to consider this question.[69] When combined with Ferrante Imperato's references to his collection in *Historia naturale*, the documents discovered by Stendardo allow us to reconstruct its nature. The collection was notable for the extraordinary quantity of stones and metals that it contained, some of which were also illustrated in

*Historia naturale*.[70] One can also grasp the character of the collection by browsing *Discorso intorno alle diverse cose naturali*. This slim volume was published in 1628 by Ferrante's son Francesco, who from the middle of the 1620s dedicated himself to publicising, enlarging and re-ordering his father's museum.[71]

Ferrante Imperato dedicated himself to gathering, cataloguing and collecting everything that would be useful to his study of the natural sciences and his pharmaceutical practice.[72] The natural orientation of Imperato's collection is revealed in a letter addressed to Ipolito Agostino and dated 25 September 1597, in which he stated that 'my theatre of nature consists of nothing other than natural things, such as minerals, animals and plants; it is a collection that now numbers several thousand artefacts, such that in my judgement it seems a wonder'.[73] As is well known, his collection is illustrated in the celebrated synoptic woodcut placed after the typographic frontispiece of the *Historia naturale*, and it was described as follows in the accompanying caption:

> Depiction of Ferrante Imperato's museum, which consists of natural plants artificially preserved and attached to the pages of books; of representations of terrestrial, flying and aquatic creatures; of classified gems, marbles and other types of stones, minerals and metals; of preserved foreign seeds and leaves; and of liquors from different lands and plants.

This image represents the oldest depiction of a natural museum. It is from this fact that it derives its importance, in so far as it allows us to understand the form and organisational principles of sixteenth-century scientific cabinets, which we can define as true and proper museum-laboratories.[74]

The section of *Historia naturale* dedicated to the discussion of the mineral kingdom runs from metals to stones. It presents a rich and variegated whole, which includes extraordinary and mysterious pieces, such as crystals, stones and a great quantity of fossils, the latter variously catalogued and described in Book 24.[75] The discussion of fossils as natural or organic entities remained in the future; nevertheless, in his work Imperato suggested that some stones 'can take their form from the petrification of things'.[76] He noted, for example, the case of

> shells, the form of which is very frequently found in stones; at times, they are from the hollow impression that the shell makes by means of its convex form, at other times, they are from the refilling of the

> convex hollow left by those shells, at yet other times, they are made through changes in the same substances transformed into stone.[77]

The deduction is clever and it complements the idea of a process completed in the 'course of time' ('corso di tempo').[78] Ferrante did not elaborate a systematic theory of the process of fossilisation, however, nor did he distance himself from a belief that a vegetative virtue or spirit was the cause of the generation of stones and metals.[79]

The task of presenting a lithological theory would fall to Ferrante's pupil, Fabio Colonna. He accomplished this in a 1606 work titled *Ekphrasism*, which was dedicated to zoological observations. In Chapter 21 he provided a list of descriptions of seashell fossils. Here he proposed that the elements earth and water were the 'material causes' and heat and coldness the 'efficient causes' of the process of fossilisation.[80] Also following in the footsteps of Imperato Ferrante, his son Francesco published *De fossilibus opusculum* (1610), in which, with regard to glossopetrae and bucardia, he continued to defend the existence of a 'vegetative spirit'.[81] In the first book of his later work *Discorso*, entitled *Intorno a gli animali, e altre cose convertite in pietre (Regarding animals and other petrified things)*, Francesco Imperato included the usual morpho-structural descriptions of fossils,[82] but alongside them there also appear observations of another kind. He noted, for example, that:

> among the many species of stones, there are those that comprise all things, that through the agent that is the petrifying power, proportionally receiving the density both on the earth and in the sea. This is particularly true for porous stones, which are more inclined to acquire such density.[83]

Here he discussed a rather diffuse theory of petrification, one that links the formation of fossils with a combination of exhalations emanating from the earth and heavens. During this encounter, the emanations provoke the coagulation of the materials that are found in the upper layers of the Earth's crust. In turn, these materials assume different forms, transforming themselves into stones, crystals, minerals or fossils.[84]

Francesco Imperato also continued to be moved by a curiosity for peculiar objects, jotting down such contributions as 'the sight of petrified sea worms is a wonder'.[85] He considered the forms of fossils to be ambiguous, writing of minerals with the appearance of animals or plants formed by 'fluid humours of varied sorts'[86] that 'take the form of the hollow place, in which they find themselves; in this manner, also soft earth is a petrifying power that takes the form of shells, woods, horns, animals and other things, and it converts itself into rock or another types of stone.'[87]

In 1637 the naturalist and member of the Lincean academy Francesco Stelluti, writing in his treatise on fossilised wood, continued to describe fossils as marvels; he defined them as a 'rare and wonderous part of nature'.[88] Stelluti wrote of stones that were special, singular and strange, which he held to be a demonstration of how nature diverted herself with the production of 'so many varieties of object, of such diverse qualities so marvellous and with such abilities, without ever using a chisel, brush or any other instrument; and therefore all the more are her works worthy of wonderment'.[89]

Did observations of this kind foreshadow modern science? On the one hand, Ferrante Imperato's observations were, one might say, of a morpho-structural nature. The Neapolitan scientist, in the guise of the petrographer, described the most relevant and significant morphological examples from his collection and examined the natural objects through the use of analogies with living organisms, whether animal or plant.[90] On the other, the principle of similitude ('similitudo') nevertheless guided and organised all of his knowledge.[91] For example, he described the form of stone as like that of 'a vegetation of stone formed of chalk in the shape of a branch'[92] and another as an 'oyster fossil, the vegetation of which imitates seashells'.[93] Imperato also shows his amazement in observing 'a haematite stone composed of multiple little blocks that resemble the brain of animals'.[94] If in stones, calcareous incrustations, fossils and shells, nature showed herself to be a sculptor,[95] in other artefacts she revealed herself as a painter. This recalled an idea proposed by Leon Battista Alberti (1404–1472), who had observed that 'nature herself appears to delight in painting, which we see in the patterns of marble when she often paints little centaurs and the faces and crests of kings'.[96]

Ferrante Imperato was inclined to regard phytomorphic and zoomorphic images that formed by chance in minerals and in stones as works conceived and accomplished by nature.[97] He believed that nature had the capacity to imprint recognisable images on stones: in *Historia naturale* they are thus described as a 'stone naturally painted with the figure of woods'[98] and 'a figurative stone shaped like corn'.[99] If it is true, as Enrica Stenardo has suggested, that Imperato's analysis has moved from the level of 'lusus naturae' to that of physical and material evidence,[100] it is also true that the 'pietra imboscata' constitutes a popular and frequent category in natural collections.[101]

The observations that Francesco Imperato set out in the *Discorsi* should also be understood in this manner. He observed, for instance, that

> to arrive at a compendium of descriptions of stones, one cannot deny that Nature produces many of those [stones], that represent the form and effigy of flying and aquatic animals, of plants and trees and similar things.[102]

Among the specimens in his collection, Imperato drew attention to 'a stone in which there appears naturally sculpted in relief, many signs, which represent human hands and feet'.[103] The principle underlying his analysis is still that of the system of resemblance among beings, derived from a system of signs furnished by nature and deciphered by humans.[104] There are numerous pieces whose description turns on this idea. In each case, one can note that the emerging impression is one of surprise, aroused by the fact that nature should choose to 'amuse' herself by deceiving the human eye with her tricks.

Many of the matters that Imperato and his son were investigating were already well researched: from the discussions relating to the toad stone[105] and the bezoar to those relating to corals.[106] It was a widely held opinion that the latter was a marine plant that acquired hardness through a process of solidification.[107] Recalling the work of Pliny,[108] Ferrante Imperato affirmed that

> corals are numbered among the plants that have life in the humours: when they are found in their natural environment, water, they remain soft, but when they are placed in the air their form hardens.[109]

Water and air: these are the two spheres to which these 'stoney plants'[110] belong. Imperato divided them into two classes, those with petrified arms and those with branches, then subdivided them further into red, black, striped, starry. Imperato was seduced by this curious natural element; he described its ornamental use as amulets[111] before proceeding to examine other types of stone. Despite the persistence of ancient beliefs, Imperato correctly recognised and described the appearance of the order of madrepore with its calcareous skeleton:

> Among those aforementioned stones are others that are recognised under the species of stony plants, as the corals are, and the species with pores, others of animal origin, like the madrepores, which

grow continually although the oldest part hardens and becomes coral.[112]

Taking into account such a conception of nature that reflected itself in itself, one should not be surprised by the considerable quantity of fossils present in the collection. In it wondrous forms are understood as works produced by nature who 'amused' herself by imitating fish, shells and plants.

Developing this theme, let us now return to the work of Francesco Imperato, who wrote:

> therefore, among the other things that one must admire is the sight of animals, whether of the earth, the sky or the water, plants, stones and minerals produced conforms to their characteristics, and also many things similarly generated by nature [that are] similar to the others, but different from their species.[113]

In the moment of drawing up this catalogue of the animal and vegetable worlds, he tied his research to all those things that fell outside the rules, demonstrating how curiosity distinguished itself as one of the supporting elements of scientific knowledge. In the company of Fabio Colonna, he was able to observe:

> A root of the said plant [mandrake], which not only displays the well-formed image of a body, legs, belly and hind quarters of a man, but also the two raised arms, above which there were the flowers, which represent the form of man, transformed into a tree, in the manner that one sees painted, and in more places sculpted, according to the inventions of the poets.[114]

Are we speaking of 'scientific pedantry'?[115] Sustained by the concept of resemblance, the concept of a 'joke of nature' could include various typologies of 'naturalia'; it is certain that the scientists of the late Renaissance, in their search for a taxonomy of knowledge capable of embracing each part of nature, privileged the irregular and the monstrous. Nourished by shared experience, Neapolitan scientists were inspired by a common method of interpreting and perceiving experimental data derived from the direct observation of nature. Although this novel method formed part of a polemical rupture with tradition, its imposition did not prevent its practitioners from maintaining a taste for 'lusus naturae', which they continued to use as a key to decipher the great book of nature. As Paula Findlen has already demonstrated, some conceptual categories (curiosity and wonder) co-existed without conflict with experimental data, the basic assumption and point of departure of the Scientific Revolution.[116]

Giuseppe Olmi has observed that to early moderns nature appeared as 'a boundless territory, an enormous reservoir of anomalies and oddities':[117] to the eyes of Imperato's contemporaries, nature retained marvellous aspects that could be discovered in its artefacts. Knowledge of nature, now ever more accessible to the human mind thanks to the fashion for producing encyclopedic knowledge inaugurated by recent geographic discoveries, prompted curiosity about monsters. This is demonstrated in the following passage of Francesco Imperato's *Discorsi* in which he wrote:

> I will add that nature can produce not only monstrous roots, but also monstrous fruits; this can be demonstrated by examining some species of cucurbits which take the shape of the container in which they are placed; some take the shape of a flask. What can one say about citrons, which arrive in Naples from the coast of Calabria and which often have monstrous shapes [...]; or about some monstrous plants such as the parietaria or the anchusa which is similar to a cornucopia. [The anchusa] has been dried and conserved in our museum and represented and described by my father.[118]

The listed monstrosities included animals 'that at times imitate in parts human form, with more of less ordinary limbs, which are often deformed'.[119] Francesco Imperato developed a kind of theory of 'biological monsters',[120] identifying the principle causes of their malformation. He variously attributed it to the lack or abundance of seed, a defect in the quality or in the place of generation and the influence of the heavenly bodies. His contention that the mother's imagination – viewed as having the power of generation – also contributed to shaping the form of the foetus. This demonstrates Francesco Imperato's inability to provide a scientific explanation of bizarre and natural phenomena that defied simple classification.[121]

The 'monsters' in Imperato's collection included a lamb with one eye, two bodies and tails and eight legs, a calf with two heads and a human body with one head and two torsos.[122] It is necessary to understand Imperato's interest in monsters in the context of a wider undertaking: monstrosity evoked fascination because through the study of anomalies and variations that fell outside the parameters of normality it was possible to reveal the mysteries of the creation.[123] This wide-ranging interest in all things that fell outside of the usual norms determined the presence of innumerable images of monstrous human beings, animals and vegetables in the iconographic apparatus contained in contemporary cosmographies, large zoological encyclopedias and medical treatises.

Among the diverse entries one finds sea monsters including whales, orcas, narwhals and cephalopods. There were also terrestrial monsters including animals with a human appearance, human monsters and 'mixed' monsters such as humans covered entirely with fur, sylvan and wild humans or people of restricted growth; prodigies, satyrs and centaurs; monstrous plants. Finally, Imperato also included serpents and dragons within this category, thereby demonstrating how in this literature the boundaries between fantasy and reality were blurred.[124] The first teratology therefore sought to describe and catalogue the monstrous, a study that included human deformities and naturally produced rarities. From these emerged the idea that nature was 'the mother of monsters', that is, of a multitude of oddities and eccentricities.[125]

## Conclusion

The scholars who were the first to dedicate themselves to the direct study of nature in the early modern period, including its most diverse manifestations, could do nothing other than accommodate all things that fell outside the normal order of things. In so doing they formulated a new body of knowledge that was in part consolidated by tradition but nonetheless subject to the scrutiny of experience. As Francesco Imperato observed:

> So delightful the knowledge of the secrets of nature [...] that has given many probable parts, parts hidden to many irrational animals, and to the parts of [them], to the plants in their roots, stem, bark, blossom, flowers and fruit; to the stones and to the minerals, that my father, Ferrante Imperato, has discussed with enthusiasm, not only has it consumed years and decades, but now almost the entire course of his life.[126]

Despite Francesco Imperato's eulogy to his father Ferrante, the problem of interpretation remains open. His natural researches were also directed towards the resolution of 'marvellous astonishment', most often rendered in observations such as this one of 'an account of a viper giving birth in a box witnessed by [Ferrante Imperato]. He killed the vipers and their mother and preserved them alongside some other things that he retained in his *studiolo*'.[127] In a similar manner Giambattista Della Porta, an emblematic figure in the efforts to comprehend the secrets of nature through the practice of natural magic, medicine and the operative arts more broadly,[128] had occasion to observe:

> A little boy from whose chest there emerged another boy, entire except for a head, because the neck of the second boy emerged from the chest of the first and they were joined in the belly, and I have seen countless other examples of such things, those with four hands, with four feet, with seven fingers per hand, [and many varieties of feet]; if you were to ask me to relate all these things, it would take a long time.[129]

Della Porta, who dedicated himself to natural magic in order to construct an ordered system of knowledge, performed experiments relating to the various 'types of monstrous chickens that I can always procure, for my pleasure, for they are born in my houseshold',[130] with the aim of discovering the secrets that lay behind the generation of similar creatures. Della Porta was closely 'tied to an idea of science that divulged secrets or marvels'.[131] By making his own observations derived from experimental verification, 'he continuously transcended the limits of a single discipline or experiment in his pursuit of the unique, the portentous or the occult',[132] convinced as he was that only the replication of proof could guarantee the correct comprehension of nature.

'[Ferrante Imperato] has made a treasury in my homeland [Naples], / by paying close attention / to the thousands of beautiful secrets of nature':[133] with these verses Giovan Battista Del Tufo summarised the intentions and emphasised the distinctive character of Ferrante's natural collection. For him it constituted a 'true and exact microcosm of nature, not entirely freed from the secrets of nature'. The sheer number and variety of the assembled objects brought together in a single room demonstrated the exceptional nature and rarity of these curiosities; moreover, the beauty, exoticism and medicinal properties of many of the exhibits displayed the real and exact 'miracle of the world' ('miracoli del mondo').[134] An embalmed pigmy, described by Francesco Imperato as a being of a height of 'less than a hand-breadth, having not yet reached its ordinary size',[135] which he paired well with 'two giant's teeth, and its petrified ulna, vertebra, which are very hard, are conserved in our museum',[136] reflected the concept – already identified in Imperato's museum by Adalgisa Lugli – of 'scarto di scala', that is wonders or forms that fall outside of expected proportions.[137] Del Tufo, making ample use of a style rooted in the rhetoric of the marvellous, conveys the sense of bewilderment that overwhelmed the visitor standing before a collection of objects so varied and heterogeneous, drawn from distant and unknown lands:

> Here one may see in a hidden space
> Chosen by prudent hand
> That which come from the east or the west:
> Here vipers, asps, water snakes and cerastes,
> Hydras, dragons, *faree*, toads and chelydridae,
> And hundreds of other [species] to see [such as] the curious
> Mountainous and venomous animals;
> So that if you admired
> The simples, the plants and the roots,
> That he [Ferrante Imperato] has often acquired
> From the Arabian Lands,
> [species] which are all well maintained and carefully displayed in locked cabinets,
> You [visitors] would be astonished and amazed.[138]

## Notes

1 'Trattasi del Mondo tutto, e di tutte quante le cose, che in esso si contengono, dico di Dio con tutte le cose intelligibili, del cielo, e di tutte le parti sue, del fuoco e sua natura; de l'aria, e sue regioni, ed altre sue pertinenze, de l'acqua con tutti i mari, e laghi, e fiumi, e fonti, ed altre cose a ciò spettanti, de la terra con tutti i paesi suoi, ed altre sue condizioni, di tutti i minerali, così pietre, come mezzi minerali, e metalli, di tutte le piante, così erbe, come arbori, di tutti gli animali, e sanguinei, e non sanguinei, e aerei, e acquatici, e terreni, de l'Huomo con tutte le parti sue del corpo, come de l'anima, insieme con tutte le operazioni naturali del corpo, e tutte le virtù, potenzi, moti, abiti, notizie, ed azion de l'anima, di tutte le scienze [...] come leggendo si può vedere', Citolini, *La tipocosmia*, lettera dedicatoria, c. 3v.

2 There is a large bibliography concerned with Citolini and his universal project; it is enough to cite Della Giustina, '*La "Tipocosmia*', 63–87; Kuwakino, *L'architettura e l'arte della memoria*, 1–15.

3 The essential works on this phenomenon are Vasoli, *L'enciclopedismo del Seicento*; Tega ed., *L'unità del sapere*; Rossi, *La memoria*, 213–337; Serrai, *Storia della bibliografia*.

4 'mondo virtuale composto solo da parole', Kuwakino, *L'architettura e l'arte della memoria*, 9.

5 C.f. Toffoli, '*La "Tipocosmia"* ', 211–26.

6 'ne la quale entrar vi si potea', Citolini, *La tipocosmia*, 549.

7 'il cielo, e nel mezzo la terra', Citolini, *La tipocosmia*, 549.

8 'cose ordinate in modo assai più grato a l'occhio del corpo, che a quello dell'intelletto', Citolini, *La tipocosmia*, 549.

9 'aperto un libro di estrema grandezza, incominciò a mostrar loro questo suo nuovo ed artificioso Mondo', Citolini, *La tipocosmia*, 549.

10 This point is noted in Bolzoni, *Il lettore creative*, 359.

11 C.f. Garin, 'La nuova scienza', 451–65.

12 On this point Foucault, *Les mots et les choses*, remains fundamental.

13 On the linguistic aspects of the recovery and circulation of ancient texts in the fifteenth and sixteenth centuries see Mottana, 'Il libro Sulle pietre di Teofrasto', 151–244; Fausti, 'Traduzione cinquecenteschi di Disoscoride', 185–205; Severi, 'Fillipo Beroaldo', 81–112; Di Teodoro, 'Vitruvio volgarizzato', 39–85; Antonella Piras, 'Le dispute filogiche', 169–76.

14 This issue is explored in Olmi, 'Per preservazione', 265–89; Olmi, *L'inventario del mondo*, 211–52.

15 Badaloni, 'Fermenti di vita intellectuale', 643–89; Galasso, 'Cultura napoletana', 105–44; Torrini, 'Dal naturalismo', 455–62; Torrini, 'Scienza e filosofia', 91–102; Torrini, 'Sciences et philosophie', 654–66; the various essays in *Napoli Viceregno Spagnolo.*

16 There is a large bibliography relating to Telesio and his work. See for example Arese, 'Il rapporto Parmenide-Telesi', 15–34; Selmi, '"Libertà dolce"', 1–74; Omodeo (ed.), *Telesio and the Natural Sciences.*

17 C.f. Galasso, 'Cultura napoletana'.

18 See for example Neviani, 'Ferrante Imperato', 57–74; Accordi, 'Ferrante Imperato', 43–56; Ciarallo, 'Ferante Imperato', 50–9; Stendardo, 'Ferrante Imperato', 43–79; Stendardo, *Ferrante Imperato*; Maio and Stendardo, 'Pioneering', 209–12.

19 On the translation of Theophrastus, see Mottola, 'Ferrante Imperato', 35–42.

20 'considerando dall'altra parte l'immensità del soggetto, e la grandiosità della natura, che quantunque pigliata a spiegare, non resta perciò o del tutto, o nella maggior parte narrata, ma lascia sempre a studiosi ampio campo di cose nuove', Imperato, *Historia naturale*, 1.

21 'cose o per l'antichità de' scrittori, e mutationi de voci già sconosciute, o pur da quelli tralasciate, o vero imperfettamente e oscuramente trattate', Imperato, *Historia naturale*, 1.

22 'la cognizione dei secreti della natura', Imperato, *Historia naturale*, lettera dedicatoria, c. 1.

23 Eamon, *Science and the Secrets of Nature*.

24 'bisogna essere ancora molto intelligente della natura de' semplici, cioè non semplice herbolajo, ma gran investigator delle piante. Né è cosa più disconveniente ad un artefice non conoscer bene gli strumenti de' quali si serve. Né men bisogna haver esatta cognitione di metalli, minerali, delle gioie e delle pietre' (Italian translation), Della Porta, *Magia naturale*, 5.

25 C.f. Muraro, *Giambattista Della Porta*, Della Porta, *Criptologia*, 11–44; Vasoli, 'L'analogia universale', 31–52; Capecchi et al. (eds), *L'Accademia dei Lincei*; Santoro (ed.), *La 'mirabile natura'*; Veradi, *La scienza*.

26 'Ho trattato in modo con il Sig. Battista Porta et il Sig. Ferrante Imperato, che son tutti miei et de' Lyncei amicissimi, et invero sono miracoli di natura, et molto più di quello che si dice; io ho imparato grandemente nel discorrere con loro et ho avuto, et *avrò bellissimi segreti*, et con questi dui ho passato buona parte del tempo in Napoli con molto utile', Gabriele, *Il carteggio linceo*, 41 (author's italics).

27 C.f. Perfetti, *L'alchimia a Napoli*, tome 1, 314–18; on Fioravanti and his place in sixteenth-century culture see Camporesi, *Camminare il mondo*.

28 The introduction to the posthumous *Secreti nuovi di meravigliosa virtù* (1567) contains evidence for the existence of a Neapolitan academy dedicated to the study of nature. The programme of activities included, other than the presence of a Prince and his associates, the knowledge of the world and natural phenomena through various practices (chemical, alchemical and distillatory). See Eamon, *Science and the Secrets of Nature*, 198 and following pages. This issue was taken up and developed in Procaccioli, 'Accademia come palestra', 214–32; Verardi, *La scienza*, 14, n. 13. This academy must not be confused with that of the Della Porta brothers. For the Della Porta brothers' academy see Badaloni, 'I fratelli Della Porta', 677–715; Eamon, *Science and the Secrets of Nature*, 198–9; Copenhaver, 'The occultist tradition', 454–512; Piccari, *Giovan Battista Della Porta*, 22–3; Verardi, *La scienza*, 14.

29 It is necessary to recall the scientific dispute over the theriac that pitted Maranta and Imperato against the Collegio medico patavino (Paduan College of Medicine), in which Stegliola intervened with the publication of *Theriace et mithridatia Nicolai Stelliolae Nolani libellus* in 1577. For more detail see Cuna, 'Editoria e testi de re medica', 59–79. See too Marra, 'La vipera e l'oppio'.

30 On Stegliola see Badaloni, 'Il programma scientifico', 161–75; Ricci, *Nicola Antonio Stegliola*; Rinaldi, *La cultura delle accademie*, 161–79; Gatto, 'Gli elementi mechanici', 262–305. On the

activities of the printer see Manzi, *La tipografia napoletana*; De Frede, 'La stampa a Napoli', 753–76; Lombardi, *Tra la pagine*, 20 and 134.

31 Imperato, *Historia naturale*, dedica al lettore, cc. 1–2.

32 Imperato, *Historia naturale*, dedica al lettore, c. 1.

33 'perlochè cominciando dalla Terra, e sue differenze, e virtù, siamo indi passati all'elemento dell'Acqua, alle varie affezioni, e impressioni che essa riceve, e dall'Acqua all'Aria, e sostanze in essa generate. Segue poi di essi la consideratione dell'operazioni naturali del caldo, e del freddo, a quali si accompagna l'artificio delle fornaci, con quali secondo le occorrenti necessità, giudichiamo l'operazioni del fuoco. A questo succede la consideration de corpi di prima generatione: dico de sali e grassezze terrene, e appresso di essi le conside a Napoli ration delle specie de metalli e pietre, con gli artifici di condurle alla loro perfezione e finalmente succede la considerazion de vegetali, e animali terrestri, e marini', Imperato, *Historia naturale*, dedica al lettore, cc. 1–2.

34 See the alchemical-Hermetic interpretation offered by Marra, *Il Pulcinella filosofo*. On the partition and alchemical symbolism necessary to interpret the text see Marra, '*Dealbate latonam*', 91–124.

35 Imperato, *Historia naturale*, 578.

36 Imperato, *Historia naturale*, 578

37 For further analysis see Stendardo, 'Ferrante Imperato', 67–79.

38 'Habbiam oltre di ciò alla dottrina detta, aggiunte figurazioni delle cose c'han certa figura, e non da altri mandate in luce, accioché quanto per noi possibile fusse ne venisse aiutata l'intelligenza del lettore', Imperato, *Historia naturale*, dedica al lettore, c.1.

39 For example, Imperato mentioned information gleaned from reports prepared by the Florentine sailor Andrea Corsali and the Portuguese sailor Odoardo Barbosa. Imperato, *Historia naturale*, 626 and 627, respectively.

40 Imperato, *Historia naturale*, dedica al lettore, c.1.

41 On the nexus knowledge-identification-words see Kristensen, 'Lo sguardo curioso', 75–89; see too Campetella, 'Note su alcuni elementi', 1–17.

42 See Motta, 'I nouvi mondi', 363–98.

43 The procedure is described in Olmi, *L'inventario del mondo*, 119–62.

44 There is an extensive bibliography concerned with iconographic apparatus in scientific treatises. These studies have emphasised the relationship between text and image, the modality of representation and the character and significance of illustrations. For general discussion, see for example Petrucci, 'I percorsi della stampa', 135–64; Olmi, 'Natura morta', 69–91; Tongiorgi Tomasi, 'L'illustrazione naturalistica', 133–52; Conigliello, 'L'illustrazione scientifica', 66–71.

45 Stenardo, 'Ferrante Imperato', 75–9.

46 Tongiorgi Tomasi, 'Arte e natura', 173.

47 This issue is addressed in its entirety with respect to the natural sciences by Foucault, *Les mots and les choses*, 144–81.

48 'Della Terra sono le gemme, e spezie di pietre più nobili, e alcune particolari nature terrene segnate o dai colori, o dalla pulitezza, e liscezza, o dalla spessezza, o da altra propria virtù', Imperato, *Historia naturale*, 582.

49 On the development of studies of the inorganic kingdom in the early modern period see Accordi and Tagliaferro, 'I lapidary', 95–100; Morello, 'Alle radice', 567–82; Morello, 'La questione della natura', 127–51; on the influence of Agricola on sixteenth- and seventeenth-century chemistry see Beretta, 'Humanism and chemistry', 17–48.

50 On Mercati see Accordi, 'Michele Mercati', 1–50.

51 On Ferrante's sources and richly stocked library see Stendardo, 'Ferrante Imperato', 53–5.

52 On philological studies of these texts in fifteenth-century Naples, see Corfiati, 'Lettori della Naturalis historia', 251–76.

53 Garin, *cultura del Rinascimento*, 130–2.

54 Olmi, 'Molti amici', 3–31.

55 Stendardo, 'Ferrante Imperato', 25–38, 56–8.

56 'dico questo, perciochè io confesso, che gli studi nostri e le cose da noi trattate, han fatto progresso dall'aiuto degli amici, che o sono concorsi come fautori in procurarmi la somministrazione delle cose venutemi da diverse parti del mondo, o sono stati come

compagni e conforti delle fatiche: a quali tutti, se nel nostro trattato è cosa alcuna di buono, si deve parte di grazia', Imperato, *Historia naturale*, dedica al lettore, c.1.

57 On the Neapolitan debate on nature useful for understanding Ferrante's role, see Stendardo, 'Ferrante Imperato', 23–49.

58 A scholar of the natural science in his own right, in 1610 he published his first study *De fossilibis pusculum*. On this work see Stendardo, 'Francesco Imperato giureconsulto'. On his links to the Linceans see Stendardo, 'Francesco Imperato e i Lincei'. On his works see Rinaldi, 'La curiosità che in me predomina'.

59 'nell'età nostra, inclinata più tosto agli ozi che alla curiosità, non ha mancato in molte città d'Europa nobilissimi ingegni, quali non solo con i loro scritti, ma con le immagini al vivo ritratte delle cose, han raccolto quanto all'umana industria è stato possibile, per illustrare così nobile professione', Imperato, *Historia naturale*, dedica al Signor Giovanni di Velasco, c. 1.

60 This was noted by Brancaccio, *Geografica*, 159.

61 Badaloni, 'Fermenti di vita intelletuale', 642–89.

62 'resta di ricordarti quel che Aristotele e altri *scrittori ingenui* nei loro scritti non son restati di confessare, che le scienze umane pigliano accrescimento dal comunicare l'un all'altro', Imperato, *Historia naturale*, dedica al lettore, c. 1 (author's italics).

63 Cherchi, *La piazza universal*, 121–2.

64 'Gli speciali sicuramente trafficano oggidì nel lor mestieri tutte le speciarie, che di Levante vengono a i paesi nostri, e per quelli son ricercati e stimati convenientemente da ciascuno. Essi speciali, ovvero Aromatici, son chiamati ministri de i Medici, perché son quelli che raccolgono i semplici, che fanno i sciropi, gli unguenti, le decottioni, gli elettuari, i violebi, i tricisci, le pillole, le bevande, e altre cose simili alla medicina pertinenti', Garzoni, *La piazza universal*, 676.

65 'Ai speciali si appartiene all'ultimo raccogliere, seccare, governare, riporre, e conservar piante, o l'atre cose e così spremer succhi, mettere in infusione, far decottioni, o simili bollimenti, tener mescolato, spumare, far conditi, comprare medicine, e cose tali', Garzoni, *La piazza universal*, 678.

66 On the development of practices of collection and display of artefacts see Findlen, 'Possedere la natura', 25–47; Findlen, *Possessing Nature*.

67 'molti spetiali / [...] / colmi d'arte e valor, di fedeltade / [...] / fan l'infusioni, / pillole, dattil, con mill'altre cose, / tutte miracolose; / preparandovi ancor confettini, / conserve sciroppate e medicine, / con che purgan le febbri e l'intestine', Del Tufo, *Ritratto*, 121.

68 'uomo litteratissimo e dottissimo [...] che ha un bellissimo Palazzo in Napoli a Santa Chiara, dove tiene una superbissima bottega di speciarie aromatiche e in esso palazzo ha uno studio con tante cose diverse, che è cosa da stupire il mondo, in vedere tante erbe, tante pietre, tanti animali che è numero senza fine', Fioravanti, *Huomini eccellentissimi*, c. 1v.

69 Stendardo, 'Ferante Imperato', 124–7.

70 Imperato, *Historia naturale*, 718, 719, 742.

71 Stendardo, 'Francesco Imperato giureconsulto', 21.

72 Olmi, 'L'arca di Noé', 53.

73 Neviani, 'Ferrante Imperato', 257.

74 Lugli, *Naturalia et mirabilia*, 257. On Imperato's museum, see Stendaro, 'Francesco Imperato gireconsulto', 81–98.

75 Imperato, *Historia naturale*, 665, 667, 669, 671.

76 'possan piglian forme dall'impetramento delle cose', Imperato, *Historia naturale*, 663.

77 'conche, la forma de quali è nelle pietre frequentissima, altre volte d'impression cava fatta dalla convessità della forma, altre volte di confessità fatta dal riempimento del cavo di esse conche, e altre volte, per commutamento dell'istesse sostanze trasmutate in pietre', Imperato, *Historia naturale*, 665.

78 Imperato, *Historia naturale*, 667.

79 Imperato, *Historia naturale*, 460–1, 587, 659, 689. On the originality of Imperato's comments on fossils, see Stendardo, *Ferrante Imperato*, 72–3.

80 Colonna was especially well known for his botanical works. On this see Ottaviani, 'La natura senza inventario'; Tognoni, 'Nature described'. He also conducted research into geology and paleontology. See Morello, 'Fabio Colonna'; Morello, *La nascita di paleontologia*; Morello, 'De glossopteris dissertatio'. See also Ottaviani, 'Fra diluvio noaico'.

81 Francesco Imperato, *De fossilibus opusculum*, 69.

82 Francesco Imperato, *Discorsi*, 2–9. On the illustration of fossilised plants see Imperato, *Historia naturale*, 668–9.

83 'tra le molte spetie di pietre, vi è quella, che comprende tutte le cose, che mediante l'agente che è il succo petrifico, proportionalmente ricevendo la condensità tanto dentro, quanto fuor della terra, e anco nel mare, e particolarmente le porose, atte a ricevere questo condensamento', Imperato, *Discorsi*, 2.

84 Morello, 'Alle radici', 570.

85 Francesco Imperato, *Discorsi*, 4. Compare with the illustrations in Imperato, *Historia naturale*, 671.

86 Francesco Imperato, *Discorsi*, 82.

87 'prendono la forma del luogo cavo, che ritrovano; così anco la terra molle, è succo petrifico piglia la forma di conca, legni, corna, animali, e altre cose, e si converte in sasso, o in altra sorte di pieta' Francesco Imperato, *Discorsi*, 82.

88 'rari e mirabili parti della natura', Stelluti, *Trattato*, 12.

89 'tante varietà di oggetti e di qualità diverse, e sì mirabili e con tanto artificio, senza che v'adopri né scalpello, né pennello, né strumento di sorta alcuna; che perciò tanto maggiormente son le opere sue degne di meraviglia', Stelluti, *Trattato*, 12. See also Morello, 'Stenone e la natura', 67–89; Morello, 'Una lis necdum decisa', 1135–58; Morello, 'Steno', 81–93; Carpita, 'Agostino Scilla', 307–84; Romano, 'Vain speculation', 1–21.

90 Imperato, *Historia naturale*, 663–4, 669, 699. On the idea of 'iconic worlds' in modern science see Long, 'Objects of art', 63–82.

91 I follow here the interpretation in Foucault, *Les mots et les choses*, 31–44.

92 'una vegetazione di pietra da gesso in forma ramosa', Imperato, *Historia naturale*, 650.

93 'ostracite le cui vegetazioni imitano le conche marine', Imperato, *Historia naturale*, 651.

94 'una pietra ematite composta da più nascimenti, che nelle superficiali impressioni imita il cervello di animali', Imperato, *Historia naturale*, 699.

95 For a historico-mythological analysis of shells and their representations in figurative arts see Ventura, 'L'ostrica', 461–500.

96 'la natura medesima pare si diletti di dipingere, quale veggiamo quanto nelle fessure de' marmi spesso dipinga ipocentauri e più facce di re e crinite', Alberti, *De statua*, 50.

97 On this tradition on interpreting images see Berra, 'Immagini casuali', 358–419.

98 'pietra naturalmente delineata di figure de boschi', Imperato, *Historia naturale*, 663.

99 'la pietra figurata di frumento', Imperato, *Historia naturale*, 663.

100 Stendardo, 'Ferrante Imperato', 73, n. 14.

101 Lugli, 'Il collezione enciclopedico', 106.

102 'per venire al ristretto della descrition delle pietre, non si può negare, che la Natura produca molte di quelle, che rappresenta la forma, e effigie de gli animali volatili, acquatili, de piante, arbori, e simili cose', Francesco Imperato, *Discorsi*, 13–14.

103 'una pietra, nella quale vi appareno naturalmente scolpite di mezzo rilievo, molti segni, che rappresentano le mani, e piedi umani', Francesco Imperato, *Discorsi*, 17.

104 Foucault, *Les mots et les choses*, 56.

105 Ferrante Imperato refuted the widely diffused idea regarding the origin of animals' skulls, an issue also taken up by Francesco Imperato in *Discorsi*, 27–31.

106 Imperato, *Discorsi*, 13–14. For a detailed examination of the properties of stones in seventeenth-century treatises, see Marra, 'La Pietra', 29–38.

107 C.f. Bredekamp, *I coralli di Darwin*, especially 93–106.

108 Pliny, *Historia naturalis*, XIII, 51, 140; XXXII, II, 21.24; XXXVII, 58, 153–4, LIX, 164.

109 'i coralli sono numerati tra le piante c'han vita nell'humore: dove naturalmente molti si ritrovano, quantunque estratti nell'aria s'induriscono in consistenza', Imperato, *Historia naturale*, 713.

110 Imperato, *Historia naturale*, 717.

111 Imperato, *Historia naturale*, 714.

112 Imperato, *Historia naturale*, 717.

113 'onde fra l'altre cose, delle quali si deve ammirare è il vedere gli animali terrestri, volatili, acquatili, piante, pietre e minerali generarsi ciascuno conforme al proprio genio, e anco mote cose similmente dalla natura generarsi simile all'altre, ma diverse dalla lor spetie', Imperato, *Discorsi*, 10.

114 Imperato, *Discorsi*, 81.

115 Findlen, '"Quanto scherzevole la natura"', 419.
116 Findlen, '"Quanto scherzevole la natura"', 413–36.
117 Olmi, '"Ricche e magnifiche curiosità"', 14.
118 'Dirò di più, che la natura può produrre non solo le radici mostruose, ma anco li frutti; l'esempio si dimostra in alcune spetie di cocozze, che pigliano la forma del luogo, nel qual con industria si rinchiudono, e prendono per accidente la forma di fiasco; che si dirà delli cedri, che vengono in Napoli dagli luoghi di Calabria esposti a mare; de quali si veggono spesso molti mostruosi [...] Delle peinate vi veggono spesso mostruose, fra le quali la parietaria, e anco l'ancusa, qual è simile a cornucopia, che essiccato si conserva nel nostro museo, e vien depinto, e descritto da mio Padre', Imperato, *Discorsi*, 81–2.
119 Imperato, *Discorsi*, 82.
120 Imperato, *Discorsi*, 82–3.
121 On the question of the monstrous in sixteenth-century botanical and zoological treatises see Tomasi, 'Scienza o Immaginario?', 385–99.
122 These examples are drawn from Cappaccio, *Il Forastiero*, 864–6, and the list reproduced in Stendardo, *Ferrante Imperato*, 125.
123 On sixteenth-century tetralogia and its chief protagonists, see Caprotti (ed.), *Mostri, draghi, e serpenti*; Baldini, 'Prodigi, simulacri e mostri', 215–43.
124 C.f. Tomasi and Tongiorgi, *Ai confine dell'immaginario*, 117–28.
125 On the configuration and generation of monsters see Guidi, 'Errata naturae', 65–105.
126 'tanto dilettevole la cognizione de secreti della natura [...] che ha dato tante parti probabili, parti occulte a molti animali irrazionali, e alle parti di quelli, alle piante nelle radici, ne tronchi, nelle cortecce, nelle frondi, ne fiori, e ne i frutti; alle pietre e ai minerali, che tratto dal gusto di quella Ferrante Imperato mio Padre, non solo vi ha consumati anni e lustri, ma oramai lo intiero corso di sua vita', Imperato, *Historia naturale*, dedica al lettore, c. 1.
127 'di una historia del parto di una vipera veduta da esso [Ferrante Imperato] in una scatola. Tutti gli ha fatti morire insieme con la madre e conditi con alcune cose che conservano nello studiolo'. Taken from: *Catalogo di cose rare che una persona teneva specialmente di Storia naturale*, in Stendardo, *Ferrante Imperato. Collezionismo e studio della natura*, 135.
128 C.f. Verardi, *La scienza*.
129 'figliolo dal cui petto ne usciva un altro intero, solo senza testa, perché dal petto del primo usciva il collo del secondo e stavano giunti nel ventre, e ne ho visti poi infiniti altri, con quattro mani, con quattro piedi, con sei diti alle mani, nei piedi e in varie forme, il che sarebbe molto lungo, se volessi narrarli', Della Porta, *Magia naturale*, 93.
130 'sorti di polli [nel senso generale di uccelli] mostruosi che sempre procuro, per mio diletto, farne nascere a casa mia', Della Porta, *Magia naturale*, 94.
131 'legato ad un'idea di scienza che divulga segreti o meraviglie', Bianchi, 'Convegno', 244.
132 'sconfinava continuamente nell'inseguimento del singolare, del portentoso, dell'occulto', Serrai, *Storia della bibliografia*, 348.
133 'Ha fatto [Ferrante Imperato] nella patria mia [Napoli] un tesoro, / ponendo attenta cura / a mille bei secreti di natura', Del Tufo, *Ritratto*, 121.
134 Imperato, *Discorsi*, dedica al lettore, c. 1v.
135 'poco meno d'un palmo, non ancora giunto alla sua ordinaria grandezza', Imperato, *Discorsi*, 32.
136 'due denti de giganti, e il suo gubito, e vertebra, che impietriti si conservano nel nostro museo quali sono di molta durezza', Imperato, *Discorsi*, 7.
137 Lugli, *Wunderkammer*, 19.
138 'Qui vedreste in disparate/ scelto da man prudente/ ciò che viene dal Levante o dal Ponente:/ qui vipere, aspidi, bisce, anguille e ceraste,/ idre, draghi, faree, rospi e chelidri,/ con cento altri a veder da' curiosi/ animalacci alpestri e velenosi;/ tal che se voi miraste/ i semplici, le piante o le radici,/ che da monti, rupi o pendici,/ dell'una e l'altra Arabia egli spesso have,/ negli armadi gentil risposte a chiave/ sempre ben custodite,/ ne rimarreste attonite e stupite', Del Tufo, *Ritratto*, 121–2.

# Bibliography

## Primary sources

Alberti, Leon Battista. *De Statua*, edited by Marco Collareta. Livorno: Sillabe, 1998.

Capaccio, Giulio Cesare. *Il Forastiero*. Naples, 1634.

Citolini, Alssandro. *La tipocosmia*. Venice, 1561.

Della Porta, Giovan Battista. *Della magia naturale*. Naples, 1611.

Della Porta, Giovan Battista. *Criptologia: Conoscenze magica e ricerca scientifica in G.B. Della Porta*, edited by Gabriella Belloni. Rome: Centro Internazionale di Studi Umanistici, 1982.

Del Tufo, Giovan Battista. *Ritratto o modello delle grandezze, delizie e meraviglie della nobilissima città di Napoli*, edited by Calogero Tagliareni. Naples: Agar, 1959.

Fioravanti, Leonardo. *Della fisica*. Venice, 1592.

Garzoni, Tomaso. *La piazza universale di tutte le professioni del mondo*. Venice, 1586.

Imperato, Ferrante. *Dell'Historia naturale [...], nella quale ordinatamente si tratta della diversa condition di miniere, e pietre, con alcune historie di piante, e animali, fin'hora non date in luce*. Naples, 1599.

Imperato, Francesco. *De fossilibus opusculum*. Naples, 1610.

Imperato, Francesco. *Discorsi intorno a diverse cose naturali*. Naples, 1628.

Imperato, Francesco. *De Fossilibus opusculum (1610)*, edited by Filippo Bartolo, Francesca Coletta, Mariantonietta Paladini, Carmela Pisaniello, Caterina Porcaro, Enrica Stendardo. Naples, Accademia Pontaniana, 2015.

Pliny. *Historia naturalis*. Turin: Einaudi, 1984.

Stelluti, Francesco. *Trattato sul legno fossile minerale*. Rome, 1637.

## Secondary sources

Accordi, Bruno. 'Michele Mercati (1541–1593) e la Metalloteca', *Geologica romana* 19 (1980): 1–50.

Accordi, Bruno. 'Ferrante Imperato (Napoli, 1550–1625) e il suo contributo alla storia della geologia', *Geologica Romana* 20 (1981): 43–56.

Accordi, Bruno and Carmina Tagliaferro. 'I Lapidari dall'antichità al Rinascimento e il concetto di classificazione del regno minerale', *Atti della Accademia Nazionale dei Lincei. Classe di Scienze Fisiche, Matematiche e Naturali. Rendiconti*, series 8, vol. 71, n. 5 (1981): 95–100.

Antonella Piras, Laura. 'Le dispute filologiche sulla Naturalis historia di Plinio il Vecchio tra Quattrocento e Cinquecento', *Schifanoia* 50–1 (2016): 169–76.

Arese, Luciano. 'Il rapporto Parmenide-Telesio, dal Persio al Maranta', *Giornale critico della filosofia italiana* 11 (1991): 15–34.

Badaloni, Nicola. 'I fratelli Della Porta e la cultura magica e astrologica a Napoli nel "500"', *Studi Storici* 1 (1959–1960): 677–715.

Badaloni, Nicola. 'Fermenti di vita intellettuale a Napoli dal 1500 alla metà del "600"'. In *Storia di Napoli, Vol. 5: Il Viceregno*, vol. 1, 643–89. Napoli: E.S.I., 1972.

Badaloni, Nicola. 'Il programma scientifico di un bruniano: Colantonio Stigliola', *Studi Storici* 26 (1985): 161–75.

Baldini, Enrico. 'Prodigi, simulacri e mostri nell'eredità botanica di Ulisse Aldrovandi'. In *Natura-Cultura. L'interpretazione del mondo fisico nei testi e nelle immagini. Atti del Convegno (Mantova, 1996)*, edited by Giuseppe Olmi, Lucia Tongiorgi Tomasi and Attilio Zanca, 215–43. Florence: Olschki, 2000.

Beretta, Marco. 'Humanism and chemistry: The spread of Georgius Agricola's metallurgical writings', *Nuncius* 12 (1997): 17–48.

Berra, Giacomo. 'Immagini casuali, figure nascoste e natura antropomorfa nell'immaginario artistico rinascimentale', *Mitteilungen des Kunsthistorischen Institutes in Florenz* 43 (1999): 358–419.

Bianchi, Lorenzo. 'Convegno Internazionale su Giovambattista Della Porta', *Nuncius* 2 (1987): 241–51.

Bolzoni, Lina. *Il lettore creativo. Percorsi cinquecenteschi tra memoria, gioco, scrittura*. Naples: Guida Editori, 2012.

Bosse, Monika and André Stoll (eds). *Napoli Viceregno Spagnolo. Una capitale della cultura alle origini dell'Europa moderna* (sec. XVI–XVII). *Atti del colloquio tenuto a Bielefeld nel 1994*. Naples: Vivarium, 2001.

Brancaccio, Giovanni. *Geografia, cartografia e storia del Mezzogiorno*. Naples: Guida Editori, 1991.

Bredekamp, Horst. *I coralli di Darwin. I primi modelli evolutivi e la tradizione della storia naturale*. Turin: Bollati Bolinghieri, 2006.

Campetella, Moreno. 'Note su alcuni elementi della lingua delle scienze naturali nei secoli XVI e XVII'. In *Le vie dell'italiano: mercanti, viaggiatori, migranti, cibernauti (e altro). Percorsi e incroci possibili tra letteratura, lingua, cultura e civiltà*. Atti del XXII Convegno Internazionale AIPI (Siena, 5–9 settembre 2018), 1–17. HAL: Archives-ouvertes, 2019.

Camporesi, Piero. *Camminare il mondo. Vita e avventure di Leonardo Fioravanti medico del Cinquecento*. Milan: Garzanti, 1997.

Caprotti, Erminio (ed.). *Mostri, draghi e serpenti nelle silografie dell'opera di Ulisse Aldrovandi e dei suoi contemporanei*. Milan: Mazzotta, 1980.

Carpita, Veronica. 'Agostino Scilla (1629–1700) a Pietro Santi Bartoli (1635–1700): il metodo scientifico applicato allo studio dei fossili e la sua trasmissione ai siti e monumenti antichi', *Rendiconti dell'Accademia Nazionale dei Lincei. Classe di scienze morali, storiche e filologiche*, s. IX, CDIII, 2006, 307–84.

Capecchi, Anna Maria, Caterini Forni Montagna, Paolo Galluzzi, Anna Nicolo, Giovanni Paoloni (eds). *L'Accademia dei Lincei e la cultura europea del XVII secolo: Manoscritti-libri-incisioni-strumenti scientifici, catalogo della mostra (Parigi, 13 dicembre 1991–8 gennaio 1992; Roma 10–25 febbraio 1992)*. Rome: Accademia Nazionale dei Lincei, 1992.

Cherchi, Paolo. 'La Piazza Universale: somma di altre somme'. In *Repertori di parole e immagini. Esperienze cinquecentesche e moderni data base*, edited by Paola Barocchi e Lina Bolzoni, 121–2. Pisa: Scuola Normale Superiore di Pisa, 1997.

Ciarallo, Anna Maria. 'Ferrante Imperato e le origini del Museo naturalistico', *Museologia* 10 (1981): 50–9.

Cinzia Marra, Antonella. 'Fossili: da oggetti magici a testimoni della storia della vita sulla Terra', *Quaderni di Rivista abruzzese* 86: 137–46.

Lucilla Conigliello. 'L'illustrazione scientifica in Italia tra Cinquecento e primo Seicento'. In *Natura morta in Italia tra Cinque e Settecento*, edited by Mina Gregori, 66–71. Milan: Electa, 2003.

Copenhaver, Brian. 'The occultist tradition and its critics'. In *The Cambridge History of Seventeenth-Century Philosophy*, edited by Daniela Garber, Michael Ayers, 454–512. Cambridge: Cambridge University Press, 1998.

Corfiati, Claudia. 'Lettori della *Naturalis Historia* di Plinio a Napoli nel Rinascimento'. In *La* Naturalis Historia *di Plinio nella tradizione medievale e umanistica*, edited by Vanna Maraglino, 251–76. Bari: Cacucci, 2012.

Cuna, Andrea. 'Editoria e testi de re medica: la controversia fra Nicola Antonio Stelliola e i medici patavini'. In Saverio Ricci, *Nicola Antonio Stigliola, enciclopedista e linceo, con l'edizione del trattato Delle apparenze celesti*, edited by Andrea Cuna, 59–79. Rome: Accademia Nazionale dei Licei, 1996.

De Frede, Carlo. 'La stampa a Napoli e la diffusione delle idee riformate'. In *La stampa in Italia nel Cinquecento. Atti del convegno (Roma, 17–21 ottobre 1989)*, edited by Marco Santoro, vol. 2, 753–76. Rome: Bulzoni, 1992.

Della Giustina, Lisa. 'La "*Tipocosmia*" di Alessandro Citolini (1561). Nuove forme di enciclopedismo nel XVI secolo', *Archivio storico italiano* 157 (1999): 63–87.

Di Teodoro, Francesco P. 'Vitruvio volgarizzato e identità locali: prime indagini'. In *Architettura e identità locali*, edited by Lucia Corrain and Francesco P. Di Teodoro, vol. 139–58. Florence: Olschki, 2013.

Eamon, William. *Science and the Secrets of Nature. Books of secrets in medieval and early modern culture*. Princeton, NJ and Oxford: Princeton University Press, 1996.

Fausti, Daniela. 'Su alcune traduzioni cinquecentesche di Dioscoride: da Ermolao Barbaro a Pietro Andrea Mattioli'. In *Sulla tradizione indiretta dei testi medici greci: le traduzioni. Atti del III Seminario Internazionale di Siena (Siena, 18–19 settembre 2009)*, edited by Ivan Garofalo, Stefania Fortuna, Alessandro Lami e Amneris Roselli, 181–205. Pisa and Rome: Fabrizio Serra Editore, 2010.

Findlen, Paula. '"Quanto scherzevole la natura". La scienza che gioco dal Rinascimento all'Illuminismo', *Intersezioni* 10 (1990): 413–46.
Findlen, Paula. *Possessing Nature: Museums, collecting, and scientific culture in early modern Italy*. Berkeley and Los Angeles: University of California Press, 1996.
Findlen, Paula. 'Possedere la natura'. In *I musei della natura. Tra storia e progetto*, edited by Luca Basso Peressut, 25–47. Bologna: CLUEB, 1997.
Foucault, Michel. *Les mots et les choses*. Paris: Éditions Gallimard, 1966.
Gabrielle, Giuseppe. 'Il carteggio linceo della vecchia Accademia di Federico Cesi (1603–1630)', *Memorie della R. Accademia Nazionale dei Lincei. Classe di scienze morali, storiche e filosofiche*, s. VI, VIII, (1938–1942): 1–1446.
Galasso, Giuseppe. 'Cultura napoletana nel tardo Rinascimento', *Archivio Storico per le Province Napoletane* 105 (1987): 105–44.
Garin, Eugenio. *La cultura filosofica del Rinascimento italiano*. Florence: Sansoni, 1961.
Garin, Eugenio. *La cultura del Rinascimento*. Milan: Il Saggiatore, 1988.
Gatto, Romano. '*Gli Elementi mechanici*, di Colantonio Stigliola, un trattato archimedeo di meccanica del XVI secolo', *Bolletino di storia delle scienze matematiche* 26 (2006): 262–305.
Guidi, Simone. 'Errata naturae. Cause prime e seconde del mostro biologico tra medioevo e età moderna', *Lo sguardo: Rivista di filosofia* 9 (2012): 65–105.
Kristensen, Jens Erik. 'Lo sguardo curioso, classificatore, biologico. L'ordine della natura e il museo di storia naturale'. In *I musei della natura. Tra storia e progetto*, edited by Luca Basso Peressut, 75–89. Bologna, CLUEB, 1997.
Kuwakino, Koji. 'L'architettura e l'arte della memoria: la fabbrica del mondo progettata nella Tipocosmia (1561) di Alessandro Citolini', *EdA. Esempi di architettura* (2018): 1–15.
Lombardi, Giovanni. *Tra le pagine di San Biagio. L'economia della stampa a Napoli in età moderna*. Naples: Edizioni Scientifiche Italiane, 2000.
Long, Pamela O. 'Objects of art/objects of nature. Visual representation and the investigation of nature'. In *Merchants and Marvels. Commerce, science and art in early modern Europe*, edited by Pamela H. Smith and Paula Findlen, 63–82. New York and London: Routledge, 2002.
Lugli, Adalgisa. *Naturalia et mirabilia. Il collezionismo enciclopedico nelle Wunderkammern d'Europa*. Milan: Mazzotta, 1983.
Lugli, Adalgisa. *Wunderkammer. La stanza delle meraviglie*. Turin: Ullemandi, 1997.
Manzi, Pietro. *La tipografia napoletana nel '500*. Florence: Olsckhi, 1972.
Marra, Massimo. *Il Pulcinella filosofo chimico si Severino Scipione: uomini e idee dell'alchimia a Napoli nel periodo del Viceregno*. Milan: Mimesis, 2000.
Marra, Massimo. 'La pietra, il belzuar e la polvere: qualche storia italiana di alchimisti, principi e truffatori nell'Italia del XVII secolo', *Anthropos & latria* 7 (2003): 29–38.
Marra, Massimo. 'La vipera e l'oppio: la teriaca di Andromaco a Napoli tra XVI e XVIII secolo', *Anthropos e Iatria* 4 (2000).
Marra, Massimo. 'Dealbate latonam et rumpite libros: come si legge un libro di alchimia'. In *Come si legge un testo antico di alchimia e astrologia medica*, edited by Paolo Aldo Rossi and Ida Li Vigni, 91–124. Castel Negrino, Viruosa: Mente, 2017.
Maio, Nicola and Enrica Stendardo. 'Pioneering herpetological researches of Ferrante Imperato', *Italian Journal of Zoology* 71: sup. 2(2004): 209–12.
Morello, Nicoletta. 'Fabio Colonna e gli inizi della paleontologia', *Physis* 19 (1977): 247–78.
Morello, Nicoletta. *La nascita della paleontologia nel Seicento: Colonna, Stenone e Scilla*. Milan: Franco Angeli, 1979.
Morello, Nicoletta. 'De glossopetris dissertatio. The demonstration by Fabio Colonna of the nature of fossilis', *Archives Internationales d'historie des sciences* 31 (1981): 63–71.
Morello, Nicoletta. 'Stenone e la natura inorganica, in Niccolò Stenone e la scienza'. In *Toscana alla fine del '600. Mostra documentaria ed iconografica*, 67–89. Florence: Biblioteca Medico-Laurenziana, 1986.
Morello, Nicoletta. 'Una lis necdum decisa: Stenone e la questione dei fossili', *Miscellanea di storia ligure* 20 (1988): 1135–158.
Morello, Nicoletta. 'Alle radici della mineralogia sistematica. Il XVI secolo', *Geologica romana* 29 (1993): 567–82.
Morello, Nicoletta. 'La questione dalla natura dei fossili nel Cinquecento e Seicento'. In *Quadricentenario della parola geologia: Unisse Aldrovandi, 1603 Bologna*, edited by Gian Battista Vai and William Cavazza, 127–51. Bologna: Minerva Edizioni, 2003.

Morello, Nicoletta. 'Steno, the fossils, the rocks and the calendar of the Earth'. In *The Origins of Geology in Italy*, edited by Gian Battista Vai, W. Glen and E. Caldwell, 81–93. Boulder, CO: Geological Society of America, 2006.

Motta, Franco. 'I nuovi mondi e la natura. Realtà e rappresentazioni nel discorso scientifico della prima età moderna'. In *Milano. L'Ambrosiana e la conoscenza dei nuovi mondi (secoli XVII–XVIII)*, edited by Michela Catto and Gianvittorio Signorotto, 363–98. Milan: Biblioteca Ambrosiana, 2015.

Mottana, Annibale 'Il libro Sulle pietre di Teofrasto. Prima traduzione italiana con un vocabolario dei termini mineralogici', *Rend. Fis. Acc. Lincei*, s. 9, v. 8: (1997): 151–244.

Mottona, Annibale. 'Ferrante Imperato, primo traduttore italiano del trattato di Teofrasto "Sulle Pietre" (Napoli, 1599)', *Atti Accademia Pontaniana* 65 (2011): 35–42.

Muraro, Luisa. *Giambattista Della Porta mago e scienziato*. Milan: Feltrinelli, 1978.

Neviani, Antonio. 'Ferrante Imperato speziale e naturalista napoletano con documenti inediti', *Atti e memorie dell'Accademia di storia dell'arte sanitaria* 2 (1936): 3–86.

Olmi, Giuseppe. '"Per presevatione, et cura degli uomini": storia naturale e farmacia nel Rinascimento', *Studi Tridentini di Scienze Storiche*, 66 (1987): 265–89.

Olmi, Giuseppe. 'Natura morta e illustrazione scientifica'. In *La natura morta in Italia*, vol. 1, 66–91. Milan: Electa, 1989.

Olmi, Giuseppe. '"Ricche e magnifiche curiosità". Collezioni e collezionisti nell'età moderna'. In *La curiosità dipinta. Elementi di collezionismo tra XVI e XVIII secolo, catalogo della mostra internazionale d'antiquariato (30 marzo-8 aprile 1990; 18 aprile–5 maggio 1990)*. Milan: Expo CT, 1990.

Olmi, Giuseppe. '"Molti amici in varii lughi". Studio della natura e rapporti epistolari nel secolo XVI', *Nuncius* 6 (1991): 3–31.

Olmi, Giuseppe. *L'inventario del mondo. Catalogazione della natura e luoghi del sapere nella prima età moderna*. Bologna: Il Mulino, 1992.

Olmi, Giuseppe. 'L'arca di Noè. La natura "in mostra" e le sue meraviglie'. In *Stanza della meraviglia. I musei della natura tra storia e progetto*, edited by Luca Basso Peressut, 48–74. Bologna. Cooperatrice Universitaria Editrice, 1997.

Omodeo, Pietro D. *Telesio and the Natural Sciences in the Renaissance*. Leiden; Boston: Brill, 2019.

Ottaviani, Alessandro. 'La natura senza inventario: aspetti della ricerca naturalistica del Linceo Fabio Colonna', *Physis* 34 1997: 31–70.

Ottaviani, Alessandro. 'Fra diluvio noaico e fuochi sotterranei. Note sulla fortuna sei-settecentesca di Fabio Colonna', *Giornale critico della filosofia italiana* 96 (2017): 272–303.

Perfetti, Amalia. 'L'alchimia a Napoli tra Cinquecento e Seicento: Leonardo Fioravanti e Giovanni Battista Della Porta'. In *Napoli Viceregno Spagnolo. Una capitale della cultura alle origini dell'Europa moderna* (sec. XVI–XVII). *Atti del colloquio tenuto a Bielefeld nel 1994*, edited by Monika Bosse and André Stoll, vol. 1, 314–18. Naples: Vivarium, 2001.

Petrucci, Armando. 'I percorsi della stampa: da Gutenberg all' "Encyclopédie"'. In *La memoria del sapere*, edited by Pietro Tosi, 135–64. Rome and Bari: Laterza, 1988.

Piccari, Paolo. *Giovan Battista della Porta. Il filosofo, il retore, lo scienziato*. Milan: Franco Angeli, 2007.

Procaccioli, Paolo. 'Accademia come palestra e come tributo: Girolamo Ruscelli sdegnato, arente, dubbioso, fratteggiano'. In *The Italian Academies 1525–1700. Networks of culture, innovation and dissent*, edited by Jane E. Everson, Denis V. Reidy and Lisa Sampson, 214–32. New York: Routledge, 2016.

Ricci, Saverio. *Nicola Antonio Stigliola, enciclopedista e linceo, con l'edizione del trattato. Delle apparenze celesti*, edited by Andrea Cuna. Rome: Accademia Nazionale dei Licei, 1996.

Rinaldi, Massimo. *L'audacia di Pythio. Filosofia, scienza e architettura in Colantonio Stigliola*. Bologna: Società Editrice Il Mulino, 1999.

Rinaldi, Massimo. '"La curiosità che in me predomina": In margine ad un opuscolo dimenticato (1606) di Francesco Imperato', *Aevum* 75 (2001): 705–13.

Rinaldi, Massimo. *La cultura delle accademie. Immaginario urbano e scienza della natura tra Cinque e Seicento*. Milan: Edizioni Unicopoli, 2005.

Romano, Marco. '"The vain speculation disillusioned by the sense": The Italian painter Agostino Scilla (1629–1700) called "The Discoloured", and the correct interpretation of fossil as lithified organism that once lived in the sea', *Historical Biology* 25 (2013): 1–21.

Rossi, Paolo. 'La memoria, le immagini, l'enciclopedia'. In *La memoria del sapere*, edited by Pietro Tosi, 213–337. Rome and Bari: Laterza, 1988.

Santoro, Marco ed. *La 'mirabile natura'. Magia e scienza in Giovan Battista delle Porta (1615–2015). Atti del Convegno Internazionale (Napoli–Vico Equense, 13–17 ottobre 2015)*. Pisa and Rome: Serra, 2016.

Selmi, Elisabetta. '"Libertà dolce alla verità impetra". La fortuna del pensiero telesiano nel dialogo con i letterati e filosofi della Studio patavino e degli ambienti settentrionali del secondo Cinquecento'. In *Il fiore delle passioni. Animo e virtù nel sistema dei saperi tra Cinque e Seicento*, with the assistance of Achille Olivieri, Cristiano Rocchio and Stefano Giazzon, 1–74. Padua: Coop. Libraria Editrice Universitaria di Padova, 2012.

Serrai, Alfredo. *Storia della bibliografia*. I *Bibliografia e cabala. Le enciclopedie rinascimentali*, edited by Maria Cochetti. Rome: Bulzoni, 1988.

Severi, Andrea. 'Filippo Beroaldo correttore di Plinio il Vecchio: La lettera a Niccolò Ravacaldo', *Schede Umanistiche* 24–5 (2010–11): 81–112.

Stendardo, Enrica. 'Ferrante Imperato. Il collezionismo naturalistico a Napoli tra '500 e '600 ed alcuni documenti inediti', *Atti e Memorie dell'Accademia Clementina* 28–9 (1991): 43–79.

Stendardo, Enrica. *Ferrante Imperato. Collezionismo e studio della natura a Napoli tra Cinque e Seicento*. Naples: Officine Grafiche Napoletane, 2001.

Stendardo, Enrica. 'Francesco Imperato e i Lincei. Con alcune lettere inedite', *Aprosiana* 9 (2001): 159–68.

Stendardo, Enrica. 'Francesco Imperato giureconsulto e naturalista napoletano'. In Francesco Imperato, *De Fossilibus opusculum (1610)*, edited by Filippo Bartolo, Francesca Coletta, Mariantonietta Paladini, Carmela Pisaniello, Caterina Porcaro and Enrica Stendardo, 11–37. Naples: Accademia Pontaniana, 2015.

Tega, Walter (ed.). *L'unità del sapere e l'ideale enciclopedico nel pensiero moderno*, edited by Walter Tega. Bologna: Il Mulino, 1983.

Toffoli, Aldo. 'La Tipocosmia di Alessandro Citolini'. In *Atti e memorie dell'Ateneo di Treviso* 18 (2000–1): 211–26.

Tognoni, Federico. 'Nature described: Fabio Colonna and natural history illustration', *Nuncius* 20 (2005): 347–70.

Tongiorgi Tomasi, Lucia. 'Arte e natura nel Giardino dei Semplici: dalle origini alla fine dell'età medicea'. In *Giardino dei semplici. L'Orto botanico di Pisa dal XVI al XX secolo*, edited by Fabio Garbari, Lucia Tongiorgi Tomasi and Alessandro Tosi, 115–212. Pisa: Pacini Editore, 1991.

Tongiorgi Tomasi, Lucia. 'L'illustrazione naturalistica: tecnica e invenzione'. In *Natura-Cultura. Interpretazione del mondo fisico nei testi e nelle immagini. Atti del Convegno Internazionale di studi* (Mantova, 1996), edited by Giuseppe Olmi, Lucia Tongiorgi Tomasi and Attilio Zanza, 133–52. Florence: Olschki, 2000.

Tongiorgi Tomasi, Lucia. 'Scienza o Immaginario? Mostri, meraviglie e prodigi nell'iconografia del tardo Rinascimento'. In *L'art de la Renaissance entre science et magie. Colloque International organisé par le Centre d'Historie de l'Art de la Renaissance* (Paris, 20–21 juin 2002), edited by Philippe Morel, 385–99. Rome: Académie de France, 2006.

Tongiorgi Tomasi, Lucia and Paolo Tongiorni. 'Ai confini dell'immaginario. I mostri: fantasia e realtà'. In *Immagine e natura. L'immagine naturalistica nei codici e libri a stampa delle Biblioteche Estense e Universitaria, sec. XV–XVII, catalogo della mostra* (Modena, 1984). Modena: Panini, 1984.

Torrini, Maurizio. 'Il topos della meraviglia'. In *Francis Bacon. Terminologia e fortuna nel XVII secolo. Atti del seminario internazionale (Roma, 11–13 marzo 1984)*, edited by Marta Fattori, 261–80. Rome: Edizioni dell'Ateneo, 1984.

Torrini, Maurizio. 'Dal naturalismo alla rivoluzione scientifica. Filosofia e scienza della natura a Napoli tra XVI e XVII secolo'. In *Storia e civiltà della Campania, Vol. 3: Il Rinascimento e l'età barocca*, edited by Giuseppe Pugliese Caratelli, 455–62. Naples: Electa, 1994.

Torrini, Maurizio. 'Scienza e filosofia nell'Italia del Seicento'. In *Il Mezzogiorno e Napoli nel Seicento Italiano. Atti del convegno Il Mezzogiorno e Napoli come motori delle rivoluzioni del Seicento italiano (Napoli, 24 maggio 2002)*, edited by Mario Tedeschi, 91–102. Naples: Rubbertino Editore, 2003.

Torrini, Maurizio. 'Sciences et philosophie à Naples de Della Porta à Vico'. In *De Florence à Venice. Études en l'honneur de Christian Bec*, edited by François Livi et Carlo Ossola, 654–66. Paris: PUPS, 2006.

Vasoli, Cesare. *L'enciclopedismo del Seicento*. Naples: Bibliopolis, 1978.

Vasoli, Cesare. 'L'analogia universale: la retorica come semiotica nell'opera del Della Porta'. In *Giovambattista Della Porta nell'Europa del suo tempo. Atti del convegno (Vico Equense, 29 settembre–3 ottobre 1986)*, edited by Maurizio Torrini, 31–52. Naples: Guida, 1990.

Veneri, Toni. 'Giovanni Battista Ramusio, molto più di uno spettatore. Le quinte di Navigationi et viaggi', *Italica* 89 (2012): 162–201.

Ventura, G. 'L'Ostrica e la Pinna: storia, leggenda e curiosità', *Memorie Descrittive della Carta Geologica D'Italia* 96 (2014): 461–500.

Verardi, Donato. *La scienza e i secreti della natura a Napoli nel Rinascimento. La magia naturale di Giovan Battista della Porta*. Florence: Firenze University Press, 2018.

Vignolo, Paolo. *Cannibali, giganti e selvaggi. Creature mostruose del Nuovo Mondo.* Milan: Pearson Paravia Bruno Mondadori, 2009.

2

# Political science in the *Settecento* University of Naples

Felix Waldmann

This chapter focuses on the status of 'political science' in the *Settecento* University of Naples. The principal context of the chapter is the tenure of Celestino Galiani (1681–1753) as the rector (or 'Cappellano Maggiore') of the University of Naples and the published work of his protégé Antonio Genovesi (1711–1769). The history of Galiani's and Genovesi's curricular reforms is well known among historians of the *Settecento* Meridione.[1] However, the place of 'political science' – that is, the teaching and study of 'politics' – in this history still awaits study. The chapter begins by surveying the institutional background to Galiani's curricular reforms as Cappellano Maggiore; it then turns to the works of Genovesi and members of the so-called 'scuola genovesiana'. Although the latter published on 'political' topics, neither Genovesi nor his students developed a curricular 'political science'. The chapter offers a conjectural explanation for this striking absence among the 'sciences' practised in eighteenth-century Naples.

## The University of Naples in the eighteenth century

In the eighteenth century the University of Naples remained the most conspicuous part of a patchwork of institutions for the education of the Kingdom's inhabitants ('regnicoli'). In the provinces, religious orders ran secondary schools, convents and monasteries, particularly within the viceroyalty's and later the Kingdom's more prominent archdioceses,[2] such as the Certosa di Padula,[3] or the many different seminaries in the city of Naples. Among these were the Seminario Diocesano and Seminario Urbano for secular clergy, the Certosa di San Martino for Carthusians, the Chiesa dei Girolamini for Oratorians, the Collegio

Massimo for Jesuits and San Domenico Maggiore for Dominicans. An efflorescence of Academic activity within the Kingdom followed the foundation of the Accademia degli Oziosi (1611–1645)[4] and continued to supplement or rival the University as educational institutions: the Accademia degli Investiganti (1663–70, 1735–7),[5] the Accademia di Medinaceli (1698–1701)[6] and the Accademia delle Scienze (1732–44) were all prominent contexts for research, debate and public education.[7] However, the University retained its significance as the principal venue for education in the higher faculties of theology, medicine and law. This was partly because of its monopolistic control of admission to the guilds, or *Collegi dei Dottori*, that could control preferment within the professions associated with theology, medicine and law.

Within the University, professorial chairs ('cattedre') were filled through public competitions ('concorsi'), adjudged by the Cappellano Maggiore and professoriate. Differences among elected professors ('ordinarii') divided those who held their chairs permanently ('primaria', noted with a **P** in the listing below) from those who were subject to quadrennial re-election through 'concorsi' ('secondaria'). A third body of teachers ('extraordinarii') was derived from public students of distinction, who were employed as 'extraordinary readers' ('lettori straordinari') or 'substitutes' ('sostituti') for infirm or absent professors ('cattedratici').[8] Many permanent professors treated their roles as sinecures or waystations for clerical preferment. A fourth body of teachers, in the form of temporary (ad interim) professors, appointed in lieu of a 'concorso', often occupied professorial chairs for these sinecurists.

The destruction of the Archivio Universitario in 1943 has complicated any reconstruction of the University's curricular requirements. However, they appear to have demanded a student's attendance in four continuous sessions running from November to September.[9] Acquaintance with philosophy (logic, metaphysics, physics and ethics) was typically expected from matriculants in the higher faculties, but students could exempt themselves from these sessions, provided they could pay the cattedratico a fee and establish their competence under examination. These requirements followed the regulations ('ordinamenti') established in 1616 by the statute *De regimine studiorum*.

Chairs ('cattedre') and chairholders ('cattedratici') within the University of Naples (1738)

1. **Civil Law** (Diritto Civile)

- Feudal Law (Diritto Feudale) **P**
  - Francesco Nicoli
- Evening Civil Law (Vespertina di Diritto Civile) **P**
  - Domenico Gentile
- Morning Civil Law (Mattutina di Diritto Civile) **P**
  - Marcello Cusani
- Law of the Kingdom (Diritto del Regno) **N**
  - Ferdinando d'Ambrosio
- Old Digest (Digesto Vecchio)
  - Andrea Caputo
- Morning Civil Institutions (Mattutina d'Istituzioni Civili)
  - Giuseppe Cirillo
- Evening Civil Institutions (Vespertina d'Istituzioni Civili)
  - Vacant

2. **Canon Law** (Diritto Canonico)
- Morning Canon Law (Primaria Mattutina di Diritto Canonico) **P**
  - Pietro Antonio de Turris
- Evening Canon Law (Primaria Vespertina di Diritto Canonico)
  - Blasio Troyse
- Morning Canonical Institutions (Mattutina d'Istituzione Canoniche)
  - Francesco de Chellis
- Evening Canonical Institutions (Vespertina d'Istituzione Canoniche)
  - Gennaro de Ferdinando
- Gratian's *Decretum* (Decreto di Graziano)
  - Vacant

3. **Theology** (Teologia)
- Sacred Scriptures (Sacra Scrittura) **P**
  - Alessio Simmaco Mazzocchi
- Scholasticism of Saint Thomas (Scolastica di San Tommaso)
  - Pio Tommaso Milante
- Theology of Duns Scotus (Teologia di Scoto)
  - Giacomo Filippo Gatti
- Moral Theology (Teologia Morale)
  - Castrensis Scaja
- Ecclesiastical History (Storia Ecclesiastica)
  - Giovanni Ruggiero

4. **Medicine** (Medicina)
- Practical Medicine (Primaria di Medicina Pratica) **P**
  - Gioacchino Poeta

- Theoretical Medicine (Primaria di Medicina Teorica) **P**
  - Giovan Battista Balbi
- Aphorisms of Hippocrates (Aforismi d'Ippocrate)
  - Giacinto Giannotti
- Practical Medicine (Secondaria di Medicina Pratica)
  - Bernardino Rossi
- Theoretical Medicine (Seconda di Medicina Teorica)
  - Francesco Serao
- Anatomy (Anatomia) **N**
  - Aniello Tirelli
- Surgery (Chirurgia)
  - Francesco de Micco
- Botany and Natural History (Botanica e Storia Naturale)
  - Orazio Biancardi

5. **Philosophy** (Filosofia)

- Ethics (Etica)
  - Isidoro Sanchez de Luna
- Physics (Fisica) **P**
  - Mario Lama
- Experimental Physics (Fisica Sperimentale)
  - Giuseppe Orlandi
- Logic and Metaphysics (Logica e Metafisica)
  - Battista Lamberti

6. **Philology** (Filologia)

- Rhetoric (Rettorica)
  - Giambattista Vico
- Greek Language (Lingua Greca)
  - Antonio Fusco
- Hebrew Language (Lingua Ebraica) **N**
  - Vacant

7. **Sciences** (Scienze)

- Geometry (Geometria) **P**
  - Nicola de Martino
- Astronomy and Nautics (Astronomia e Nautica) **N**
  - Pietro de Martino

A series of failed reforms – aimed at remedying a perceived decline in educational standards – had marked the years between 1703 and Celestino Galiani's appointment as Cappellano Maggiore in 1731.[10] Yet within three years of his arrival Galiani had succeeded in introducing several new chairs to the University (noted with an **N** in the listing above), as well as reforming perceived abuses among the professoriate. These reforms affected the study of 'politics' only negatively. The curricular structure of the professoriate – that is, the roles nominally performed on the 'ruolo degli insegnanti' in the listing above – did not include 'politics' except in the form of 'ethics', the 'cattedra' grouped within the Faculty of Philosophy. Politics formed one part of the tripartite division of ethics, which consisted of moral philosophy, politics and economics. The last referred restrictedly to the sense of Oeconomics or 'household management' embodied by the work of that title attributed to Aristotle.

In 1739 the cattedratico in ethics was Isidoro Sanchez de Luna (1705–1786), later the Archbishop of Salerno, who proved to be a characteristic example of a sinecurist anticipating clerical preferment. Sanchez de Luna published nothing relating to any of the three component parts of curricular ethics – a distinction he shared with several of his predecessors, as we will see. It was this tendency which underpinned Galiani's campaign between 1732 and 1734 to abolish the cattedra of ethics entirely.

## The reforms of Celestino Galiani

This decision is partly explicable by reference to Galiani's own intellectual sympathies, which were aligned with innovations in the natural sciences to the detriment of scholastic theology and Aristotelianism. Galiani was born in Foggia in 1681; in *c.*1711 he transferred to Rome and trained among the Celestines.[11] In 1723 he was appointed the Procurator General of the Order. In 1728 he was promoted to General, the highest position in the Order, before he returned to Austrian Naples as Cappellano Maggiore in 1731. Following his appointment as Cappellano Maggiore, Galiani swiftly undertook a series of reforms relating to the natural sciences. He promoted his fellow Celestine, Giuseppe Orlandi, to a professorship of experimental physics, petitioned for the creation of a chair in astronomy and campaigned to abolish the chair in Scotist theology.[12]

The decision to abolish the 'cattedra' of ethics coincided with Galiani's decision in June 1732 to create a chair in the 'law of nature and nations'.[13] The teaching of ethics in the university had reportedly languished for years. In early 1714 the Municipal Electors of Naples wrote to the Cappellano Maggiore, Diego Vincencio de Vidania (1644–1732). They complained expressly about the teaching of the subject and the unintelligible accent of the cattedratico Diego de Loya (d.1720) when speaking in Latin. De Loya was the cattedra's second incumbent (1705–20). He succeeded the inaugural professor Ottavio Santoro (1703–5), appointed following the creation of the cattedra in 1703 by the statute *De regimine studiorum III*. De Loya's successor ad interim, Arcangelo Maria Ciccarelli, would lose a concorso in 1721 to Niccolò Crescenzo, formerly a professor of logic. Crescenzo went on to hold the cattedra from 1721 until his death in 1734, when the medical doctor Giacinto Giannotti requested it 'per l'interim'. In 1734 Galiani questioned Giannotti's credentials for the post, 'even for the time being [*etiam ad tempus*]',[14] and suggested an alternative to Giannotti's appointment:

> On this occasion, I ought to mention that two years ago, when lessons in the law of nature and nations were introduced into the University, a subject which contains the true principles of ethics […] it was suggested that the chair of ethics […] be suppressed in order to create some other chair, of which the University had a pressing need.

Galiani submitted that the chair of ethics should be 'abolished' and that the 'professor of the law of nature and nations' could be 'obliged […] to teach the material that the professor of ethics had taught'.[15] Galiani's attempt to abolish the chair of ethics was unsuccessful.[16] The cattedra of ethics remained a fixed component of teaching in the university for the remainder of the eighteenth century; indeed, one of the great ironies of Galiani's contempt for the cattedra is that it would become the principal vehicle for the promotion of Antonio Genovesi.

Galiani's claim that the teaching of 'ethics' could be entrusted to the professor of the 'law of nature and nations' fitted within a general tendency to associate teaching in moral philosophy in the century after the publication of Hugo Grotius's *De Jure Belli ac Pacis* (1625) with the study of the 'jus naturae et gentium'. This is a complex subject which I have discussed elsewhere.[17] The most important consideration for Galiani was that Aristotelian 'ethics' was insufficiently concerned with the question of 'obligation'; instead, it focused principally on the definition of 'virtue', without apprising students of the moral necessity to practise it. In the language typically associated with this development, Aristotelian moral philosophy was considered to be 'indicative', while the species of moral philosophy associated with the post-Grotian natural lawyers – Samuel Pufendorf (1632–1694), Christian Thomasius (1655–1728), and Christian Wolff (1679–1754) – was considered to be 'imperative'.

In the 1720s Galiani had authored a manuscript *Ricerche intorno alle prime origini della scienza morale*, which was animated by a fear of the resurgence of 'scepticism' in matters of morality and suffused with the tropes of post-Grotian natural jurisprudence. By 1725 Galiani was ordering works by Pufendorf from the Lausannois bookdealer Marc-Michel Bousquet.[18] In the following year, he was reportedly offering private tutelage in 'Moral Philosophy and the Principles of the Law of Nature'.[19] When he sought to abolish the teaching of 'ethics' in 1734, Galiani was thus pressed by an abiding concern about the adequacy of Aristotelian moral philosophy to confront the challenge of scepticism in a curricular context by providing students with a clear explanation of the imperative character of morality. (A separate and significant impetus for the concerns of Galiani and others about the urgency of finding an 'imperative' moral philosophy related to a perceived decline in the status of moral theology, which many believed to be excessively preoccupied with the debate over 'probabilism' – a topic that cannot detain us here.[20])

Yet an obvious difficulty accompanied Galiani's campaign to abolish the cattedra of 'ethics' on account of its 'indicative' moral philosophy: what about the teaching of 'politics'? This issue is absent from Galiani's surviving manuscripts and from the surviving correspondence pertaining to Galiani's campaign: correspondents sympathetic and antipathetic to his request either felt that natural jurisprudence was a suitable alternative to Aristotelian ethics or they did not, but they never once refer to the adequacy or inadequacy of natural jurisprudence as a surrogate for the teaching of 'politics'. In April 1733 Pietro Giannone, writing in exile from Vienna, observed that Galiani's campaign to abolish the cattedra of ethics was objectionable, without any reference to the fate of curricular 'politics'. Giannone notably described the professorship as the 'cattedra di etica, o sia morale', reaffirming the restricted synonymy of 'ethics' with moral philosophy:

> The chair of Ethics, or rather Morality [*Etica, o sia Morale*], should not be suppressed, since the professor of the Law of Nature and Nations has a different concern, nor ought he to discuss moral virtue [*virtú morale*] and the regulation of behaviour [*costumi*]. Morality should be the aim of every preceding philosophical study; and likewise the speculations of the most serious philosophers, both ancient and modern, have been addressed, ultimately, to the issue of Morality. If the class of Theology considers Morality, then it is unreasonable that the class of Philosophy should abandon it, since Morality is the noblest part of Philosophy.[21]

The surviving publications of the cattedratici in 'ethics' and 'concorrenti' for the chair – de Loya's eccentric *Quinque porticus morales ad probaticam piscinam mysticae sanitatis* (1717), Giuseppe Maria Amati's *Ethica ex-tempore concinnata in publica universitate neapolitana* (1721) and Crescenzo's *Pro morali philosophia* (1733) – do not refer to the teaching of 'politics'. Indeed, only one extant manuscript course in 'ethics', which I have located in a repository in the territories of the former viceroyalty or Kingdom of Naples, preserves a 'politics': MS Branc. IV F 6, now held in the Biblioteca Nazionale di Napoli, includes a *Tractatus politicus*. As the subtitle to the tract notes, the course is one

> in which it is shown in what ways society, instead of labouring under tyranny, is where monarchical power has its place, as well as where it ought to be established that the best men rule, such that the inviolate peace and liberty of the citizens may endure.

That is, it considers precisely the topics one would expect from an early modern course in 'political science'.[22]

These courses, typically taught by professors of 'ethics', have received significant attention from scholars of political thought in early modern England and Germany.[23] The focus of this work is the tradition of commentary on Aristotle's *Politics* and *Nicomachean Ethics*. This was not the only place in which the study of 'politics' was pursued: writing on 'rhetoric', 'mirror for princes' literature and post-Machiavellian ruminations on 'reason of state' all provided a context for the development of 'political thought'. Yet the treatment of politics as a 'science' ('scientia') in England, Germany and elsewhere in Europe was particularly associated with the teaching of Aristotelian 'ethics' – and it is here that one confronts an absence in early modern Naples. The reason for this absence is difficult to identify except by surmise, but a reasonable surmise is that the curricular discussion of 'political science' or publication on the subject was viewed with suspicion by the succession of monarchical authorities that controlled the university.

It is important to emphasise that the Aristotelian commentary tradition was not concerned necessarily, or even typically, with questions that a monarch might regard as potentially subversive, such as inquiries into the sources of political obligation, which were the defining characteristic of later seventeenth-century Anglophone political thought. Commentaries would rather often focus on demonstrating the 'properties' that are shared by any conceivable instance of a 'res publica' or on providing an overview of Aristotle's taxonomy of constitutions.[24] However, one can only assume the discussion of Aristotle's *Politics* was considered by the presiding authorities in early modern Naples – Spanish, Austrian and Neapolitan – as an unwanted invitation to contemplate the merits of monarchy against alternative political arrangements. This surmise is not tantamount to the suggestion that the presiding authorities in England or Germany were congenial to these contemplations, nor that Naples was uniquely beset by demotic uprisings such as the revolt of Masaniello, but 'ex hypothesi' the restrictions in Naples were more effectual in repressing the development of curricular 'political science'.

Different evidence could be advanced in support of this proposition, but two eighteenth-century examples merit attention. First, there is the judgement of Giuseppe Pasquale Cirillo (1709–1776) – a professor of civil law who served, after 1747, as the ex officio professor of the law of nature and nations in the university[25] – on whether natural jurisprudence comprehended the teaching of 'politics'. In an undated tract on the 'jus naturae et gentium', Cirillo identified the distinguishable foci of 'natural law' and 'ethics', which he impliedly equated with moral philosophy tout court:

> The purest definition of natural law is the will of God, promulgated to the human race through right reason, commanding certain things before any action is performed, other things after any action is performed, prohibiting some actions with prospective rewards, and others with attached punishments, and leaving the remainder to man's free will. Natural jurisprudence is an art teaching the rules through which human reason may perceive the enjoined will of God, and accommodate every part of life to it. Now it must readily be understood, that the matter of natural jurisprudence and ethics differ. Both concern the good, but ethics concerns the good that perfects man and truly brings happiness, whereas natural jurisprudence concerns that good that is just, or rather what is endorsed by the law. We are obliged by the latter good, but not the former.[26]

The tendency to associate the 'law' with the will of God, and natural jurisprudence as the mode through which one could understand this law, divorced the subject from any inquiry into the status of positive law – that is, whether positive law could be justifiably violated by its subjects. Cirillo associates positive law – 'qua Lege sancitum est', without a clear distinction between 'lex naturalis' or 'lex positiva' – with divine authority, in such a way that pre-empted any discussion of the 'authority' of a human lawmaker. One could conclude that Galiani viewed the teaching of the 'jus naturae et gentium' in a similar way: a discipline as far removed from any inquiry into the sources of political authority as it was from the taxonomic concerns characteristic of the tradition of commentary on Aristotle's *Politics*. By substituting Aristotelian ethics with teaching in the law of nature and nations, Galiani did not risk the intrusion of 'political' science into the curriculum.

The second example to consider is the absence of printed works on 'political science'. The books of Paolo Mattia Doria (1667–1746) are outliers in this respect: Doria's *Vita Civile*, published under a false imprint in 1710, is a prominent instance of an early *Settecento* work published in Naples that nominally concerned itself with 'political' matters.[27] Yet Doria's interventions in this sphere were highly unusual. They were predictably capped by the so-called 'rogo posthumo' that accompanied the suppressed publication of his *Idea di una Perfetta Repubblica* (*c.* 1741) in 1753, a manuscript in which Doria expressly condoned a quasi-republican constitutional arrangement.[28] There is every indication, in other words, that 'political science' was deprecated by the authorities in eighteenth-century Naples. It was in this context that Antonio Genovesi commenced teaching 'ethics' in the University of Naples, following his appointment in 1746 to the chair ad interim.

## Antonio Genovesi and curricular ethics

In addition to its oversight of the university, the Cappellania Maggiore co-administered the Kingdom's system of press censorship – providing Galiani with the means to promote authors sympathetic to his intellectual commitments and curricular reforms. Every work published within the Kingdom required evidence of censorial approval. Publishers applied to the Cappellano Maggiore and the Archbishop of Naples, who subsequently appointed civil and vicarial censors, drawn from the ranks of the professoriate in the university or the clergy. An imprimatur, transcribing the censor's approval of the book or any particular opinion of its merits, was usually affixed to each volume.[29] Notwithstanding this tortuous process, the Kingdom's streets were awash with print, from gazettes of court business to ponderous works of theology and civil law to posters and pamphlets, chronicling disputations in seminaries and legal proceedings.[30] In the seventeenth century alone, more than 2,700 publications were published within the city of Naples itself.[31] The absence of works nominally devoted to 'politics' cannot therefore be attributed to the inadequacies of the city's or the Kingdom's publishers. Instead, it stemmed from a persistent apprehension that publishing a work objectionable to the court could have severe repercussions for the publisher, the printer and the author – wholly apart from the threat of intercession by the Congregation of the Index in Rome.

Genovesi's career had an inauspicious start in this respect. After migrating to the city of Naples from the region of Salerno, where he was born in 1713, Genovesi published the first volume of an *Elementa metaphysicae* in 1743 without vicarial approval. He was promptly hauled before the Archbishop of Naples, Giuseppe Spinelli (1694–1763), and condemned for his insufficient confutation of the views of the many 'heretics' that the *Elementa* had compendiously summarised. He published a contrite *Appendix* to the *Elementa* in 1744, but remained under clerical suspicion for the rest of his career.

Shortly after this episode, Genovesi secured the patronage of Galiani, who sponsored his employment as the 'cattedratico' of ethics ad interim in the room of Isidoro Sanchez de Luna, who was selected as the Bishop of Ariano in 1748. In a typically self-regarding passage of his *Autobiografia* (*c.*1748), Genovesi claimed to have devised a system of 'ethics' that superseded the teachings of his predecessor 'cattedratici', 'who did not possess those cognitions, and that eloquence, which their profession required'.[32] As Genovesi boasted, his 'system' would commence with a study of the 'physiology of man', or our 'inclinations, passions, virtues and vices' (Book 1), before demonstrating the existence of God and the law of nature, as well as our 'need of a regulatory law to live well' (Book 2). His work would then summarise 'the principal systems of great men who have discussed the law of nature' (Book 3) and 'the various duties of men' (Book 4).[33]

Portions of Book 1 were later discussed in chapters of Genovesi's *Psychesophia* (1747), the second volume of his *Elementa Metaphysicae*, and portions of Books 2, 3 and 4 were published in his *De Legibus Naturae* (1752), the fourth volume of the *Elementa Metaphysicae*. In 1765 Genovesi reworked *De Legibus Naturae* into *De Jure et Officiis in Usum Tironum*, a work that he subsequently reissued in Italian, in two volumes, as the *Diceosina, o sia, della Filosofia del Giusto e dell'Onesto* (1766–71). None of these instalments were nominally 'political'. Moreover, when Genovesi was appointed as the university's first professor of commerce and mechanics in 1754, he acted consciously to distinguish the curricular focus of his previous work in the 'cattedra' of ethics from his present and prospective work as an author on 'economics'. In this way, Genovesi attended to two parts of the tripartite Aristotelian curriculum in ethics – moral philosophy (in the second and third volumes of the *Elementa*, in the *De Jure et Officiis* and the *Diceosina*) and economics (in his several publications on the topic, culminating in his *Lezioni di Commercio* of 1765–7). Yet 'politics' was pointedly omitted as the nominal focus of any of his publications, which were uniformly produced *in usum tironum* or for curricular use.

One might respond that concentrating on the 'nominal' focus of Genovesi's works detracts from their implied or incidental focus on concerns that fall within any reasonable conception of 'politics'. In *De Legibus Naturae*, for example, Genovesi discussed how Grotius, Hobbes and Pufendorf, among others, had resolved the question of our inclination to obey natural law and co-exist in civil society. As the title of 'De Jure et Officiis' intimated, with its allusion to the works of Cicero and Pufendorf on 'duties', Genovesi was concerned with our conduct inside the context of civil society, in lieu of a focus on our 'duties' to God in a context abstracted from the proprieties of 'civiltà'.

Furthermore, as Sophus Reinert has incisively shown, Genovesi's writings on commerce fit within a genre of writing on reformative political economy that was at once continuous with the anxieties of seventeenth-century reason of state, in obedience to the logic of a rivalrous European states system, and a contribution to the emergence of eighteenth-century cameralism – celebrating the 'rationalising' tendencies of absolutist monarchies to concentrate jurisdiction in their own hands against the pretensions of a self-policing feudal aristocracy.[34] In these guises, Genovesi arguably introduced a form of 'political science' into the curriculum of the University of Naples and nurtured a generation of students who would publish on related topics.

Yet it must be emphasised that each of these students published their 'political' works typically – if not necessarily – outside a curricular context. Clear evidence of a disinclination to introduce the curricular study of political science in the University of Naples is provided by the later history of the cattedra in ethics. In 1769, after Genovesi's successor Gaetano Maria Capece (1720–94) was selected as the Archbishop of Trani, the university staged a concorso for the post, in which it invited concorrenti to discuss a locus on 'friendship' from Aristotle's *Nicomachean Ethics*.[35] This reversion to 'indicative' Aristotelianism is surprising, given Capece's recorded self-description as the cattedratico of 'ethics, and natural law'[36] and the evidence of his curriculum from an *Album Professorum Regii Archi-Gymnasii Neapolitani* of 1761–2, in which Capece is described as teaching the 'laws of nature, or the duties of man'.[37] Yet the authorities responsible for the concorso evidently saw nothing anachronistic in reviving the conceit that the teaching of ethics consisted of indicative Aristotelianism.

A controvertible exception is provided by the Royal Military Academy or *Nunziatella*, which developed its own curriculum and advertised a concorso for its chair of ethics in 1769–70.[38] One of the concorrenti was Genovesi's student Francesco Mario Pagano (1748–99), who circulated a short *Disegno del Sistema della Scienza degli Ufizj* in anticipation of the concorso, which he lost to Trojano Odazi (1741–94).[39] As its title suggests, Pagano's work mirrored Genovesi's *De Jure et Officiis* in concentrating on the 'imperative' dimension of the subject, to the detriment of indicative Aristotelianism.

Whether this was considered a fault by his assessors is impossible to establish. However, one of Odazi's successors in the post, Michelangelo Grisolia (1751–94), provides a surprising instance of a substantive engagement with the study of politics in an eighteenth-century curricular context. Grisolia's *Ragionamento sul Sistema dell'Origine della Sovranità* (1783) was an analytical discussion of 'sovereignty'. It was accompanied by other works that Grisolia issued on the topic of the authority of the monarchy in Naples, including a *De Principe Liber Unus* (1783), dedicated to Maria Carolina.[40] The paratext of *De Principe* reveals that its publication occurred only after Grisolia had navigated the sensitivities of the court. In their 'parere' for the volume, two civil censors noted that they had asked Grisolia to amend his original title – *De Principe, sive de Ortu, et Progressu Regiae Potestatis* – to '*De Principe* senza altr'aggiunzione'. Moreover, they had demanded that Grisolia clarify a passage in the manuscript, in which he had failed adequately to report that only 'Monarchomachs, or enemies of royal power' believed that the latter originated in a delegation 'from the people ['a populo']', instead of 'directly from God ['αμέσως a Deo']'.[41]

## Conclusion

In the light of this minatory experience, it is unsurprising that 'political science' failed to find a curricular exponent in *Settecento* Naples: even a strident monarchist could risk censure for their reportage of a theory – framed, as Grisolia protested, by 'doubts [*dubia*]' – that monarchical sovereignty stemmed from popular delegation. Yet there is every indication that an appetite for discussion of 'political science' existed within the Kingdom, in spite of the overhanging threat of censorship for discussing it in print. One of the two censors who compelled Grisolia to amend his manuscript, Giacinto Dragonetti (1738–1818), would serve as an enthusiastic republican in 1799.[42] Joining him as a leading republican was Pagano, whose *Saggi Politici* (1783 and 1792) provide the fullest example of a nominally 'political' work produced in the later eighteenth-century Kingdom. Pagano's *Saggi* were *Politici* because they examined the history of civil society, in a manner expressly indebted to Giambattista Vico's *Scienza Nuova* (1725, 1730, 1744). However, the *Saggi* did not include a systematic investigation of different constitutional forms, and they only tentatively embraced the possibility of a reformed monarchical government for Naples – potentially assisted by an ephorate.[43]

Other members of the so-called 'scuola genovesiana' ventured criticisms of the Kingdom's baronial classes in defence of a centripetal monarchy, assisted by a *ceto* of modernising bureaucrats. However, their criticisms were typically couched in historical synopses of the development of feudalism or in chorographic overviews of the Kingdom's feudal jurisdictions.[44] Were these authors unfamiliar with 'political science'? Did its absence in the curricula that they would have encountered in their schooling or at university deprive them of access to texts in which it was discussed? The answer to both of these questions must be 'no' and the supporting evidence is *reading*.

Each educational institution throughout the Kingdom typically maintained a semi-public library, where matriculants, inmates and inquisitive locals could read or borrow its holdings.[45] Within the capital, surviving catalogues for the Carthusians, Jesuits and Oratorians report vast holdings, rivalling the Biblioteca Brancacciana, the University's de facto library.[46] In addition to the Royal Library, newly enriched by Carlo Borbone's inheritance of Parma's Farnese collection,[47] the later eighteenth-century capital was served by a private but accessible library within the Palazzo Spinelli di Tarsia.[48] Each of these libraries provided their readers with access to 'political science', in its ancient and most modern forms. Indeed, as early 1672, the librarians in the Collegio Massimo recorded the accession of a copy of Thomas Hobbes's *Opera philosophica* (1668), containing within it his *De Cive*, the work in which he purported to invent 'civil science'.[49] The Biblioteca Brancacciana possessed a copy *De Cive* as well, albeit in a section labelled 'libri proibiti'.[50] In this respect, as ever, the pursuit of a 'science' was extra-curricular.

## Notes

1 For the better-known studies see Nicolini, *Un grande educatore italiano*; Ferrone, *Scienza, natura, religione*; Zambelli, *La formazione filosofica*.

2 For provincial seminaries see d'Andrea, *I frati minori napoletani*, 143–55; di Biase, 'La soppressione dei Monasteri'; Amato, 'La formazione'.

3 For the *Certosa di Padula* see Guerrieri, *Per il recupero del patrimonio*.

4 For the Oziosi see Riccio, *Cenno storico*; Comparato, 'Società civile'; De Miranda, *Una quiete operose*.

5 For the Investiganti see Torrini, 'L'Accademia degli Investigati'.

6 For the Medinacoeli see Ricuperati, 'A proposito dell'Accademia Medina Coeli'; Torrini, 'Antonio Monforte'; Conti, 'Paolo Mattia Doria'.

7 For the Accademia delle Scienze see Nicolini, 'Della società di scienze', 8–16; Nardella, 'Celestino Galiani'; Borrelli, 'Istituzioni e attrezzature scientifiche'.

8 For the eighteenth-century university see Beltrani, 'Contributo alla storia della Università'; Schipa, 'Il settimo decimottavo'; Scandone, *L'Università degli Studi in Napoli*; Trifone, *L'Università degli Studi di Napoli*; Sarubbi, 'Lo studio napoletano'; Ascione, *Seminarium doctrinarum*; Cammisa, *L'Università di Napoli*.

9 For attempts to reconstruct the curriculum see Azzinnari, 'Un contributo'.

10 For these reforms see Luongo (ed.), *All'alba dell'Illuminismo*.

11 For Galiani see Nicolini, *Un grande educatore italiano* and Ferrone, *Scienza, natura, religione*.

12 Palladino, 'La riforma'.

13 Amodeo, 'Le riforme universitarie', 4 and 8.

14 'essendo [...] D[ottor]e Giannotti attualm[en]te Professore di Medicina, n[on] istimo utile dell'Un[iversi]tà, che gli si faccia insegnare, etia[m] ad tempus, l'Etica, la Politica e la Morale, [*deleted*: perché queste non son] discipline affatto disparate dalla professione sua', Naples, Archivio di Stato, *Cappellano Maggiore*, *Relazioni*, 718/VII, fo. 92$^{r}$.

15 'Con tal occas[io]ne poi debbo riferire a M[aestà] S[ua] Ill[ustrissi]ma, che da due anni a questa parte essendosi introdotta in questi Studj pubblici la lezione del diritto della natura e delle genti, che contiene i veri principj [*deleted*: dell'Etica, della Morale] dell'Etica [...] si pensava supprimere la [...] Cattedra di Etica [...] p[er] fondarne qualche altra, di cui avesse maggior bisogno l'Un[iversi]tà [...] io sarei di parere [...] di estinguere la d[etta]a cattedra di Etica, con imporre al professore del diritto della natura e delle genti, che nelle sue lez[io]ni insegnasse anche le materie, che in q[ue]lla si trattavano', Naples, Archivio di Stato, *Cappellano Maggiore*, *Relazioni*, 718/VII, fos. 92$^{v}$–3$^{r}$.

16 For a discussion of this episode see Waldmann, 'Natural law'.

17 Waldmann, 'Natural law', 61–3.

18 Naples, Società Napoletana di Storia Patria, ms. XXXI A 2, fos. 86$^{r}$–96$^{v}$.

19 Natale, 'Eclecticismo teoretico', 216–17.

20 For this development see Waldmann, 'Natural law', 66–7.

21 Turin, Archivio di Stato, *Carte Giannone*, mazzo I, ins. 7: Pietro Giannone, 'Parere intorno alla riforma de' Regij Studj di Napoli', printed in Luongo (ed.), *All'alba*, 150: 'La cattedra d'Etica, o sia Morale, non conviene supprimersi perché il professore del Jus di natura e delle genti tien altra incumbenza di questa, né dee trattar di virtú morale e di regolamento di costumi. La Morale dee esser il fine di tutti i precedenti studj filosofici ed i piú gravi filosofi, cosí antichi come moderni, le speculazioni sopra tutta la natura le drizzavano finalmente alla Morale. E se nella classe della Teologia si ritiene la Morale, non è ragionevole che in quella di Filosofia si tralasci questa ch'è la piú nobile sua parte'.

22 'in quo demonstratur, quomodo societas, ubi imperium monarchium locum habet sicut et ea, ubi optimi imperant debet institui, ne in tyrannidem labatur, et ut pax libertasque civium inviolata maneat', Naples, Biblioteca Nazionale di Napoli, MS Branc. IV F 6, title page.

23 For an overview see Smith, 'The language of "political science"'.

24 Smith, 'The language of "political science"', 217.

25 A separate chair in the discipline was only created in 1777; the law of nature and nations was an ex officio curricular addition to the course offered by the morning professor of civil law, rewarded with an annual gratuity of 100 ducati, see Cammisa, *L'Università*, 304 and 331.

26 'Ecce nunc tibi absolutissimam juris naturalis definitionem. Naturale jus est voluntas Dei per rectam rationem humano generi promulgata, quaedam nullo praeeunte hominis facto,

quaedam posito facto hominis jubens, vetansve propositis praemiis, adjectisque poenis, reliqua omnia libero hominis arbitrio permittens. Ecce etiam tibi definitionem Jurisprudentiae naturalis. Ars est regulas tradens, quibus humana ratio praecipientem Dei voluntatem cognoscat, accommodatque ad omnem vitae partem. Jam vero facile intellectu est, qua re naturalis Jurisprudentia, atque Ethica differant inter se. Utraque circa bonum versatur; sed haec bonum spectat, qua hominem perficit, ac vere felicem reddit: illa bonum, qua justum est, seu qua Lege sancitum est: proinde ad posterius hoc bonum obligamur: non item ad prius illud', Cirillo, 'Juris naturalis', 277.

27 For an incisive overview of Doria's *Vita Civile* see Robertson, *The Case for the Enlightenment*, 191–7.

28 Zambelli, 'Il rogo postumo'.

29 For censorship in early modern Naples see Maria Monti, *Dal Duecento al Settecento*; Lopez, *Inquisizione, stampa e censura*; Napoli, *Letture proibite*; Sabato, *Poteri censori. Disciplina*; Sabato, *Il sapere che brucia*.

30 For the book trade in this period see Croce, 'Stampatori e librai in Napoli', reprinted in Croce, *Stampatori e librai*; Rao (ed.), *Editoria e cultura*.

31 For these works see Santoro (ed.), *Le secentine napoletane*. For works produced in Puglia see Tafuri and Virno, 'Edizioni pugliesi'.

32 'Che non avevano [*sc.* i maestri d'etica] quelle cognizioni e quella eloquenza che questa professione ricerca. Per la qual cosa io formai un nuovo piano d'etica', Genovesi, *Autobiografia*, 18–19.

33 Genovesi, *Autobiografia*, 18–19.

34 Reinert, *Translating Empire*.

35 For the identifiable 'concorrenti' see Naples, Archivio di Stato, *Cappellano Maggiore, Consulte*, 751, fos. 398$^{v}$–9$^{v}$ (3 Feb. 1770); Nicola Barone, 'Alessio Aurelio Pelliccia', 4, 17 n. 8; De Luca, *Per la morte*, 28–9; Pagano, *Gli esuli Tebani* [1782] in Pagano et al., *Teatrali contese*, 3.

36 Maria Capece, *Opuscula nunc prima edita*.

37 *Album Professorum Regii Archi-Gymnasii neapolitani* (Naples, 1761–2), preserved as an excerpt in de Felice, *Excerptum totius italicae*, 273.

38 For the *Nunziatella c.*1770 see Martullo Arpago (ed.), *L'Accademia militare*, ix; Rao, 'Esercito e società', 644–5.

39 Naples, Società Napoletana di Storia Patria, II Stanza 01.B. 3/1: Francesco Mario Pagano, *Disegno del sistema della scienza degli ufizj* ([Naples, 1769]), reprinted in *Archivio Storico del Sannio*. For the appointment of Odazi see Di Leonardo, *L'illuminista abruzzese*, 69–71.

40 For Grisolia see Napolitano, 'Il problema della sovranità', 65–79.

41 Michelangelo Grisolia, *De principe liber unus* (Naples, 1783), sig. b2$^{r}$.

42 For Dragonetti see Bruni, 'Su *Delle virtù e de' premi*'.

43 Pagano, *Saggi Politici* [1791–2], 171, 176.

44 For treatments of this literature see Calaresu, 'Images of Ancient Rome'; Naddeo, 'A cosmopolitan in the provinces'.

45 For libraries within the Kingdom of Naples see Trombetta, *Storia e cultura*; Sabato, 'Libri e frati cappuccino'.

46 For these libraries see Padiglione, *La Biblioteca del Museo Nazionale*; Trombetta, 'Viaggiatori stranieri'; Trombetta, 'La libraria di S. Angelo a Nido'; Trombetta, *Storia della biblioteca*; Trombetta, 'Erudizione e bibliofilia'; Trombetta, 'La libreria del Collegio dei Nobili'; Russo, *Storia della Biblioteca*.

47 For the Royal Library see Lusignani, *Catalogus bibliothecae*; Castellano Lanzara, *La Real Biblioteca*; Guerrieri, *La biblioteca nazionale*; Dallasta, *Eredità di carta*.

48 For the Spinelli di Tarsia see Rizzo, *Ferdinando Vincenzo Spinelli di Tarsia*.

49 For this copy see Naples, Biblioteca Nazionale, Sala Farn. 24. C 0025. The copy is inscribed 'Bibliotheca Coll[egii]. Neap[olis]. Soc[ietatis]. Jesu catal[ogus] inscriptus die 28 aug 1672'.

50 Naples, Biblioteca Nazionale, MS Branc. II G 14, *Inventario de' libri prohibiti della libraria Brancaccio*.

# Bibliography

## Primary sources

Naples, Archivio di Stato, *Cappellano Maggiore*, *Relazioni*, 718/VII.
Naples, Archivio di Stato, *Cappellano Maggiore*, *Consulte*, 751.
Naples, Biblioteca Nazionale di Napoli, MS Branc. IV F 6.
Naples, Biblioteca Nazionale, Sala Farn. 24. C 0025.
Naples, Biblioteca Nazionale, MS Branc. II G 14.
Naples, Società Napoletana di Storia Patria, ms. XXXI A 2.
Naples, Società Napoletana di Storia Patria, II Stanza 01.B. 3/1: Francesco Mario Pagano, *Disegno del sistema della scienza degli ufizj* ([Naples, 1769]).
Turin, Archivio di Stato, *Carte Giannone*, mazzo I, ins. 7.

## Secondary sources

Amato, Domenico. 'La formazione del clero e l'opera del Seminario a Molfetta agli inizi del Settecento'. In *Studi in onore di Mons. Antonio Bello*, edited by Luigi Michele De Palma, 255–82. Molfetta: Mezzina 1992.
Amodeo, Federico. 'Le riforme universitarie di Carlo III e di Ferdinando IV Borbone', *Atti dell'Accademia Pontaniana* 32 (1902): 1–30.
Ascione, Imma. *Seminarium doctrinarum. L'Università di Napoli nei documenti del '700, 1690–1734*. Naples: Edizioni scientifiche italiane, 1997.
Azzinnari, Marina. 'Un contributo allo studio del sistema scolastico nel secondo Settecento napoletano. Le fonti documentarie dell'Archivio di Stato di Napoli'. In *Sulle vie della scrittura. Alfabetizzazione, cultura scritta e istituzioni in età moderna*, edited by Maria Rosaria Pelizzari, 297–312. Naples: Edizioni scientifiche italiane, 1989.
Barone, Nicola. 'Alessio Aurelio Pelliccia cattedratico di diplomatica nella R. Università degli studi di Napoli nel primo quarto del sec. 19', *Atti dell'Accademia Pontaniana* 35 (1905): 3–24.
Beltrani, Giovanni. 'Contributo alla storia della Università degli Studi di Napoli durante la seconda metà del secolo XVIII', *Atti dell'Accademia Pontaniana* 32 (1902): 1–88.
Borrelli, Antonio. 'Istituzioni e attrezzature scientifiche a Napoli nell'età dei Lumi', *Archivio Storico per le Province Napoletane* 114 (1996): 131–83.
Bruni, Luigino. 'Su *Delle virtù e de' premi* di Giacinto Dragonetti (e una polemica di Benedetto Croce)', *History of Economic Thought and Policy* 1 (2010): 33–49.
Calaresu, Melissa. 'Images of Ancient Rome in late eighteenth-century Neapolitan historiography', *Journal of the History of Ideas* 58 (1997): 641–61.
Cammisa, Francesco. *L'Università di Napoli nella seconda metà del '700. Documenti e profilo delle riforme*. Naples: Jovene, 2001.
Capece, Gaetano Maria. *Opuscula nunc prima edita*, 2 vols. Naples, 1785–90.
Castellano Lanzara, Maria Giuseppina. *La Real Biblioteca di Carlo di Borbone e il suo primo bibliotecario Matteo Egizio*. Naples: A. Miccoli, 1942.
Cirillo, Giuseppe Pasquale. 'Juris naturalis ac gentium. Prolegomena'. In idem, *Opera omnia*, 2 vols. Naples, 1781–83.
Comparato, Vittor Ivo. 'Società civile e società letteraria nel primo Seicento: l'Accademia degli Oziosi', *Quaderni Storici* 23 (1973): 360–88.
Conti, Vittorio. 'Paolo Mattia Doria e l'Accademia di Medinacoeli', *Il Pensiero Politico* 7 (1975): 204–14.
Croce, Benedetto. 'Stampatori e librai in Napoli nella prima metà del Settecento', *La Strenna della R. Tipografia Giannini* 4 (1892): 137–46.
Croce, Benedetto. *Stampatori e librai nella prima metà del Settecento*. Macerata: Biblohaus, 2010.
d'Andrea, Gioacchino. *I frati minori napoletani nel loro sviluppo storico*. Naples: Laurenziana, 1967.
de Felice, Fortunato Bartolomeo. *Excerptum totius italicae nec non helveticae literaturae*. Bern: October–December 1761.
Dallasta, Federica. *Eredità di carta. Biblioteche private e circolazione libraria nella Parma farnesiana (1545–1731)*. Milan: Franco Angeli, 2010.

De Luca, Giovanni. *Per la morte di monsignor d. Luca Nicola De Luca*. Naples: Angelo Coda, 1827.

De Miranda, Girolamo. *Una quiete operose: forma e pratiche dell'Accademia napoletana degli Oziosi, 1611–1645*. Naples: Federiciana Editrice Universitaria, 2000.

di Biase, Pietro. 'La soppressione dei Monasteri di Terra di Bari nel decennio francese', *Archivio Storico Pugliese* 44 (1991): 135–62.

Di Leonardo, Giovanni. *L'illuminista abruzzese Don Trojano Odazj*. Mosciano Sant'Angelo: Mosciano S.A., 2003.

Ferrone, Vincenzo. *Scienza, natura, religione: Mondo newtoniano e cultura italiana nel primo Settecento*. Naples: Jovene Editore, 1982.

Genovesi, Antonio. *Autobiografia, lettere e altri scritti*, edited by Gennaro Savarese. Milan: Feltrinelli editoria, 1962.

Grisolia, Michelangelo. *De principe liber unus*. Naples, 1783.

Guerrieri, Guerriera. *Per il recupero del patrimonio bibliografico, archivistico, artistico e sacro della Certosa di Padula disperso nell'Ottocento*. Salerno: Centro Raffaele Guariglia di Studi salertini, 1974.

Guerrieri, Guerriera. *La biblioteca nazionale "Vittorio Emanuele III" di Napoli*. Milan: Riccardo Ricciardi, 1974.

Lopez, Pasquale. *Inquisizione, stampa e censura nel Regno di Napoli tra '500 e '600*. Naples: Edizione del Delfino, 1974.

Luongo, Dario (ed.). *All'alba dell'Illuminismo. Cultura e pubblico studio nella Napoli austriaca: Contegna, Vidania, Caravita, Giannone*. Naples: A. Guida, 1997.

Lusignani, Benedetto. *Catalogus bibliothecae farnesianae parmensis*. Parma, 1689–95.

Martullo Arpago, Maria Antonietta (ed.). *L'Accademia militare della Nunziatella dalle origini al 1860*. Naples: Regione Campania, 1987.

Monti, Gennaro Maria. *Dal Duecento al Settecento. Studi storico-giuridici*. Naples: I.T.E.A Editrice, 1923.

Naddeo, Barbara. 'A cosmopolitan in the provinces: G. M. Galanti, geography and Enlightenment Europe', *Modern Intellectual History* 10 (2013): 1–26.

Napoli, Maria Consiglia. *Letture proibite. La censura dei libri nel Regno di Napoli in età borbonica*. Milan: F. Angeli, 2002.

Napolitano, Saverio. 'Il problema della sovranità nella cultura napoletana di fine Settecento: il contributo del mormannese Michelangelo Grisolia'. In *Tra Calabria e Mezzogiorno. Studi storici in memoria di Tobia Cornacchioli*, edited by Giuseppe Masi, 65–79. Cosenza: Pellegrini, 2007.

Nardella, Tommaso. 'Celestino Galiani e l'Accademia degli Illuminati', *Archivio Storico Pugliese* 35 (1982): 453–8.

Natale, Maria. 'Ecletticismo teoretico e pragmatismo, alle origini delle riforme illuministiche. L'autobiografia di Celestino Galiani', *Frontiera d'Europa* 1 (2002): 115–219.

Nicolini, Fausto. *Un grande educatore italiano: Celestino Galiani*. Naples: F. Giannini, 1951.

Nicolini, Fausto. 'Della società di scienze, lettere e arti di Napoli e di altre accademie che la precederono', *Accademie e Biblioteche d'Italia* 25 (1957): 3–33.

Padiglione, Carlo. *La Biblioteca del Museo Nazionale nella Certosa di S. Martino in Napoli ed i suoi manoscritti*. Naples: F. Giannini, 1876.

'Pagano Francesco Mario. Disegno del sistema della scienza degli ufizj ([Naples, 1769])', edited by Francesco Berti. In *Archivio Storico del Sannio* 15 (2010): 153–74.

Pagano, Francesco Mario. *Gli esuli Tebani* [1782]. In *Teatrali contese*, edited by Alberto Granese, 1–70. Salerno: EdiSud, 1999.

Pagano, Francesco Mario. *Saggi Politici [1791–2]*, 2 vols, edited by Luigi Firpo and Laura Salvetti Firpo. Naples: Vivarium, 1993–2004.

Palladino, Franco. 'La riforma dello studio napoletano nella prima metà del Settecento e l'istituzione della cattedra di fisica sperimentale', *Rendiconti della Accademia Nazionale delle Scienze detta dei XL* 9 (1985): 333–6.

Rao, Anna Maria. 'Esercito e società a Napoli nelle riforme del secondo Settecento', *Studi Storici* 28 (1987): 623–77.

Rao, Anna Maria (ed.). *Editoria e cultura a Napoli nel 18. secolo*. Naples: Liguori, 1998.

Reinert, Sophus A. *Translating Empire: Emulation and the origins of political economy*. Cambridge, MA: Harvard University Press, 2011.

Riccio, Camillo Minieri. *Cenno storico intorno all'Accademia degli Oziosi in Napoli*. Naples: Stamperia della R. Università, 1862.
Ricuperati, Giuseppe. 'A proposito dell'Accademia Medina Coeli', *Rivista Storica Italiana* 84 (1972): 57–79.
Rizzo, Vincenzo. *Ferdinando Vincenzo Spinelli di Tarsia. Un principe napoletano di respiro europeo 1685–1753*. Aversa: Macchione, 1997.
Robertson, John. *The Case for the Enlightenment*. Cambridge: Cambridge University Press, 2005.
Russo, Francesco. *Storia della Biblioteca teologica "S. Tommaso" di Napoli*. Florence: Olshki, 1980.
Sabato, Milena. *Poteri censori. Disciplina e circolazione libraria nel Regno di Napoli fra '700 e '800*. Galatina: Congedo, 2007.
Sabato, Milena. *Il sapere che brucia. Libri, censure e rapporti Stato-Chiesa nel Regno di Napoli fra '500 e '600*. Galatina: Congedo, 2009.
Sabato, Milena. 'Libri e frati cappuccino di Puglia in età moderna: un'indagine tra proibito e permesso'. In *La provincia dei frati minori Cappuccini di Puglia. Cinque secoli di storia*, edited by Rosa Anna Savoia, 37–58. Lecce: Edizione Grifo, 2010.
Santoro, Marco (ed.). *Le secentine napoletane della Biblioteca nazionale di Napoli*. Rome: Instituto poligrafico e zecca dello stato, 1986.
Sarubbi, Antonio. 'Lo studio napoletano nella cultura meridionale del Sei-Settecento', *Archivio Storico per le Province Napoletane* 16 (1978): 231–45.
Scandone, Francesco. *L'Università degli Studi in Napoli nel Settecento: ordinamento-concorsi-locali*. Santa Maria Capua Vetere: Tip. Cavotta, 1927.
Schipa, Michelangelo. 'Il settimo decimottavo'. In *Storia della Università di Napoli*, edited by Nino Cortese, Riccardo Filangieri di Candida, Gennaro Maria Monti, Luigi Russo, Michelangelo Schipa, Francesco Torraca and Alfredo Zazo, 433–66. Naples: Riccardo Ricciardi, 1924.
Smith, Sophie. 'The language of "political science" in early modern Europe', *Journal of the History of Ideas* 80 (2019): 203–26.
Tafuri, Maria Teresa and Maria Virno. 'Edizioni pugliesi dei secoli xvi–xviii nella Biblioteca Nazionale di Bari', *Archivio Storico Pugliese* 36 (1983): 55–96.
Torrini, Maurizio. 'Antonio Monforte, uno scienziato napoletano tra l'Accademia degli Investiganti e quella palatina di Medinaceli'. In *Ricerche sulla cultura dell'Italia moderna*, edited by Paola Zambelli, 96–146. Bari: Editori Laterza, 1973.
Torrini, Maurizio. 'L'Accademia degli Investigati. Napoli 1663–1670', *Quaderni Storici* 48 (1981): 845–83.
Trifone, Romualdo. *L'Università degli Studi di Napoli dalla fondazione ai giorni nostri*. Naples: Università di Napoli, 1954.
Trombetta, Vincenzo. 'Viaggiatori stranieri nelle biblioteche napoletane del Settecento', *Rivista Italiana di Studi Napoleonici* 31 (1994): 143–68.
Trombetta, Vincenzo. 'La libraria di S. Angelo a Nido', *Accademie e Biblioteche d'Italia* 62 (1994): 11–43.
Trombetta, Vincenzo. *Storia della biblioteca universitaria di Napoli dal Viceregno spagnolo all'Unità d'Italia*. Naples: Vivarium, 1995.
Trombetta, Vincenzo. 'Erudizione e bibliofilia a Napoli nella prima metà del XVIII secolo: la biblioteca di Domenico Greco', *Rara Volumina* 4 (1997): 59–91.
Trombetta, Vincenzo. *Storia e cultura delle biblioteche napoletane*. Naples: Vivarium, 2002.
Trombetta, Vincenzo. 'La libreria del Collegio dei Nobili e le biblioteche dei Gesuiti, a Napoli, tra Sette e Ottocento'. In *Educare la nobiltà*, edited by Gianfranco Tortorelli, 123–63. Bologna: Pendragon, 2005.
Waldmann, Felix. 'Natural law and the Chair of Ethics in the University of Naples, 1703–1769', *Modern Intellectual History* 19 (2022): 54–80.
Zambelli, Paola. *La formazione filosofica di Antonio Genovesi*. Naples: Morano, 1972.
Zambelli, Paola. 'Il rogo postumo di Paolo Mattia Doria'. In *Ricerche sulla cultura dell'Italia moderna*, edited by Paola Zambelli, 149–201. Bari: Editori Laterza, 1973.

# Part II

# Environment, disaster and disease in the Bay of Naples

# 3

# Giambattista Della Porta's *De aëris transmutationibus*: natural philosophy and the Earth sciences

Alfonso Paolella

> They divide Magic into two parts, the first they call infamous, as it is unclean and besmirched by unworldly spirits and commerce with demons [...] The other is natural, it is honoured and revered as the active part, the principal part of philosophy, which produces its marvellous effects [...] you should not believe that magic is anything other than the works of nature herself.[1]

## Introduction

In 1604 Federico Cesi (1585–1630), the youthful founder of the Accademia dei Lincei, arrived in the city of Naples, drawn by Giambattista Della Porta's reputation for producing scientific knowledge. He wanted to expand his institution within the large and cultured city of Naples. Della Porta eventually joined the academy in July of 1610, just a few months prior to Galileo Galilei. Della Porta's membership of the academy was mutually advantageous. The title 'Academico linceo' conferred dignity and honour upon the Neapolitan scholar, while his renown throughout Europe as a celebrated scientist lent enormous prestige and lustre to the nascent institution.[2] It was intended that the Neapolitan branch of the Lincei would be called 'Liceo' and Della Porta would be awarded the title of 'Viceprincipe', but after a series of negotiations it never opened.

Della Porta's reputation rested above all on two works that enjoyed success up until the beginning of the Enlightenment, namely the *Magia naturalis* and *Humana physiognomia*. Editions and translation of these two works circulated across Europe, reaching France, Germany, the

Low Countries and England. This chapter will consider one of his lesser-known works, *De aëris transmutationibus*. The work was published in 1610, at the behest of Federico Cesi, who wanted Della Porta to provide an original work that could provide a critique of Aristotle's *Meteorologica*. It was to be the last of Della Porta's works.

## The intellectual context of Della Porta's works

During the sixteenth century, philosophers and scientists began to set the Renaissance on a new course. Following the enthusiasm for translating works that existed between the fifteenth and sixteenth centuries, a new conception of nature began to emerge from the middle of the century which sought to free itself from the 'ipse dixit' of Aristotle. There was a growing recognition in many places that the conclusions drawn by the Stagirite no longer held up in the face of experimental evidence, and that one should only accept that which had been established and proved.

During this crisis Aristotle's works of logic, rhetoric, poetry and metaphysics – cultivated especially in Padua by such scholars as Nicoletto Vernia (d.1499), Agostino Nifo (1470–1538), Pietro Pomponazzi (1462–1525), Francesco Piccolomini (1523–1607) and Jacopo Zarabella (1535–1589) – continued for many years to be an object of often harsh discussion. On the other hand, Florentine Neoplatonism – whose exponents included Marsilio Ficino (1433–1499), Giovanni Pico della Mirandola (1463–1494), Angelo Poliziano (1454–1494), Nicolas of Cusa (1401–1464), Leon Battista Alberti (1404–1472), Bartolomeo Scala (1493–1497) and Cristoforo Landino (1424–1498) – set out a formidable doctrinal apparatus. They sought to decipher nature by using the magico-hermetic and Neoplatonic traditions, which regarded it as a manifestation of the Plotinian 'One'. For this reason, during the Renaissance understanding the relationship between the individual and the cosmos constituted an essential part of philosophico-scientific debate.

These debates raised important questions about the relationship between human knowledge and revealed faith. From the medieval period, the central problem had been the co-existence of a science of nature with Christian theology. The latter discipline presented a quite different vision of the world, both in terms of its content and in the methods used to apprehend nature. Nature and divine will, a theory of natural causes and miraculous explanation, faith and argumentation – none of these elements could be easily reconciled, even if Alan of Lille

(1128–1202/03) could observe: 'Every creature of the world/ as if a book or picture/ is to us a faithful mirror/ of our life/ our death/ our status/ our fate'.[3] Certainly, already in the ambit of the twelfth-century School of Chartres, and above all between the thirteenth and fourteenth centuries, a model for their relationship was elaborated in which the fields of natural science and revealed theology could be balanced, not distinguished by repurposing and renewing the concept of the Anima mundi taken from Plato's *Timaeus*.

This idea was in turn taken up once more by Ficino.[4] Admitting the possibility of a direct intervention in natural affairs on the part of God, the absolute master of the world, the practitioners of 'physics' recognised the limitations of their arguments – but, contemporaneously, sought their autonomy. Founded on human natural reason, physics could not go beyond its limited capacity. However, by exercising this capacity within this field philosophers could, without compromising theologians' claims to superior knowledge, dismiss all external interference.

Della Porta's scientific activity was ostensibly inspired by the Hermetic-Neoplatonic tradition, but it seems that this philosophy was only a façade that he had chosen. His work was characterised by a constant preoccupation with offering precise verifications of his research into the philosophy of nature. This he did by observing natural phenomena that produced 'marvellous effects' and seeking out their causes. This restless desire to understand is present above all in those works deemed to be at the boundaries of science and Christian dogma and which possibly verged on heresy, such as the *Magia naturalis* and *Human physiognomia*. In almost all of those works, Della Porta started, as in the *Magia naturalis* (1589),[5] with methodological self-awareness. When describing phenomena, he would first enunciate the opinions of the ancients and then, having submitted them to the scrutiny of experience, affirm whether they were true or false.

Magic was the dominant form of thought throughout his life. The natural magician was someone who, through his skilful understanding of nature, could distinguish between 'miracles' (in the sense of wonders, spells or marvels) and the most occult and recondite aspects of nature. The magus was thus the only person in a position to reveal and comprehend nature's secrets. Everything was explicable according to the laws of nature, including supernatural and preternatural phenomena. Della Porta tended to dissimulate his opinions, but he sometimes revealed them in an indirect manner. For example, although he noted that one must accept the existence of witches as a matter of faith, he suggested that in natural reality devils and witches did not exist. Everything that was attributed to the actions of witches or devils could, in fact, be explained and understood by natural means. The witches' salve, for example, contained various hallucinogenic substances that acted upon the minds of those poor women who had confessed to unspeakable acts.[6] If his true intention was to disenchant the magical, that is, to interpret magical traditions as natural practices freed from associations with superstition or demonism, in popular opinion his reputation occupied an ambiguous position between a natural and ceremonial magician.[7] A semantic ambiguity remained rooted in the term 'magician'.

The constant presence of Della Porta's elder brother Giovan Vincenzo, a scholar with a deep and precise knowledge of ancient sources, including the rarest texts, allowed him to order his materials in a logical and reasoned manner according to the schemes used in the works that inspired him.[8] The result was a formidable and authoritative encyclopaedic *summa*, in the medieval sense, treating a single theme. When researching sources, Della Porta tended to downplay the significance of more recent examples, possibly because he believed that they lacked the 'authoritative' force of those authors regarded as 'auctoritates'. All scientific works, whether written in Greek, Latin or Arabic, were systematically searched and the list of works consulted was extensive.

It is enough to consult the sources referenced in the volumes published in the 'Edizione Nazionale delle Opere di Giovan Battista Della Porta'[9] to realise the great number of citations and to imagine the quantity of volumes that he must have held in his library. This also helps to explain why the library, among other things, tempted Cesi. Following Della Porta's death, the Roman prince wanted to relocate the library to the palace being built by the Neapolitan 'Liceo'. We possess no knowledge of the fate of this vast library, but it was probably widely dispersed.

For Della Porta, the subject matter treated in *De aëris transmutationibus*[10] was most suitable and appropriate for realising and proving his programme of research. That is, it offered the opportunity to demonstrate the marvels of nature either as a methodology in the modern sense, namely the study of such phenomena as the dynamics of winds, rains, storms, lightning and meteors, or as an Earth science studying volcanoes, earthquakes, formation of seas and thermal waters according to the scheme that structured and composed Aristotle's *Meterologica*.

Although *De aëris* enjoyed only a limited circulation, it assumed its place in a long and rich line of research on nature that reached back to the age of the so-called pre-Socratics. This research had also interested poets such as Lucretius in *De rerum natura* and Pontano in *Meteorum libri* (1490), who had treated the material in a manner that straddled the scientific and the imaginative. The major authorities in this field included Plato, Aristotle, Seneca, Pliny the Elder and Theophrastus. Considered to be of no less importance were Isidore of Seville's *Etimologiarum libri*, Bartolomeo Anglico's *De proprietate rerum* and Albertus Magnus's *Librum meteorum*, as well as the numerous commentaries on Aristotle's *Meteorology* produced by Alexander of Afrodisias. The translation of Aristotle's natural works from the mid-twelfth century opened new perspectives; however, they appeared to medieval scholars as an organically structured body of knowledge that would enable them to analyse the physical reality of the world in its entirety.

In his research, Della Porta departed from existing traditions of cosmological thought embodied in the work of scholars such as Nicholas of Cusa and Ficino, who believed in the unity of the cosmos from a religious or theological perspective. Instead he followed the example of Pico. In his condemnation of divinatory astrology *Disputationes adversus astrologiam divinatricem* (1494), Pico had separated the spiritual world of man from the natural. Already in this work astrology assumed the task of integrating two hitherto separate fields and, most importantly, seeking to offer a naturally determinist account of the whole human world from an act of will to fate.

In his writings, Della Porta divided the cosmos into two regions: the astral and the sublunary. In *Coelestis physiognomia* (1603) he continued to talk of both, but in *De aëris*, written seven years later, he completely ignored the astral region. Although Della Porta did discuss falling stars and meteors, he did so because he believed that they belonged to the sublunary realm (III, 3). Moreover, from the start of the sixteenth century, following the discovery of the New World, Aristotelian doctrine was in severe crisis; limited in scope to that which was known in the ancient world, it could not be used to explain natural phenomena observed in these new lands. While it was not possible to recover experimental proofs, it was possible to critique Aristotle's use of deductive reason – such as his claims that in the southern hemisphere humans must be standing upside down, or that on account of the great heat in regions around the equator, it was likely these areas would be completely uninhabited.

The Aristotelian system was nevertheless reasonably flexible: it was capable of accommodating new experiences and Aristotle had already admitted that the sublunary realm could be known with increasing degrees of certainty, as one reads in *Meteorologica*. As we shall see, this was in marked contrast to the manner in which Aristotelianism had become restricted by Renaissance hermeneutics. Contemporaries maintained that while geometry could describe the motion of the stars and the planets with precision, because their movements were regular and eternal, the sublunary realm offered only disordered and episodic change. Knowledge of this realm was consequently hypothetical and approximate; above all, it was impossible to predict phenomena that would occur there.[11] Precipitation, earthquakes, volcanic eruptions and the origins and movements of the seas, like the origin of rivers and underground springs, were all unforeseeable and inexplicable. It was therefore impossible to provide an account of the true cause of natural effects; instead, one could offer only provisional knowledge established through conjecture of the material and efficient causes of a given phenomenon.

According to the obstinate dogmatism of many Aristotelian commentators, especially during the sixteenth century, applying the rules of logic to 'materia naturale' signified, in the first place, reading and interpreting the fundamental texts of a single science, reducing them where possible to the form of a syllogism. This involved investing them with a unity and a systematicity far removed from Aristotle's own philosophy. This methodological system thus ended up reducing the Aristotelian theoretical framework to a cage without an exit, leaving behind the dynamic capabilities of the Stagirite's method. It was precisely this hermeneutic 'sickness' within Aristotelianism that was so conducive to the anti-Aristotelian revolution. Della Porta aligned himself with these critics of Aristotelianism, and in almost all his works one finds chapters that attempt to demolish Aristotelian texts: *Aristotelis confutationes*.

During the Renaissance *Meteorologia* was considered as the fourth part of Aristotle's philosophy after the *De physica*, *De caelo* and *De generatione et corruption animalium*; all featured in the curriculum of university students. According to the statute of the University of Bologna, for example, students studied these texts in the second year of their course. Some commentators were also studying meteorology in order to sustain the view that all of natural philosophy was a conjectural science.

This point of view emerges clearly in the works of Agostino Nifo (1473–1538), who adopted the so-called method of *regressus/ progressus* when practising natural philosophy. It was theorised by Paduan Aristotelians, in particular Jacopo Zarabella (1533–1589),[12] and it allowed for a greater logical certainty when investigating the variability and uncertainty of sublunary phenomena. The correctness of inductive and deductive reasoning is proved by the evidence of the consequent results. One can proceed 'from the effect to the cause' or 'from the cause to the effect'. In other words, the theory sustained that through the perception of the senses it was possible to establish by means of induction the 'quid est' (what the thing is) and therefore fix empirical truth on the basis of natural observations. Deduction, by contrast, explains the causes – that is, the reason *why* a given thing exists. This method, also adopted by Della Porta, is considered to be the root of the 'scientific method' that entailed searching for the cause of phenomena from within the system of nature itself.

In particular, the aforementioned *Commentaria* on *Meteorologica* of Agostino Nifo and Francesco Vimercati confirm this concrete methodology that placed in doubt the complex edifice of Aristotle's thought. This was especially the case in light of the experience of sixteenth-century explorers such as Christopher Columbus, Giovanni Cabato, Amerigo Vespucci, Filippo Sassetti, Ferdinand Magellan and Giovanni da Verrazzano. Travellers' accounts and empirical evidence demonstrated that – contrary to Aristotelian logical deductions which he had used to establish that the coast of Africa and the equatorial regions must be deserts on account of the excessive heat of this torrid zone – were in fact inhabited.[13]

The impotent Aristotelian system endured, however, in the sense that his works both continued to provide a model for presenting materials relating to a particular theme and treating them as part of a single coherent system, in the imposition of an irrefutable demonstrative method, or in the resolution of problems that started with sensible experience. In any case, Aristotelianism's total collapse, which would arrive with the rise of enlightened rationality, remained in the future. The problem that contemporaries faced was, above all, how to make nature co-exist and agree with theology, as Cusanus had sought to do. How could one reconcile divine will with the theory of natural causes, that is, faith and philosophy, and above all provide an epistemological justification for a philosophy of nature in which meteorology could be included?

For these reasons, Pietro Pomponazzi, his adhesion to the thought of Aristotle notwithstanding, actively encouraged his Paduan students to doubt claims of established and absolute knowledge. His doubts were above all born of his awareness of the contradictions between texts and the diverse interpretations placed on them, logic and experience. Diverse fields of study required differing methods and conferred corresponding levels of certainty. Deductive arguments, for example, could clarify the nature of mathematical entities, such as lines and forms, but they could not be applied to politics – a sphere that, although ordered to a certain degree, could not be understood with the same degree of certainty as mathematics.

Della Porta was assailed by the same doubts. In the preface to *De aëris* he clearly affirmed that he would prefer to risk making mistakes by using only experimental knowledge of phenomena than to appear learned by following the false doctrine of *auctoritates*.[14] It was this methodological freedom that allowed Della Porta to develop his encyclopaedic interests, leading him to invent or intuit the principle theories and practices of some objects that feature today in our daily lives.[15] Let us now turn to consider *De aëris transmutationibus*.

## *De aëris transmutationibus*

For Aristotle, meteorology was not a predictive but a descriptive discipline. The themes included in this field were precipitation, rainbows, comets and meteors (which he held to be created in the sublunary region), earthquakes, movements of the seas and rivers and underground springs. As in all fields of natural philosophy, the object of meteorology was to establish the causes of phenomena. According to Aristotle, the four elements (water, air, earth and fire) made up two exhalations, one wet and vaporous, the other hot and smoky; these exhalations circulated between the surface of the Earth and the moon, that is, the threshold between the terrestrial and celestial regions. The elements and their exhalations were the material causes of meteorological phenomena. The efficient causes were provided by the motions of the celestial bodies, in particular the sun, which drove the exhalations and caused their transformations.

The fact that Della Porta divided his treatise into four books, and followed the order and structure of Aristotle's *Meteorologica*, almost gives it the appearance of a commentary on this text. It is not possible in this chapter to discuss all of Della Porta's arguments, but one can note that the structure of each argument was consistent. As we have already seen, for the most part they followed the methods of the medieval 'Summa' – that is, it presented an exposition of the theories of diverse 'auctores', then raised objections (often not sustained by robust argumentation), before finally providing an account of the correct theory. In many cases, these explanations did not greatly differ from those offered by the 'auctoritates'. The analysis that I offer here will be solely concerned with a limited range of examples.

The first book of *De aëris* addresses the problem of defining air and its properties, vapours, the effects of solar heating, the classification and movements of the winds, generation and corruption of the airs and the diseases associated with the air such as plague and other epidemic illnesses.

By addressing these themes, Della Porta was participating in a contemporary debate about the heat of the sun. Other thinkers who had engaged with this question included Girolamo Fracastoro (1467/8–1553) in his *Homocentricorum sive De stellis liber unus* (1538) and Bernardino Telesio (1509–1588) in *De rerum natura iuxta propria principia* (1570). The question they addressed was whether heat, which is a property of bodies, was derived from light. Since lunar light did not produce heat, as Francis Bacon observed again in 1620 in the 'tabulae absentiae' of the *Novum organum* (II, XII),[16] one had to conclude that heat and light are two different entities. Although light was often associated with heat, they were not necessarily linked.

The problem of the heat of the sun could be resolved, Della Porta maintained, by using principles of refraction and reflection derived from optics. He believed that the ideas sustained by Neoplatonists such as Proclus and Iamblichus, namely that light was a body, were not correct. Aristotle's belief that the sun heated the sublunary world by means of the speed of its movement towards the Earth was no more accurate (*Meteorologia*, 341.a). The capacity of the rays of the sun to warm the Earth, Della Porta instead argued, were affected by the manner in which they reflected on the concave surface of the Earth. If the sun was in a position perpendicular to the Earth, as when it was found in Cancer, for example, the rays did not reflect anywhere; they therefore produced an intense heat. When the sun gradually moved into a more oblique position relative to the Earth, however, the rays became dispersed through refraction, reducing the concentration of heat and provoking the cold of winter (I, 7).

According to Aristotle, the material cause of winds should be attributed to exhalations, whether wet and vaporous or warm and smoky, that circulated in the sublunary space. For Della Porta (1, 16), winds were instead caused by humid air being heated and rising, thus leaving behind an empty space into which new air would be drawn. This idea was proved by the fact that in summer there was less circulation of air because, since the air was all heated, it could not induce any movement in the atmosphere. The concept of atmospheric pressure had not yet been identified – it was indeed discovered by Evangelista Torricelli (1608–1647) some 30 years after the publication of Della Porta's work – but he was nonetheless close to this idea. The wind, therefore, did not derive from exhalations but rather from evaporation.

He also maintained that this was the cause of secondary winds such as the Etesian and 'orniti', as well as hurricanes and typhoons. Among the other properties of the wind is that typical of the 'foehn' (I, 20) – a wind local to Lisbon (or ancient Olisipo) that, according to ancient sources including Columella, Pliny, Solinus and Virgil, had the power to make horses and vultures fertile. This legend is still alive in Portugal today.[17]

Della Porta also argued that if evaporation became corrupted, it would also corrupt the airs and, at times, produce epidemics. The putrefaction of the air was caused by excessive humidity being transported by winds that blew from pestiferous locations such as infected marshes, lakes and ponds. The regions worst afflicted could be found above all close to Sardis and Hierapolis in Turkey or in Egypt, from where the Athenian plague of the fourth century BC originated. Della Porta also listed the signs that presaged the outbreak of plague; these included, for example, a large presence of flies, birds abandoning their nests in order to avoid foetid air, bread quickly becoming stale and eggs suddenly going bad. To avoid the plague, it was enough to burn odoriferous wood such as spruce or pine, as Guy de Chauliac (1300–1368), physician to the papal court in Avignon had done during the Black Death of 1348 to protect Pope Clement VI (1291–1352), or to sprinkle perfumed unguents (1.30).

The second book of *De aëris* described atmospheric phenomena such as rain, snow, hail, dew and frost. Following the customary model found in translated commentaries of Aristotle, he dedicated entire chapters to prodigious events such as rains of blood (II, 20). This was caused by raindrops being saturated with yellow sand from the Sahara or red earth from Armenia; when they fell, they appeared to resemble drops of blood. Other examples included rains of frogs, fish or small animals (II, 21) – that is, a phenomenon in which these creatures fell from a great height after a whirlwind had lifted them from the surface of a pond or marsh. He also maintained that frogs, and also some types of fish, could be born from the mud that formed after rains. Della Porta, like all scientists of his age, believed in spontaneous generation.

The third book of *De aëris* was dedicated to thunder and lightning, but it also treated falling stars and comets. For Aristotle, lightning was the result of a dry exhalation that was released from clouds following the condensation of air into water. This exhalation, Aristotle argued, was expelled from the densest part of the cloud towards the ground, just as seeds squirt through fingers when one tries to crush them. According to Aristotle, thunder was caused by impact of the dry exhalation striking against the cloud. Della Porta objected to this by posing the question of why, if heat tended to rise, did lightning, which is hot, descend towards the ground? He instead argued that lightning in fact descends from the clouds because each body tends to move towards the centre of the Earth (III, 11). The problem of thunder and lightning must have fascinated Della Porta because he dedicated no less than 11 chapters of this book to this subject. He discussed the various types of lightning, their consequences and the fear that they produced, but refuted the ancient Etruscan and Roman belief that lightning bolts could be interpreted as portents.

Finally, the fourth book of the work discussed the seas: their salinity, the motion of the waves, lunar influxes, rivers, springs and thermals. An interesting question addressed by many of Aristotle's commentators was from where did the seas originate and why were they salty? Many, taking inspiration from Homer and Hesiod, maintained that the sea had always existed and that the land emerged as the waters dried out. Della Porta objected: if this were true, over time the oceans and the seas would, through a continued process of drying, simply disappear. Plato for his part argued in *Phaedo* (112a) that the sea was produced by subterranean waters, which also produced rivers. Aristotle rejected this idea, noting that water, like all heavy bodies, tended to descend towards the centre of the Earth and therefore could not rise.

Having reviewed the opinion of all the various philosophers, Della Porta prudently sought refuge in the explanation offered in Genesis (9, 1–4), in which God gathered the waters together in one place in order to make dry land appear. Telesio had offered a different explanation. The sea, he argued, could not be derived from rivers for their waters were not salty. He also rejected Aristotle's idea that the seas' salinity depended on evaporation, maintaining instead that the salinity came from salty soils that formed due to the excessive heat of the sun. When the heat of the sun is intense, it liquefies the ground, which is initially dense. It passes through various forms, successively being transformed into soft, gelatinous, viscous fluid and then into vaporous states. Vapour, raised skywards by the winds, forms clouds. The denser parts of these clouds produce hailstorms and snow; the less dense part produces rain. This created a perennial cycle. Perhaps intentionally, Della Porta ignored this explanation (the first edition of Telesio's *De rerum natura* was published in 1570). He argued instead that if the salinity of the sea was generated by the rocks and the naturally salty terrain, the depths of the sea could be too.

Della Porta then moved on to address the question of volcanic eruptions and earthquakes. Like Lucretius, he maintained that these were caused by underground winds that circulated in caves, heating the internal walls of the volcano.[18] For this reason, Della Porta distinguished between two 'airs': one that circulated on the surface and the other underground. As we have seen, he had already discussed the surface area in Book I. In the subsoil there existed another air that could penetrate through the caverns and the gaps in the earth; this could be variously expanded or compressed by cold and heat. Such a phenomenon was affected by the transition from day and night and by the changing seasons. The idea that winds played a preponderant role in the causation of volcanic eruptions would endure until the end of the eighteenth century.

What was the cause of volcanic eruptions? Underground heat and coldness were often held to be the cause of both eruptions and earthquakes. According to Plato, there existed subterranean rivers that were both hot and cold and a great central river of fire, the *Piriflegetonte*, that nourished all volcanoes. For his part, Aristotle believed that fire was created by friction, caused by winds passing through the fractures in subterranean rocks. Georgius Agricola (1494–1555), the father of mineralogy, proposed that eruptions were caused by water vapours under pressure, thereby incorporating elements of mineralogy into his explanation.

Johannes Kepler (1571–1630) developed his seismic doctrine after having read Julius Cesare Scaliger's *Exercationes exotericae* (1557) – that is, his commentary on Girolamo Cardano's *De subtilitate* (1550). In his *Mysterium cosmographicum* (1596), Kepler theorised the presence of an *anima* similar to the force that guided planetary movement. He used the same metaphor in his *Harmonices mundi* (1619) to explain the phenomenon of earthquakes. A review of treatises on earthquakes produced in the fifteenth and sixteenth centuries, such as *De terraemotu*, composed in 1456 by Giannozzo Manetti (1396–1459), *De his qui in aere fiunt et de terrae motibus*, a minor work written by Telesio at some point after 1586, Pirro Ligorio's 1571 work, *Il libro dei diversi terremoti* and the works of various ancient authors produced no new theories.

With regard to seismology, Della Porta did not deviate far from contemporary authors. He maintained that the causes of earthquakes were similar to those of volcanoes, arguing that the bowels of the Earth contained an enormous amount of bitumen which at times produced earthquakes at other volcanoes. Posidonius had already acknowledged the existence of two types of earthquake. He described them, employing terminology still in use today, as 'shaking' and 'undulating':

> One, I say, is the shaking motion, when the earth is shaken and it moves up and down; the other is undulating motion, during which

> the earth folds itself from its flanks with alternating movements similar to those of a boat.[19]

Della Porta believed the cause of earthquakes to be a thunder of the subsoil, because each earthquake was preceded by roaring noises just as thunderclaps are the earthquakes of the air that shake the clouds. The earthquake was, in fact, caused by fire in the subsoil that mixed with bitumen and sulphur, an opinion that, as he noted, Apollonius of Tyana had also maintained:

> Apollonius of Tyana speaking of this matter with regard to Etna, said the earth mixed with bitumen and sulphur burns, although occasionally they are not emitted because if the earth contains caverns as soon as some spirit (air) penetrates through it, it brings a flame, which increases. Flowing down from the mountain, it spreads in the field.[20]

In Della Porta's view, the fire heated the air in the underground caves to the point that they sought an aperture through which they could escape, thereby shaking the earth all around. For this reason, the odour of bitumen and sulphur can be smelled if the earthquake precedes a volcanic eruption, as Pliny had observed occurred prior to the eruption of Vesuvius in AD 79. The same phenomenon, he observed, had also occurred in the area around Pozzuoli, on Mount Etna and Stromboli (IV, 3). Della Porta also rejected Aristotle's opinion that earthquakes occurred in autumn because the Earth, experiencing the first spells of cold weather, reacted by shivering (IV, 46).

Yet of all the earthquakes and volcanic eruptions, the one that amazed Della Porta the most was the sudden and terrifying earthquake that occurred between 29 September and 6 October 1583 in the region of Pozzuoli. This quake caused the ground to rise up and form Monte Nuovo (New Mountain). For some strange reason, the testimony included in this work on the appearance of this mountain has never been considered by historians of Campi Flegrei, a region that has been celebrated and much visited since antiquity on account of its natural beauty and thermal springs. Throughout the medieval period these springs were developed by the creation of establishments, hospitals and even the creation of Pozzuoli's own medical school.

The phenomenon of the appearance of the Monte Nuovo, Della Porta wrote, had already begun two years earlier, with the whole region being 'shaken by perpetual earthquakes' (*perpetuis terraemotibus conquassata*) (IV, 44. 6). On Saturday 22 September a rapid bradyseismic movement caused the sea to recede by around 350 metres. On the following day the surface of the soil suddenly dropped, creating a depression. The earth began to swell in the centre of this depression, until it formed crevasses. This swelling, which measured some 130 metres, eventually collapsed, in turn forming a crater from which stones, fire, smoke and dry, muddy ashes erupted. The eruption was accompanied by fierce explosions that could be heard as far away as Naples.

During the night, the eruption destroyed Monticello del Pericolo, along with the *Balneum Tripergulae* situated in the adjacent village. Also lost were the springs *Balneum Circeronis*, *Arcus e Reynerii*; the *Balneum de Scrofa*; the *Balneum de Sancta Lucia* and the *Balneum de Cruce*, which were known to Pietro da Eboli. The spring known as *Balneum Ciceronis* (or *Bagno del Prato*) disappeared, and in addition to the remains of Cicero's villa, known as *Cumanun* (or *Academia*), the eruption destroyed also the villa of Faustus Cornelius Silla, nephew of the dictator.[21] Even Santa Marta's Hospital and its adjoining church, founded by Charles III of Anjou at the end of the thirteenth century for the purpose of offering hydrothermal cures to the poor, was destroyed. A further five thermal baths, two guesthouses for rich patients and a Roman building with a cupola of remarkable style and proportion designed by Giuliano Sangallo (1445–1516) were also reduced to rubble.

In the following days periods of dormancy alternated with phases of intense volcanic activity. This culminated in a sudden final eruption that killed 24 imprudent observers who had travelled from Naples to witness the events. The chronicles recount that on the second day of the eruption the inhabitants invoked the assistance of Saint Januarius by undertaking a penitential procession bearing the silver bust of the saint that to this day contains his skull; the procession departed from Naples and concluded in Pozzuoli. The devotion of the faithful notwithstanding, seismic activity continued in the area for a further five days. In Della Porta's account, the absence of any suggestion of divine intervention, whether for protection or punishment of humans, is striking.

There were diverse interpretations of these events. Many of them took their lead from Camillo Porzio,[22] a Neapolitan physician and scientist. A follower of Aristotle, he had attended the lectures of Pomponazzi at the University of Padua. In his own account of the earthquake, contained in *De aëris*, Della Porta argued against his ideas.

Porzio maintained that the sun, in heating up the dry earth, caused exhalations, which in turn made it easier for things to catch fire. If the fumes of the exhalations built up in caves, only some of them escaped; others remained trapped there. Pozzuoli was a region subject to earthquakes because the earth was sponge-like and full of deep caves in which exhalations gathered; they were unable to escape because they were blocked by the sea. At length, the exhalations with their rapid movements inflamed the material that could be found in the caverns and created sinkholes. The sea entered the caves, trapping the exhalations. The waters then compressed the exhalations, causing an earthquake. He explained the dry and muddy ashes by the fact that ash emerges dry, but if it reaches the clouds where there is humidity they become wet, then fall to earth in a muddy form.

Della Porta rejected this hypothesis, noting that if the hypothesis depends on the exhalations being compressed in underground caves, why does the same phenomenon not occur on the very similar coasts surrounding Sorrento and Salerno? With regard to the collapse of the land, he believed that it happened because accumulated ashes and stones had obstructed the cracks that led outside, not because the earth had swallowed seawater.

As we have seen, Della Porta maintained that the land around Napoli and Pozzuoli, as well as Ischia and those areas around Etna, Lipar, Vulcano, Stromboli and Abano Terme, near Padua, had different cracks ('rugosa'); through these winds could enter and be heated by underground fires. The fire was formed by a mixture of bitumen, sulphur, oils and other inflammable materials that easily caught alight. The wind mixed with water, and when it came into contact with fire it boiled, until it was either blown towards the surface or caused the ground to explode, thus allowing the incandescent material to escape. These subterranean explosions constituted the thunder that accompanied earthquakes and the eruption of volcanoes.

According to Della Porta, this was the geological constitution of Solfatara ('Leucogeo puteolano'), the volcano near to Pozzuoli. He maintained that here an underground fire burned constantly and had done so from time immemorial ('ignis semper ardet exarsitque ab initio'). The thermal waters present in the Phlegraean region were also created by the subterranean winds mixing with water and sulphur. The Bagni di Pozzuoli, the healing properties of which varied according to the particular type of spring from which they emerged, provided a clear example of this phenomenon.

## Conclusion

Della Porta's purpose in *De aëris* was to describe phenomena and to explain them through plausible causes that could be reconciled with the laws of physics. However, his relationship to existing philosophical systems was ambiguous. On the one hand, Della Porta appeared to possess a vision of nature that conceived the natural order of living organisms in which each individual part played a role in the functioning of the whole. This was similar to ideas found in the works of other authors such as Telesio, Campanella and Bruno when they spoke of the general theory of the 'anima mundi' and the 'melothesìa'. The idea of underground fires that informed Della Porta's account of hot springs and the phenomenon of volcanic eruptions was analogous to that of internal heat within the body.

However, he did not push this idea as far as other thinkers. Bruno, for example, had extended the theory of the Earth as a living body to the extent that it included all celestial bodies in the infinite universe. Della Porta also overlooked some of the most important innovations of the period. When he entered the cosmological debate, first with *Coelestis physiognonomia* in 1603 and then in 1610 with *De aëris*, 67 years had passed since the publication of Nicolaus Copernicus's *De revolutionibus orbium coelestium*. His ideas nevertheless remained firmly rooted within an indisputably Ptolemaic model.[23] Indeed, in order to resolve any astronomical doubts, he actually preferred to retranslate Ptolemy's *Almagest*.[24]

The knowledge that he was subject to the surveillance of both the Neapolitan and Roman Inquisitions may have guided Della Porta's choices. His use of Lucretius may be instructive. Criticising Aristotle was a constant feature of philosophical debate, but one should remember that Francesco Patrizi's *Nova de universis philosophia* (1591) had been placed on the Index of Forbidden Books, and that both Giordano Bruno and Tommaso Campanella were in prison. Della Porta made sober and sensible use of Lucretius, citing only Books V and VI. This may be explained by the fact that in Book I Lucretius enunciated an atomistic theory that he applied in the following three books. Corpuscular and atomistic theory only entered mainstream scientific debate in the following century with the work of Pierre Gassendi. Della Porta may have realised that it was too dangerous to discuss atomistic ideas. Their materialist implications and the potential to incur a charge of atheism might expose him to further investigations and condemnations from the ecclesiastical tribunal. His own trial at the hands of the Inquisition and the prohibition of the Italian translation of his work *Humana Physiognomia* had shown him the tribunal's power. It remains unclear whether Della Porta's failure to engage with these beliefs was informed by a genuine faith or whether he was dissimulating his true beliefs.

He must have taken account of the difficult intellectual climate of the post-Tridentine era and attempted to unify the differing and opposing theories, allowing first one to prevail and then the other. In any case, his lingering objective was to free himself from the tutelage and burden of the Aristotelian system – without ever explicitly proposing an alternative solution, whether rooted in Platonism or the magico-hermetic tradition, such as the idea of the 'anima mundi'. In his own work only the vestiges of this concept remained. In truth Della Porta was not a philosopher; the appeal to magical, hermetic and Neoplatonic theory provided him with ideological and philosophical cover that allowed him to convey his scientific knowledge while remaining under the inquisitor's watchful eye. One should remember that *Coelestis Physiognomonia*, written in 1594, remained in the possession of the inquisitor until 1603, while *De aëris* was published soon after, possibly thanks to the intervention of Federico Cesi.

The enormous quantity of citations and the methods of research based on those used in medieval *Summae* showed that Della Porta retained aspects of an Aristotelian method, although he did not always support Aristotelian conclusions. From this combination, the 'new science' of the seventeenth century was born. It was necessary to reserve honour and reverence for ancient and medieval knowledge, but their models were now in decline because they could no longer answer the questions posed by a changing society.

## Notes

1 'La Magia la dividono in due parti, l'una chiamano infame, come sporca et imbrattata di spiriti immondi e di commercii di demonii [...]. L'altra è naturale, l'honora, e la riverisce come [...] la parte attiva, e più principalissima della Filosofia, e che produce i suoi maravigliosi effetti [...] non crediate che altro sia la Magia, che l'istesse opere della natura', Giambattista Della Porta, *Della magia naturale*, 1611, libro I, 2–4.

2 One gains an idea of Della Porta's reputation from the numerous editions of his published works. *Magia naturalis* was published in Antwerp (1560, 1561, 1562, 1564, 1576 and 1585), Lyon (1561), Cologne (1562) and Frankfurt (1591, 1597 and 1607). It was also translated into French, with editions published in Lyon (1565) and Paris (1570). An edition of *De furtivis litterarum notis* was published in London (1591); an edition of *De humana physiognomonia* in Hanau, a city in Hesse (1593, 1601); an edition of *Coelestis Physiognomia* in Strasburg (1606).

3 'Omnis mundi creatura / quasi liber et pictura / nobis est in speculum / nostrae vitae, nostrae mortis / nostri status, nostrae sortis / fidele signaculum', Alan of Lille (Alanus ab Insulis), *Rhytmus alter*, P.L. 210, col. 579.

4 'L'Anima del Mondo è un'energia naturale degli esseri per cui alcuni hanno soltanto la capacità di muoversi, altri di crescere, altri di percepire attraverso i sensi, altri di giudicare. [...] Ci si chiede cosa sia quell'energia. Ma, come mi sembra, quell'energia naturale è lo Spirito Santo cioè una divina e benigna armonia che è ciò da cui tutte le realtà hanno l'essere, il muoversi, il crescere, il sentire, il vivere, il giudicare', William of Conches, *Glosse al Timeo di Platone*, cited in Gregory, *Anima mundi*. According to Ficino, '*L'Anima mundi* infatti, secondo i Platonici più antichi, per mezzo delle sue *ragioni* ha costruito in cielo, oltre gli astri, le figure astrali e le parti delle figure, tali che esse stesse diventano figure; ed ha impresso in tutte queste figure determinate proprietà. ... E precisamente essa ha posto in cielo quarantotto figure universali, cioè, dodici nello Zodiaco trentasei fuori dello Zodiaco', Ficino, *De vita*, 207.

5 'Noi, scrivendo, porremo prima l'opinione de' nostri antichi e de' moderni, poi scriveremo appresso quando l'abbiamo esperimentate, se l'abbiamo ritrovate vere e false appresso l'invenzioni nostre, acciò veggano gli uomini dotti quanto la nostra età avanza quella de gli antichi perché molto di loro han scritto cose che giamai videro o sperimentaro, né mai conobbero gli ingredienti della composizione', Della Porta, *Della magia naturale*, 12.

6 Della Porta described the recipe for the witches' salve in the first edition of *Magia Naturalis*, see Della Porta, *Magiae naturalis sive de miraculis rerum naturalium libri IIII* (Naples: 1558), 102. See also Buccolini, *Il diavolo nel Malleus maleficarum*; Perifano, 'Introduction', 14; Valente, *Della Porta e l'Inquisizione*; and also Valente, *Johann Wier*. The efficacy of this recipe was tested in 1960, see Camilla, *Allucinogeni vegetali*, 138.

7 See for example Michaela Valente, 'Streghe o povere vecchiette?'

8 Fulco, *Per il 'Museo' dei fratelli Della Porta*, 105–75, now also in Fulco, *La 'meravigliosa passione'*, 251–325.

9 See for example Della Porta, *Ars reminiscendi*; Della Porta, *De Munitione*; Della Porta, *Humana Physiognomonia*; Della Porta, *Coelestis Physiognomonia*; Della Porta, *De aëris transmutationibus*.

10 Della Porta, *De aëris transmutationibus*.

11 Aristotle, *Meteorologicorum*, 338a, 25–338b, 2; Aristotele, *De Partibus animalium*, 641b, 19.

12 Zabarella, *De methodis libri quatuor*.

13 Martin, 'Experience of the New World'; Martin, *Renaissance Meteorology*.

14 'Annisus sum ea scribere quae visus et experientia, sensu non blandiente, me docuit; et o utinam, ita omnes facerent; nam errando, corrigendo et investigando, veritas eruitur. Bene igitur nobiscum actum erit si, ex multis quae scribimus, aliquid proficiat lector; et si absolutum opus assequi non licuerit, hoc tamen unum assequemur nec me poenitebit ex meis erroribus alios praestantioris doctrinae et ingenii doctiores factos et ansam praebuerim meliora et cultiora cum tempore proferantur', *Prooem*, 56–63.

15 He invented the telescope – for which his friend and competitor Galileo claimed credit – through the construction of a tube. He perfected the 'camera obscura' (a precursor of the photographic camera), discovered the principle that led to the invention of the telephone to speak 'from miles of distance', discovered a rudimentary code to transmit light signals using torches and constructed a parabolic mirror, with which he experimented in Venice with Paolo Sarpi. Additionally, Della Porta invented the siphon and new forms of alembic for use in distillation which are still in use. Della Porta also created an incubator for chickens that enabled 94 per cent of eggs to hatch; discovered the principle of air conditioning by devising a system for the refrigeration of air and developed the principle behind cluster munitions, used during the Vietnam War and, more recently, in the war in Ukraine.

16 Bacon, *Novum organum*.

17 Collumellae, *Res Rustica* VI, 27, 5–10; Pliny the elder, *Naturalis Historiae*, IV, 116, 4–5 and VIII, 166, 1–4; Solini 23, 7–8; Virgil, *Georgica* III, 272–9.

18 Lucretius, *De Rerum Natura*, VI, 680–702.

19 Della Porta, *De aëris*, IV, 48, 11–13.

20 Flavii Philostrati, *Vita Apollonii*, 5, 17, 2–7.

21 Petrus de Ebulo, *I bagni di Puzzuoli*; Pietro da Eboli, *Le acque cumane*.

22 Porzio, *De conflagratione*.

23 Della Porta, *Claudii Ptolemaei Magnae*.

24 Paolella, '*Il cielo stellato*'.

# Bibliography


Alan of Lille (Alanus ab Insulis), 'De planctu naturae'. In *Patrologia Latina*, edited by J. P. Migne. Paris: Excudebatur apud Migne, 1844–.

Aristotle. *De Partibus animalium*. Edited by I. Bywater. Oxford: Clarendon Press, 1894.

Aristotle. *Meteorologicorum libri quattuor*. Edited by F. H. Fobes. Cambridge, MA: Harvard University Press, 1919.

Bacon, Francis. *Novum organum*. Edited by E. de Mas. Bari: Laterza, 1968.

Buccolini, Claudio. 'Il diavolo nel *Malleus maleficarum*'. In *Il diavolo nel Medioevo, Atti del XLIX Convegno storico internazionale, Todi, 14–17 ottobre 2012, Centro Italiano di Studi sul Basso Medioevo*, 519–51. Spoleto: Accademia Tudertina e del Centro di Studi sulla Spiritualità Medievale, 2013.

Camilla, Gilberto. *Allucinogeni vegetali: Culto antico e moderno, uso ricreazionale*. Turin: Edizioni del Cortile, 1982.

Columellae, L. 'Iunii moderati'. In *Res rustica*, edited and translated by E. S. Forster and E. H. Heffner. London: Heinemann, 1955.

Della Porta, Giambattista. *Magiae naturalis sive de miraculis rerum naturalium libri IIII*. Naples, 1558.

Della Porta, Giambattista. *Della magia naturale*. Naples: Gio. Giacomo Carlino and Costantino Vitale, 1611.

Della Porta, Giambattista. *Ars reminiscendi*. Edited by R. Sirri. Naples: Edizioni Scientifiche Italiane, 1996.

Della Porta, Giambattista. *Claudii Ptolemaei Magnae constructionis liber primus*. Edited by Raffaele. De Vivo. Naples: Edizioni Scientifiche Italiane, 2000.

Della Porta, Giambattista. *De aëris mutationibus*. Edited by Alfonso Paolella. Naples: Edizioni Scientifiche Italiane, 2000.

Della Porta, Giambattista. *Villae libri XII*, 4 vols. Edited by F. Tateo. Naples: Edizioni Scientifiche Italiane, 2010–21.

Della Porta, Giambattista. *De Munitione libri tres*. Edited by R. De Vivo. Naples: Edizioni Scientifiche Italiane, 2010.

Ficino, Marsilio. *De vita*. Pordenone: Biblioteca dell'Immagine, 1991.

Fulco, Giorgio. 'Per il "Museo" dei fratelli Della Porta'. In *Rinascimento meridionale e altri studi in onore di Mario Santoro*, edited by M. C. Cafisse, F. D'Episcopo, V. Dolla, T. Fiorino and L. Miele, 105–75. Naples: Società Editrice Napoletana 1987.

Fulco, Giorgio. *La 'meravigliosa passione': Studi sul barocco tra letteratura ed arte*. Rome: Salerno, 2001.

Gabrieli, G. *Il carteggio linceo della vecchia Accademia di Federico Cesi*. Rome: G. Bardi, 1939–41.

Gregory, Tullio. *Anima mundi. La filosofia di Guglielmo di Conches e la scuola di Chartres*. Florence: Sansoni, 1955.

Lucretius Carus, T. *De Rerum Natura Libri Sex*. Edited by J. Martin. Leipzig: Teubner, 1957.

Martin, Craig. 'Experience of the New World and Aristotelian revisions of the Earth's climates during the Renaissance', *History of Meteorology* 3 (2006): 1–16.

Martin, Craig. *Renaissance Meteorology: Pomponazzi to Descartes*. Baltimore: Johns Hopkins University Press, 2011.

Paolella, Alfonso. 'Il cielo stellato in Copernico e nella Coelestis Physiognomonia di G. B. Della Porta'. In *La diffusione del Copernicanesimo in Italia, 1543–1610*, edited by Massimo. Bucciantini and Maurizio. Torrini, 189–202. Florence: Olschki, 1997.

Patrizi, Francesco. *Nova de universis philosophia in qua aristotelica methodo, non per motum, sed per lucem et lumina ad primam causam ascenditur*. Ferrara: 1591.

Perifano, Alfredo. 'Introduction' to Jean-François Pic de la Mirandole, *La Sorcière. Dialogue en trois livres sur la tromperie des demons*. Translated and edited by A. Perifano, 5–33. Turnhout: Brepols, 2007.

Petrus de Ebulo, *I bagni di Pozzuoli: poemetto napolitano del secolo 14*. Edited by Erasmo. Percopo. Naples: Federico Furchheim, 1887.

Philostratus, Flavius. 'Vita Apollonii'. In *Flavii Philostrati Opere*. Edited by Carl Ludwig Kayser, Leipzig: Teubner, 1870.

Pietro da Eboli, *Le acque cumane ovvero la fortuna delle terme di Pozzuoli nel Medioevo*. Edited by Teofilo De Angelis. Florence: Sismel Edizioni del Galluzzo, 2022.

Plinius C. Secundus. *Naturalis Historia Libri XXXVII*. Edited by Karl Mayhoff. Leipzig: Teubner, 1870–98.

Porzio, Simone. *De conflagratione agri puteolani Simonis Portii neapolitani epistola*. Naples: 1538(?).

Solinus, Caius Iulius. *Collectanea rerum memorabilium*. Venice: 1498.

Valente, Michaela. 'Della Porta e l'Inquisizione. Nuovi documenti dell'Archivio del Sant'Uffizio', *Bruniana e Campanelliana* 5 (1999): 415–34.

Valente, Michaela. *Johann Wier. Agli albori della critica razionale dell'occulto e del demoniaco nell'Europa del Cinquecento*. Florence: Olsckhi, 2003.

Valente, Michaela. 'Streghe o povere vecchiette? Il medico Johann Wier (1515–1588) contro la caccia alle streghe', *Storicamente* 4 (2008): 1–12. Accessed on 2 April 2023. valente.pdf (storicamente.org).

Vergilii, P. Maronis. *Georgica*. Edited by Roger Aubrey Baskerville Mynors. Oxford: Clarendon Press, 1972.

Zabarella, J. *De methodis libri quatuor: Liber de regressu*. Venice: 1578.

4

# 'Miserable land, burnt by fire and submerged in water': the 1631 eruption of Vesuvius

Lorenza Gianfrancesco

## Introduction

The violent eruption of Vesuvius that occurred between 15 December 1631 and early January 1632 marks an important event in the history of the Bay of Naples. The scale of destruction and damage that this explosive and violent eruption caused provides substantial evidence for this being the most catastrophic natural event in seventeenth-century Europe.[1] In the last decade this eruption has received increasing attention from historians.[2] Moreover, within the multidisciplinary field of Environmental Humanities, this event constitutes a major case study to assess the geological, climatic and human impact of early modern natural disasters.

The aim of this chapter is to contribute to this broad scholarly discussion by analysing the environmental impact that the 1631 eruption of Vesuvius had on Naples and on some of the most affected areas in the volcano's vicinity. These included important coastal settlements such as Portici, Torre del Greco and Torre Annunziata, together with the inland villages of Somma Vesuviana and Ottaviano, both located to the east of Vesuvius.

This study is also concerned with human responses to the catastrophe. Analysis of an array of manuscript and printed accounts – some of which are here brought to scholarly attention for the first time – will deepen our knowledge of how individuals from different backgrounds made sense of this tragic event. Astounded friars and desperate civilians left emotional accounts of the eruption; travellers visited devastated areas to assess the scale of destruction; physicians compared a fiery Vesuvius to a bursting pustule; astrologers interpreted the eruption as the result of a ruinous planetary conjunction; scientists merged classical natural philosophy with new hypotheses in their explanations of the disaster. Indeed, one of the effects of the eruption was the debate that it generated. Scholars engaged in discussions about the causes and nature of this event, now contained in treatises, accounts, correspondence and unpublished academic speeches.

In the first section of this chapter we will briefly present the landscape of the Vesuvian area, its biodiversity and economic importance prior to the eruption. Contemporary sources written by scholars, travellers and merchants will offer a reconstruction of individual encounters with Vesuvius, then highly regarded for its beauty, fertility and salubrious climate. Next we will discuss the phases of the eruption; its geographical spread; its impact on the climate and the sea of the Bay of Naples; the damage that the eruption caused in urban and rural areas located around the volcano; and the extent to which this event generated a new environmental awareness about the destructive – and unpredictable – nature of Vesuvius. Finally we will analyse contemporary interpretations of the 1631 eruption with a focus on the Neapolitan scientific debate on earth sciences that developed in the aftermath of this calamitous event.

## A land of plenty

*Mountains come first*[3]

After a journey from the Papal States that took him to Caianello and then onto the Appian Way, the protonotary apostolic Giulio Cesare Braccini reached Naples on 15 December 1631.[4] The following night he witnessed the catastrophic eruption of Mount Vesuvius that would continue until mid-January 1632. On 23 December, while the volcano was still erupting, Braccini published in Naples a short account of that event, which he dedicated to Cardinal Gerolamo Colonna.[5] In April 1632 Braccini published a second text, *Dell'Incendio*, in Naples that was concerned with the history of Vesuvius from antiquity to his own time.[6] As an eyewitness to the eruption, he also devoted sections of his writings to analysing the causes and impact of the 1631 disaster on areas located along the Bay of Naples. To provide a vivid picture of the damage caused by the eruption, in chapter 4 of his *Dell'Incendio* (1632) Braccini described the landscape and geomorphology of the land surrounding Vesuvius prior to the 1631 disaster. He portrayed the area as a fertile land filled with 'flowery meadows and a variety of trees marvellously aligned, perfectly pruned and laden with fruit'.[7] This bucolic image of Vesuvius was derived from Braccini's experience some 20 years earlier when he had ascended Vesuvius, accompanied by some locals. During this ascent, he observed that the steep path to descend into the crater had a thick vegetation with 'trees that were very close to one another'.[8] This area of the crater was lit by the sun and rich in grass, oaks, euonymus trees, privets, 'genisteae' and even 'simples' that physicians would use to produce medical remedies.[9] As Antonio Nazzaro has recently discussed, Braccini's observations indicate that during periods of dormancy moderate masses of lava flows had been released through vents located at the bottom of Vesuvius; these explained the presence of undamaged vegetation at the top of the crater.[10]

Braccini was not alone in celebrating the fertility and botanical variety of the land around the volcano. Giulio Cesare Recupito, a Jesuit and Ozioso academician, stated that 'Mother Nature has trusted one mountain to provide for our lives'; he wrote that the woods and twigs located at the summit of Vesuvius were cut and collected to build ships, light fires and bake bread.[11] Similarly, in his assessment of the environmental damage caused by the eruption, the physician and scholar Antonio Santorelli remembered the fields of citrus fruits and vineyards that grew in the area of Vesuvius.[12] In a similar vein, Giulio Amodio and the Ozioso academician Giulio Cesare Capaccio commended 'the wines of Nola, Vico [Equense] and Sorrento which are born from that delicate soil' and praised the 'excellent wines of Torre del Greco'[13] – a village that 'on one side overlooks the sea, while on the other faces that dangerous Mountain'.[14] Recupito too considered the wines of Vesuvius delightful and noted that the white Grechi and the red Lagrime 'are highly valued in all nations'.[15] So famous were the wines produced in the Bay of Naples that in a book of his travels to Italy the Antwerpian jurist Franz Schott wrote that 'Neapolitan' wines such as the 'Greco di Somma' and the 'Lacrima Christi' were mostly consumed in Rome.[16]

Prior to the 1631 eruption, agricultural activities in the Vesuvian area had represented an important sector in the economy of the Kingdom of Naples. Products such as flour, walnuts and wine were widely exported to Europe through major Mediterranean ports.[17] In his travels to southern Europe, the British merchant Lewes Roberts noted that Naples and its vicinity abounded 'in several commodities for merchandising', including metals, wine, saffron, silk, oil, brimstone, aniseeds and argol.[18] Moreover, Roberts praised the richness and layout of the local agricultural crops, with one field

> yeelding at one and the same time three severall crops, the ground bearing *Corne*, having *Mulberry trees* intermixed and *Vines* planted at the foote of each *Mulberry*, which have made excellent *Wines*.[19]

Sustainable farming and agriculture in an area that stretched from Torre Annunziata to Portici were particularly thriving and lucrative, with crops harvested 'three or four times per year'.[20] Giuseppe Mormile described those villages as the land of plenty; he specifically praised the abundant production of textiles such as linen, hemp and silk.[21] So lucrative was the Vesuvian primary sector that, according to Francesco Ceraso, in the early seventeenth century the annual revenues generated by the selling of fruits, wines, vegetables and wood from the area amounted to 'two million of gold'.[22] Ceraso also admired the beauty of the villages located around Vesuvius, all enriched with the 'splendid gardens and houses of Neapolitan gentlemen who lived there with their families'.[23]

Besides the existence of a profitable agrarian system, the biodiversity of the land surrounding Vesuvius extended to its rich fauna, which included wild mammals, insects, amphibians, birds and reptiles. Hares, deer, fallow deer and a variety of birds are some of the species listed in early modern texts on Vesuvius.[24] In his account of the eruption, Francesco Bernaudo mentioned how some of the species populating the area coped with the challenges of human presence and 'the assaults of daring hunters' by hiding in secure shelters located along the mountain.[25] Besides wildlife, sheep, pigs, chickens, birds and cows were bred in the villages located around Vesuvius. Intense livestock farming supplied the city of Naples and was a contributing factor to the economic growth of the area. Local producers would make daily trips to the city carrying baskets of meat and produce, to be sold in the markets located near the port.[26] This microcosm of sustainability, where humans lived in a symbiotic relationship with the ecosystem of Vesuvius and where the fulfilment of human needs did not disrupt the cycles of nature, ended abruptly in December 1631.

## The eruption

When Vesuvius suddenly awoke after a long period of modest activity, the impact was catastrophic and the environmental disaster it caused of an unprecedented scale. The eruption became the subject of numerous accounts, some of which were printed and sold cheaply as the volcano was still erupting.[27] This large corpus of contemporary material offers historians valuable information to reconstruct the typology of this eruption and the emergence of a scientific debate on volcanology that soon spread beyond a local context. The 1631 disaster was a Plinian-type explosive eruption – albeit on a smaller scale than some previous eruptions of Vesuvius, such as those that had occurred in AD 79 and in AD 472.[28] Moreover, the 1631 eruption had been preceded and followed by other destructive natural events that caused damage to the environment.[29] These included severe seismic activities and plumes of smoke emanating from Vesuvius prior to the eruption's explosive phase.

On 15 December 1631 some Neapolitans, alerted by a frightening rumble, quickly gathered at the city's shore; others climbed to the roofs of their houses to witness a terrifying scene. They saw that the western flank of the volcano had cracked and a dense and caliginous cloud was billowing from it.[30] Compared to the shape of 'a tall and large pine tree', an eruption column of dark smoke grew so violently that it reached Naples at dawn on 16 December. In less than three hours that dark and polluting substance, along with the ashes travelling in the air, eclipsed the sun and left the city in darkness.[31] So copious was the release of ashes that in some areas of Naples the ground was covered in a carpet of dust measuring up to half a palma (13 cm).[32]

Together with the roar and the effusive release of gases and smoke, sources describe the severe earthquakes that preceded and followed the explosions from the cone of Vesuvius. The violent and rapid movement of masses underground was held responsible for the eruption, during which the volcano 'spewed out fiery stones in the air, lightnings and very dense clouds whose weight generated the falling of ashes'.[33] The series of earthquakes that followed hit Naples for several days, with the most violent tremor being experienced at midday on 23 December.[34] So strong were these earthquakes that people abandoned their homes. Some escaped the city entirely; those who remained found refuge in their coaches or built wooden shelters in squares and open spaces.[35] In the crowded area of Castle Square, near the port, the Spanish troops in the viceroy's service guarded the wooden shelters in an attempt to maintain public order.[36] Groups of Neapolitans also convened in the open areas of the Chiaia district, the closest to the city's seashore.[37]

During the explosive phase that followed, Vesuvius ejected stones, pumice, lava and bitumen which quickly descended to the valley from the volcano's caldera ('atrio').[38] The mass of stones that crashed onto the ground differed in size. In his letter to Nicolas-Claude Fabri de Peiresc, a Carthusian friar in Naples named Dom Severo (Severo Trafaglione) reported that in the village of Somma Vesuviana one stone measured

45 palmi (11.88 m) in circumference, 20 palmi (5.28 m) in height and 20 palmi (5.28 m) in width.[39] Together with explosions, the cone released large quantities of watery mud and stones.

In highlighting the destructive nature of Vesuvius's exhalations, Braccini recounted that the viscous mixture flowing from the volcano 'was so ardent that it resembled melted lead or tin'.[40] Others believed that the flammable nature of these liquids was partly due to a certain oil released by Vesuvius, known locally as 'olio petronico'. This bituminous mixture was held responsible for the catastrophic burning of the surrounding environment and local settlements.[41] Moreover, torrents of water gushed down the slopes of Vesuvius and grew due to incessant rain.[42] Braccini reported that this aquatic avalanche flowed in three streams, moving at an alarming rate. Debris and rocks borne along by the water crashed at violent speed into the villages dotting the valley along the coast (from east to west), partly destroying them.[43] The final encounter of water and incandescent matter with the sea proved equally catastrophic.

## Soapy waters and strong winds

On 17 December a terrifying earthquake hit Vesuvius and caused the collapse of its central cone.[44] On that day it was reported that Vesuvius looked as though its summit had been mutilated ('nunc capite mutilatus').[45] As Antonio Nazzaro has recently established, the central cone dropped by 481 metres in height.[46] So violent was the tremor that had preceded the partial destruction of the volcano's summit that the sea near the village of Ottaviano 'moved with great force'.[47] A similar phenomenon happened in Naples. Due to the sudden movement of water, 'galleons and vessels were at risk of being submerged, although they had been moored with thick ropes'.[48] Shortly after the sea receded, leaving behind 'a large quantity of fish'.[49] Masino reported that a similar event occurred in Sorrento and Nisida, where the sea receded by more than 20 metres ('tre picche').[50] Around the same time the sea temperature in Amalfi rose dramatically; it was compared to a boiling cauldron.[51] In Naples the sea receded three times, putting the galleons and boats anchored within the city's port at great risk.[52] In his account, Oliva reported that the sea in Naples receded by about 60 metres ('più di trenta passi').[53] Soon after, the destructive force of the waves reached the seashore, causing coastal flooding that hit nearby urban areas such as the Chiaia district.[54]

Oliva observed that the burning of volcanic substances continued even after they had entered the sea. He also reported that some friars trying to escape to Salerno by sea had their felucca overthrown by the violent movement of debris and volcanic matter. Following the capsizing of their boat, due to the scorching temperature of the seashore, some of the friars scalded their feet and calves.[55] In interpreting this occurrence – the description of which resembles a phenomenon known in modern science as volcanic laze – some Neapolitan scholars hypothesised that the tall columns of smoke and fire emerging on the sea surface originated from chasms that the mass of incandescent pyroclastic flow 'had opened in the sea'.[56] On 18 December the Spanish viceroy in Naples, Manuel de Acevedo y Zúñiga, count of Monterrey, deployed about a hundred sappers ('guastatori') on two galleons to patrol the coastal area that was being hit by the eruption. Upon disembarking in Torre del Greco, the soldiers had to place wooden planks on the seashore as 'they could not touch the burning ashes with their feet'.[57]

This unusual heat was not the only thing that affected the sea in the Bay of Naples. Volcanic substances pushed into the water caused a type of pollution that may have been triggered by a high concentration of magmatic silicate. In early January 1632 it was observed that the seawater in some areas was viscous, as though some soap had been melted in it. Indeed, it was remarked upon that people who had touched the sea with their hands noticed that when they dried them their fingers remained sticky. They also noticed that the water was rather hot, as though one had boiled lime mineral ('calcina') in it.[58]

The environmental impact of the eruption influenced the climate too. Famously described as the land of eternal spring where roses blossomed throughout the year, the Bay of Naples had a temperate air which was praised for its 'great benefit to the sick and even to the kings of Naples who had often stayed there'.[59] During the months that preceded and followed the eruption, however, people noticed a strange meteorological pattern.[60] A sudden drop of temperature hit the area with unusually cold weather that lasted from October 1631 to March 1632.[61] In his letter to Peiresc, friar Dom Severo wrote about some extreme weather events. He reported that in October 1631 a tramontana (a type of wind) blew so strongly that it seemed as though winter had arrived abruptly instead of autumn. He added that the summit of Vesuvius appeared unusually covered in snow and frost – which, as the friar noted, was 'contrary to the customs of our climate'.[62] On 14 December another, more severe tramontana hit the city again. Dom Severo reported that the extreme weather made the friars in his convent unable to walk in the cloister as 'the wind blew the habits on their faces'.[63] On 17 December the strong wind contained large quantities of sulphur that were being released from the volcano, and which caused severe air pollution across the region.[64]

Interestingly, some of the details contained in historical accounts of the 1631 eruption are relevant to recent studies on volcanology. Besides providing explanations of the dynamics of eruptions, new research has variously focused on the climatic and polluting impact of large volcanic conflagrations. Scientists have highlighted, for instance, that dust veils and the release of volcanic sulphur gases form aerosol particles; these 'absorb incoming radiation from the sun' and, in turn, generate a cooling effect on both the Earth's surface and the lowest atmosphere.[65]

The cooling effect caused by violent eruptions has also been variously linked to changes in seasonal weather and rainfall patterns. In that sense, Dom Severo's letters and other sources containing observations of unfamiliar climatological events constitute valuable material for reconstructing the phases and impact of past volcanic eruptions. As we have seen, certain historical sources also provide detailed accounts of other disastrous events connected to the eruption. These include sea disturbance, destruction of natural and urban areas, mortality and the impact of pollution on public health. The last was a major concern in the aftermath of the 1631 eruption.

Indeed, the conflicting debate that assessed the consequences of pollution on public health led some Neapolitan physicians to fear that the contaminated air might lead to an outbreak of plague.[66] The physician Cesare De Martino, a defender of the miasmatic theory, wrote that earthquakes and the burning of venomous minerals in the Earth's crevices 'infect the air that we breathe, causing malignant and pestilential diseases'.[67] Relying upon Galenic medical knowledge, he advised people to purify domestic spaces from bad smells by preparing 'perfumed balls of rosed vinegar and orange blossom' mixed with cinnamon, amber and 'other powdered odorous substances'.[68] In order to 'purify the blood and clean the stomach' from polluted air, De Martino recommended drinking 'lemon juice and citron [...] mixed with theriac, bezoar and Armenian bole'.[69] By contrast Antonio Santorelli, who at the time of the eruption held the Chair in Medicine and Philosophy at the University of Naples, firmly discarded De Martino's theory of air pollution. He asserted instead that contaminated waters and unburied decomposing corpses (rather than venomous minerals) might cause an outbreak of infectious disease. Santorelli considered an imminent upsurge of plague to be a remote possibility, explaining that 'pestilence occurs in summer when the heath easily contaminates waters and cadavers, both of which release venomous vapours that pollute the air'.[70]

Moreover, people feared the threats of an uncommon wet season. Indeed, the areas that had been spared by the fire of Vesuvius were instead hit by 'torrential and incessant rains [...] which inundated the plains of Campania north of Vesuvius that stretched from Nola to Acerra', as Dom Severo wrote.[71] In Naples, Masino reported that as late as the beginning of February 1632 the city continued to be hit by bad weather. On 4 February, he observed, 'this city was struck by the roar of thunders and lightning, thick hale and snow fall'.[72] Bad weather, earthquakes and rains of ash and water also hit territories to the north and south of Vesuvius, stretching as far as the east of the Mediterranean and central Europe. Seismic activities were reported around Mount Maiella in Abruzzo, Spoleto, Perugia, Ancona, Recanati and Loreto, where people compared the terrifying roar caused by earthquakes to the noise of mass artillery.[73] South of Vesuvius, ashes covered large areas of Basilicata, Calabria and Apulia. There the environmental and economic damage was severe in many areas, including the cities of Bari and Lecce.[74]

In December 1631 the governor of Foggia, Francesco De Ocampo, reported that a heavy rain of ash had fallen in the area for eight hours.[75] On 22 December that year a letter from Barletta reached the Spanish viceroy in Naples.[76] It reported heavy rains of ash and sand, the impact of which was deemed more serious than damage caused by the snow that had fallen earlier in the month. The letter also voiced the concerns of farmers. So large was the quantity of ash and sand left on the terrain that sheep could not graze. This resulted in unprecedented livestock mortality, not to mention the damage caused to the agricultural sector.[77] Ashes reached land on the eastern Adriatic too, with areas such as Kotor (*Cattaro*) left in darkness; mountains, houses and streets were covered in them.[78] Rains of ash and sand were reported to have reached as far as Hungary, Corfu and Istanbul, as well as the ports of Volos and St. John of Acre.[79] The eruption that had begun in 1631 lasted for months. Sources variously reported that Vesuvius continued to release smoke as late as March 1632. Dom Severo noted that in the spring of that year

> The Mount continues to billow smoke and sometimes to release a large quantity of stones. These obstruct the narrow mouth of Vesuvius and cause earthquakes that are still felt in close areas and distant lands.[80]

## A spectacle of destruction and an environmental disaster

*The most beautiful windows became doors*[81]

On 24 May 1632 an anonymous gentleman in Naples, accompanied by four travellers from Perugia, embarked on a felucca bound for Torre del Greco. With the calm sea and a light wind, they travelled leisurely and reached their destination in a short time. Once they touched the soil, the devastating scene that met their eyes made the pleasurable view of the Bay of Naples a distant memory. A once populous and thriving village located on the slopes of Vesuvius, previously known for its fertility, Torre del Greco was in a ruinous state; it stood as a poignant reminder of what the eruption had left behind. The village appeared half buried 'not so much by the diluvium of ashes and fiery stones that the mountain had vomited, as by the torrents of water, fire and bitumen'.[82] The devastation of the urban area of Torre del Greco and other villages located nearby was accentuated by unprecedented damage to the local ecosystem. In assessing the impact of the eruption, Bernaudo wrote that

> not only did the mount cause a disaster with its fire [...] but it also released a fast-flowing river of bituminous and sulphurous waters, [which] opened new paths; levelled out valleys; eradicated the vegetation; uprooted woods; destroyed buildings, sheds, flocks, forage, plants, bushes, roofs and foundations [...] with debris pulled towards the sea.[83]

The first green area to be erased was the thick vegetation that had grown inside the volcano's crater and which had long been used for pasture and collecting wood.[84] The quantity of burnt wood was such that 'it was estimated that it would have supplied the city of Naples for ten years'.[85] Destruction of vegetation was also caused by the torrential rain that could not be absorbed by the mountain due to the hardening of lava ('dura crosta').[86] Together with rain and fire, the strong wind was held responsible for the devastation of the tops of many trees.[87] Similarly, masses of fallen oaks and beech trees were carried by the storm towards the sea.[88] Entire fields of olive and fruit trees, so important to the local economy, were described as being incinerated.[89]

Incalculable was the damage to the soil and the forage it produced, as both the mountain and the surrounding land were covered in ashes.

It was estimated that in some villages the ash covered the terrain to a depth of more than 5 metres.[90] Moreover, liquified metals contained in the pyroclastic flow polluted the soil. Sources report that large quantities of sulphur and bitumen had covered the terrain and polluted the air which became 'unbreathable and intolerably malodorous'.[91] Reserves of wine, grain and oil were lost, and the damage to the agricultural sector was such that the land – it was speculated – would not be suitable for arable use for years.[92] Damage inflicted upon the fauna was equally severe. Fish, cattle, sheep, oxen, birds, horses, foxes, hares, wolves and unnamed creatures populating the area were decimated as much as humans were.[93] Due to the movement of the sea and the sudden rise of marine temperature caused by earthquakes and the pyroclastic flow, dead fish covered the seashore along the volcano's coastal path.[94] Gregorio Carafa reported that in Naples a copious amount of fish that had been burnt and killed simply floated on the water, or else lay along the seashore ('in arena iacebat, & mox in aqua piscium necatorum, exustorumquè copia fluitabat').[95]

Moreover, the heavy loss of livestock contributed to the collapse of the farming economy. In his account to Peiresc, Dom Severo wrote that the destruction of the Carthusian monastic grange located in Minardo (Somma Vesuviana) caused the loss of six oxen and about 400 sheep. He also reported that by late December 1631 more than 20,000 sheep had been killed.[96] By early January 1632 De Martino stated that an additional 15,000 livestock had been lost.[97] In assessing the reason why many sheep had allegedly been found dead with their forelegs contracted, Santorelli explained that they had assumed this position as a defensive posture against the unbearable heat of the terrain which, he noted, caused an irreversible contraction of the sheep's muscles.[98] Sources also contain accounts of the heavy losses of wild fauna and birds, with descriptions of dead creatures everywhere.[99] Indeed, the number of decomposing carcasses was so great that it was deemed necessary 'to inter the animals to avoid their stench corrupting the air'.[100] The pervasive resonance of this catastrophe – from natural environment to the urban landscape – filled the pages of a large corpus of material on this event that circulated both as manuscript and in print. It is therefore not surprising to read that many people travelled to the Vesuvian area to observe the aftermath of the eruption.

Accompanied by three locals, the five travellers that had reached Torre del Greco in May 1632 decided to climb Vesuvius. They journeyed for 4 miles, undertaking the first 2 miles on horseback but completing the remaining distance to the summit on foot. During their journey they noted that underneath the blanket of ash the soil was still hot.[101] As the group made their way up, two members remained behind, fearful of the smoke still being emitted by some openings in the rock and the roar that could be heard from different parts of the mountain.[102] The steep path of the last mile forced the group to climb on all fours; when they reached the summit they had to take their shirts off, as they were soaking in sweat.[103]

Upon observing Vesuvius and Mount Somma located next to it, the anonymous writer of this account calculated that the central cone of Vesuvius had collapsed by about a quarter of a mile.[104] Interestingly, this figure matches, as discussed earlier, Nazzaro's calculation.[105] Holding his body up by his arms, his feet raised from the ground and his head positioned over the lip of the crater, the anonymous writer recounted his view of the mountain's interior. He saw smoke coming out from seven points and a waterfall in the opposite side of the crater which, upon touching the hot volcanic matter, caused reverberation and a dense column of smoke 'that is still visible from Naples and its vicinities'.[106]

A few hours later, the group descended what was described as a mountain stripped of its vegetation. In the evening they reached Torre del Greco, where they ate in the courtyard of a heavily damaged inn and slept in one room with their mattresses on the floor. The following morning they left 'astonished' and returned to Naples by boat.[107] What these visitors perceived as 'the uncontrollable wonder of nature' encapsulated their new environmental awareness, indeed the emotional impact of seeing what the force of Vesuvius had left behind: a devastated landscape and the destruction of human activities.

Although the eruption was often interpreted as a manifestation of God's wrath against human misconduct, in some cases scholars advanced ideas that located this disastrous event within a discourse on earth sciences and the environment, drawing mostly from classical natural philosophy and partly from new hypotheses about natural phenomena that reconsidered the centrality of religion. The new scholarly interest in disciplines such as volcanology that were sparked by the Neapolitan discourse on Vesuvius would soon acquire a European dimension. The study of Vesuvius and the geomorphology of the Bay of Naples that developed within the Royal Society is a well-known example. However, it is to the Neapolitan scholarly debate stemming from the 1631 eruption of Vesuvius that we now turn.

## Assessing disasters, debating nature. Vesuvius and scholarship

In 1632 Giovanni Tommaso Giovino delivered an academic lecture at the Infuriati academy in Naples.[108] A doctor of philosophy and medicine, he was also affiliated to the Oziosi academy under whose auspices some of the texts concerned with the eruption were published. In his manuscript lecture Giovino discussed the natural causes of the recent earthquakes and conflagration of Vesuvius, although he did acknowledge that 'sometimes our irate and righteous God uses these tools (natural disasters) to punish human misconduct'.[109] He drew from a tradition that dated back to the branch of ancient natural philosophy concerned with seismology and volcanology: a corpus of knowledge that was receiving increasing attention among his contemporaries.[110]

In analysing the origins of earthquakes, Giovino distanced himself from pre-Socratic natural philosophy. He labelled as 'discordant and unfounded' the views of Thales, Anaximenes, Anaxagoras and Democritus, according to which seismic activities are essentially caused by the violent movement of aether, excessive water and the collapse of 'caverns' (layers of rock) situated in the substratum of the Earth.[111] Instead Giovino embraced the theory of exhalations advanced by Aristotle in *Meteorologica* (Book II) to explain the recent earthquakes caused by the eruption.

Giovino stated that the movement of 'hot or cold' air attains its most destructive force when it becomes denser and 'in attempting to find a passage from the cavernous bowels of the Earth, the air generates a violent motion that causes an earthquake'.[112] This view was shared by many of Giovino's fellow scholars. Examples include the physician Cesare De Martino, who employed the Aristotelian theory of seismology to explain that the recent earthquakes had been caused by the violent motion of the air trying to overcome obstacles during its passage underground.[113] These impediments 'made the earth shake and collapse; mountains move and in the end the pressure caused the opening of chasms'.[114] The Augustinian friar Agnello Santamaria reiterated the validity of this interpretation by demonstrating that the nature of Vesuvius's fire – which was partly responsible for the earthquakes that had occurred concomitantly to the eruption – was not 'infernal' (i.e. related to hell) and should therefore be analysed as a natural event.[115]

Moreover, Santamaria highlighted the role of two additional factors in explaining the causes of the recent earthquakes. First, he considered the strong tramontana wind and the snow that had covered Vesuvius as events that caused 'the pores of the Earth to shrink', which subsequently 'trapped the dry aether underneath the Earth's crust'.[116] Second, he regarded the action of the sun as a key factor to explain the time during which earthquakes had occurred. Santamaria argued that when the sun reaches its highest point at noon it causes the exhalations and the winds positioned in the substratum to move upwards. Earthquakes occur, therefore, when these elements push aside obstacles that they encounter. Likewise, he asserted that the cold generated by the absence of sunlight during the night 'affects the pores of the Earth, causing the dry aether to move violently and to shake the bowels of Mother Earth'.[117]

The scholarly debate that sparked from the 1631 eruption was not solely concerned with Aristotelian theories of seismology as the only scholarly model to explain the phenomenology of earthquakes. Braccini, for instance, doubted that the recent seismic activities had been triggered by obstacles encountered by the air in finding an exit from the substratum of the Earth. Instead, he hypothesised that tremors and rumbles linked to the eruption of Vesuvius had been generated by 'the bituminous and sulphuric matter clashing with the narrow passage from which it was supposed to make its way out from the roots of Vesuvius'.[118] This debate also encouraged some scholars to classify earthquakes. Drawing from Posidonius of Rhodes (*c*.135 BC–*c*.50 BC), Capaccio identified two types of tremors; he believed that the first type was caused by waves 'moving in succession upwards', while the second type was generated by violent oscillations.[119] Likewise, Giulio Amodio defined the tremors that had preceded and followed the eruption as '*collesivi*' – namely 'two hard bodies' colliding in the substratum, which caused 'rumbles and a violent ejection of stones that has been seen from our Vesuvius'.[120]

Besides discussing the nature and causes of earthquakes, scholars analysed the hazards and risks linked to the recent eruption of Vesuvius. Again they did so by combining Aristotelian seismology with reference to other scholarly models. For instance, Santorelli deemed the occurrence of future tremors in the Bay of Naples as a remote possibility. He resorted to Seneca's *Natural Questions* to substantiate his argument, which he based on two points. First, he believed that the newly formed crevices of the Earth's surface would ensure the harmless escape of the air. Second, he explained that the production of water resulting from the mix of air and vapour would not trap the air in the substratum.[121] In line with Santorelli's views, some scholars considered not the earthquakes but rather the release of a large quantity of stones, bitumen and liquified minerals to be the most hazardous elements of the recent eruption. Friar Santamaria followed this interpretation.

He relied upon Seneca's analysis of the eruption of Vesuvius that had occurred at the time of Emperor Nero, and more specifically on the Roman philosopher's thesis that 'these types of torrents are the most dangerous of all'.[122]

In assessing the violence and extent of the volcano's ejecta, Giovino compared Vesuvius to Stromboli and Etna (known as *Mongibello*).[123] Here he explained that the more effusive and less dangerous exhalations released by those two Sicilian volcanos were linked to two factors, namely their height and the frequency of their eruptions. Giovino argued that the lower the height of a volcano – as was the case of Stromboli – the less violent its ejecta, since they were not expelled outwards 'with great force'.[124] The frequency of volcanic explosions also dictated the magnitude of eruptions. Giovino explained this concept by comparing Vesuvius to Etna and asserting that, although the two volcanos were of a similar height, Etna's frequent explosions made them less forceful. By contrast he declared that Vesuvius 'impregnates itself with such a large quantity of exhalations for centuries and then explodes with horrendous and astonishing effects'.[125] These phenomena, Giovino argued, explained the rare yet violent eruptions of the Neapolitan volcano.

During its long periods of apparent inactivity, however, Vesuvius was not dormant. On the contrary: its force was periodically cushioned by an underground network that enabled exhalations and the 'matter' positioned at the 'roots of Vesuvius' to move across rather than push upwards. As a result, Giovino speculated, this matter moved within a network of communicating openings that extended from Mount Somma to the Phlegraean area of Pozzuoli. Interestingly, he compared these subterranean apertures to 'the veins and arteries that spread in the human body', even as the Ozioso academician Francesco de Pietri had compared the process of volcanic eruptions to the inflammation of the veins caused by imbalanced humours.[126]

It is within this framework that Giovino also explained the reason why Naples had not been destroyed by the recent eruption of Vesuvius. He advanced some uncensored hypotheses that the confined space of an academic gathering and the manuscript form of his lecture may have enabled him to express more freely. Although he briefly stated that the survival of Naples resulted from divine protection, Giovino also considered the city's geographical location in analysing why it had not been destroyed by the eruption. Here he explained that Naples was positioned between 'the two big exhalation exits positioned between Mount Somma and Pozzuoli. These exits had enabled air and matter to get past Naples underground while travelling towards the escape points'.[127]

Giovino was not alone in thinking of Vesuvius as a live volcano, albeit one with periods of seemingly undetectable activities. The physician and astrologer Giovanni Francesco Porrata Spinola stated that 'this mountain hides within itself a large quantity of water, bitumen and sulphur which is the cause of everlasting fires, although these are not always visible'.[128] Here Porrata Spinola drew from Giambattista Della Porta, who in *De Transmutationibus Aëris* had argued that the ever-burning fire within Vesuvius 'transforms its matter into smoke and ashes and at times ejects stones'.[129]

Examination of the causes and dynamics of the 1631 eruption was also concerned with the violent sea disturbance that, as discussed in previous pages, had hit the Bay of Naples. In analysing this aspect of the eruption, Giovino drew upon Agostino Nifo to explain that the recession of the coastline and the rise of the sea level had been triggered by 'exhalations ejected from underwater'. These were considered to be responsible for the elevation by more than 5 metres (*venti palmi*) of a large cliff located in the coastal area of Vico Equense, near Sorrento.[130] Similarly, Santorelli analysed the impact of the eruption on the sea and compared the 1631 disaster to the volcanic activities that had hit the area of Pozzuoli in 1538. He challenged the theory about coastal recession contained in Simone Porzio's *De Conflagratione Agri Puteolani*, published in 1551.[131] In this work Porzio wrote that in 1538 the sea in Pozzuoli had receded by about 360 m ('mare passib[us] fere. CC. recessit').[132] He explained that this phenomenon had been caused by exhalations desiccating the soil. The dried-out soil had thus absorbed seawater more readily.[133]

Santorelli dismissed Porzio's theory by arguing that the soil's absorption of water is a slow process that diverges from 'the velocity with which the coastal line has receded due to the eruptive force of Vesuvius'. He explained that the recent coastline recession in Torre del Greco and Naples had instead been caused by 'the large quantity of matter that entered the sea, similarly to what happens in Egypt when the Nile carries masses of soil into the sea'.[134]

Like Santorelli, Porrata Spinola rejected Porzio's theory, asserting instead that the underwater chasms produced by earthquakes caused 'the seawater to recede violently, leaving the port of Naples dry'.[135] Capaccio followed this line of thinking and argued that the seawater was absorbed by crevices in the substratum that caused the sea to recede. Moreover, he linked this phenomenon to volcanic eruptions, stating that the masses of water underground moved; when they reached the caves at the roots of Vesuvius 'by the vehement force of the vapour they were subsequently pushed upwards and finally ejected with fire'.[136]

The ideas presented in this section have highlighted the multidisciplinary nature of the Neapolitan debate on the 1631 eruption. The renewed scholarly interest in earth sciences was coupled with an understanding of nature as being governed by the forces of the macrocosm. It is the analysis of the connection between the Earth and the heavens, of the inner relation of the parts to the whole, that we will discuss next.

## Vesuvius and the macrocosm

In his academic speech on the eruption, Giovino argued that 'earthquakes are predicted by eclipses and the great conjunctions of the Planets'.[137] He explained that astrological transits 'generate exhalations' from the Earth, whereas the darkness resulting from eclipses 'cools and closes the pores of the Earth'. Earthquakes are thus caused by the obstacles encountered by the exhalations 'which make the Earth shake'.[138] Giovino also considered comets and portents that had appeared in previous years to be premonitory signs of misfortune and wars that 'we hear occur in many parts of the world'.[139] Specifically, he viewed the recent eruption of Vesuvius as the manifestation on Earth of the 'conjunction of Saturn and Jupiter in the house of Sagittarius, which is the strongest zodiacal sign of this igneous triplicity'.[140]

In a similar vein, Cesare De Martino argued that some of the explanations for the recent eruption lay in the influence of celestial bodies on the Earth. 'Inferiors are governed by superiors ('Inferiora à Superioribus reguntur'),' he stated in accordance with Aristotle's *Meteorologica* (Book 1), thereby maintaining that there existed a correlation between the inner workings of the Earth and the heavens.[141] De Martino relied upon astrology to explain the eclipses of the sun and the moon that had darkened the Neapolitan sky in October and November 1631 respectively. Considered to be harmful manifestations of a cosmic disorder, those two eclipses had been caused by the transit of the sun in the house of Leo and, successively, by the passage of the moon in the house of Taurus.[142] These conjunctions, De Martino noted, were premonitory signs of an imminent catastrophe 'as Leo dominates Italy and Taurus dominates Campania', meaning that 'in Italy & in *Campania Felix* damage was about to happen'.[143] He added that this adverse planetary conjunction had been worsened by 'the appearance of a star of the greatest apparent magnitude in the same point where earthquakes occurred'.[144]

Porrata Spinola shared De Martino's view and stated that the order of nature is based on superior celestial motions.[145] Moreover, he followed Ptolemy and explained that by nature lunar eclipses are not harbingers of imminent disasters. Rather, their negative effects must be analysed according to planetary conjunctions. Porrata Spinola thus believed that the ruinous nature of the lunar eclipse that preceded the eruption had been caused by Mars touching the moon during its rotation in the 14th grade of Leo. This transit, he explained, increased the fiery power of Mars. By contrast, the position of the moon in the 15th grade of Taurus caused the eclipse. The concomitant opposition of Jupiter and Venus, which was 'outside its orbit', also contributed to a broader negative celestial conjunction and featured in the threatening astrological chart that Porrata Spinola had calculated for the area of Naples in late November 1631.[146]

There was agreement among astrologers and physicians about the negative effects caused by the position of Mars. In January 1632 Filippo Finella, a follower of Della Porta's empiricism and a member of the Incauti academy of Naples, published his astrological prognostications about the eruption under the pseudonym of Lanelfi.[147] Finella stated that in October 1631 he had produced two astrological charts concerned with the year 1632. There he calculated that Mars would remain in Leo from 10 October 1631 until 31 May 1632, and considered this unusually long transit to be 'the efficient cause of the portents' that were connected to the eruption and other calamitous events.[148] It was this ominous planetary conjunction that caused 'the vapours to remain blocked in the Earth', Finella wrote. He also believed the fiery nature of both Mars and Leo to be one of the causes of the eruption of Vesuvius. In discussing the danger of this unfavourable conjunction to Naples, Finella argued that the city 'is governed by the sign of Aries which is in the house of Mars, whose heat brings drought, fire, wars, pestilence and death'.[149]

Nevertheless, he ended his astrological discussion of the recent calamitous events on a positive note. Like Giovino, Finella labelled as 'an unfounded speculation' the conjecture that the passage of bituminous matter beneath Naples would cause the city to suffer from a future catastrophe similar to that caused by the eruption of Vesuvius. He justified his views by explaining that the city's foundations lay upon 'a certain stone called tuff'. Tuff, he wrote, is 'dry, insulating and impossible to ignite', thus making Naples 'together with the grace of God, exempt from this type of misfortune'.[150]

Within the bustling urban sphere of what was at the time the most populous city in Europe and one of the largest ports in the

Mediterranean, astrological interpretations of the 1631 disaster dominated the Neapolitan public debate. 'Many have written about this occurrence, and I hear public discussions that this eruption has been caused by the stars,' exclaimed the foreigner in Capaccio's *Forastiero*.[151] Capaccio's rejection of all the astrological theories that were publicly and dangerously circulating in the city proved that the local debate on the eruption was marked by contrasting views. In distancing himself from what he defined as the debate of charlatans 'who claim to know what is only manifest to God', Capaccio equated astrologers to itinerant fortune-tellers.[152] Instead, he invited Neapolitans to think in a religious way about the relationship between the heavens and the Earth, and to dismiss the groundless stance of some modern astrologers whose pernicious ideas were being disseminated among Neapolitans from all walks of life. Capaccio's fierce criticism was directed against the likes of 'Copernicus and Tico Brahe', about whom he wrote that

> they would have given more comfort to their souls if, in their futile efforts and nocturnal observations of the planets, they would instead have thought of ways to unite themselves with God.[153]

## Conclusion

The eruption of Vesuvius that occurred between December 1631 and January 1632 is an important case study for reconstructing the environmental, scientific and cultural history of the Bay of Naples. The range of sources consulted for this chapter has presented the landscape of Vesuvius and the impact of its catastrophic eruption on its natural and human environment. Within a scholarly context, the event generated a lively debate that resulted in the production of academic speeches, treatises, short accounts and manuscript sources. Astrologers, physicians, academicians and erudite ecclesiastics variously interpreted that calamitous event within a multidisciplinary debate. As we have seen, the language of medicine saw Vesuvius as an anthropomorphised entity, while astrology explained the volcano's violent awakening as being caused by a ruinous planetary conjunction.

In his interesting study of meteorology in early modern Europe, Craig Martin has highlighted that 'the emphasis on causation did not end with the rejection of Aristotelian thought'.[154] Within a Neapolitan scholarly context, the Aristotelian framework – although still central – no longer provided answers to explain the complexity of natural disasters.

An empirical approach that located the eruption within the geomorphology of the Bay of Naples discussed the validity of classical natural philosophy that paved the way for the development of a European debate on earth sciences, within which the study of Vesuvius would emerge as a discipline.[155] In assessing the impact of the eruption on Naples, scholars such as Giovino and Finella advanced new hypotheses to explain the reasons for which the city had been spared from destruction. Whether it was the tuff stone upon which the city rested or the series of underground cavities that enabled air and matter to travel from Mount Somma to Pozzuoli, empirical observation strengthened an understanding of the link between Vesuvian eruptions and the geological structure of the region that partly questioned – albeit clandestinely – the centrality of religious explanations of natural events.

Nonetheless, beyond the factual evidence they provide, the sources analysed in this chapter also offer poignant narrations of human responses to the eruption. Indeed, descriptions of the volcano's terrifying sounds and its overflowing crater erupting fiery matter; of the ashes being spread in such a large quantity to obscure the sun; of the valley's vegetation being destroyed by the crushing amount of lava; or of the catastrophic damage to the marine habitat were paired with equally intense descriptions of human emotions. Horror in witnessing the uncontrollable violence of Vesuvius; astonishment at seeing the volcano billowing smoke; fear in coping with Naples being hit by eclipses, cold wind, smoke and earthquakes; trepidation in hearing the terrifying sound of howling dogs that resembled human crying; desperation in asking God for forgiveness; resilience in coping with losses and immense suffering make the 1631 eruption of Vesuvius as much of a human story as it is that of a natural disaster.[156] It is therefore not surprising to read that in a letter sent to his brother in Rome in December 1631, the Jesuit father Ascanio Capece encapsulated the desperation of Neapolitans when he wrote: 'my brother Antonio, one cannot see Naples and not cry constantly'.[157]

## Notes

1 See the quotation that forms the title of this chapter: 'Misere terre arse dal fuoco e sommerse nell'acqua', Bernaudo, *L'Incendio del Monte Vesuvio*, 15.

2 Recent publications on early modern Vesuvius include Everson, 'The Melting Pot'; Cocco, *Watching Vesuvius*; Cocco, 'Locating the Natural Sciences'; Calaresu and Hills (eds.), *New Approaches to Naples*; Casapullo and Gianfrancesco (eds.), *Napoli e il Gigante*; Tortora, *L'eruzione Vesuviana del 1631*; Tortora, Cassano & Cocco (eds.), *L'Europa moderna*; Cecere,

De Caprio, Gianfrancesco and Palmieri (eds.), *Disaster Narratives*. For recent projects in the field of digital humanities which contain material on the history of Vesuvius, see 'The Italian Academies Database' (hereafter IAD) (www.bl.uk/catalogues/ItalianAcademies/Default.aspx) and 'Discompose' (Disasters, Communication and Politics in Southwestern Europe – DISCOMPOSE – ERC StG2017 n° 759829 (unina.it).

3 Braudel, *The Mediterranean*, vol, 1, 25.

4 Braccini, *Relazione dell'Incendio*, 10.

5 The dedication to Cardinal Colonna, dated 23 December 1632, is in Braccini, *Relazione dell'Incendio*, 2.

6 Braccini, *Dell'Incendio*.

7 'prati fioriti e una varietà di alberi [...] maravigliosamente disposti, artificiosamente potati, e carichi di frutti', Braccini, *Dell'Incendio*, 3.

8 'Ma che gli arbori erano tanto l'uno all'altro vicino', Braccini, *Dell'Incendio*, 25.

9 Braccini, *Dell'Incendio*, 25–6. In highlighting the health benefits of Vesuvius, on the same page Braccini also wrote that the heat emitted in some areas of the crater was used by the locals to detoxify the body through sweating. See also Cocco, *Watching Vesuvius*, 56–9.

10 Nazzaro, 'Some considerations', 561–2.

11 'Così gran sostegno del viver nostro hà riposto la madre natura in una sola montagna [...]', Recupito, *Avviso Dell'Incendio*, 144–54. See also Giuliani, *Trattato del Monte Vesuvio*, 14–17. For entries on Recupito and Giuliani see IAD.

12 Santorelli, *Discorsi, Accidenti, e Pronostici*, 30.

13 'i Vini di Nola, d'Ischia, di Vico, e di Surrento che prorompono da quel delicato terreno', Capaccio, *Il Forastiero*, 939 and 1011.

14 '[Torre del Greco] è da una parte bagnata dal mare, dall'altra riguarda il troppo temerario monte', Amodio, *Breve Trattato del Terremoto*, 24. An entry on Capaccio is in IAD.

15 'appo tutte le nationi sono tenuti in pregio', Recupito, *Avviso*, 141.

16 Schottus, *Itinerarii Italiae Rerumq[ue]*, 571–3.

17 Capaccio, *Descrizione Di Napoli*, 45; De Pietri, *Dell'Historia Napoletana*, 26; Tortora, 'Il Vesuvio ed il suo territorio', 124. A reference to the large production of walnuts is in Mormile, *Descrittione*, 72.

18 Roberts, *The Merchants Mappe*, 52.

19 Roberts, *The Merchants Mappe*, 52.

20 Braccini, *Dell'Incendio*, 3.

21 Mormile, *Descrittione*, 71.

22 Cesaro, *L'Opre Stupende*, Sig A4$^{r}$.

23 'Vedesi per tutto ripieno di delitiosissimi giardini [...] erano habitationi di Cavalieri Napolitani, che in esse trasferiteve le loro case, e famiglie vivevano', Cesaro, *L'Opre Stupende*, Sig A4$^{r}$.

24 See for instance Pollera, *Relatione dell'Incendio*, Sig. A3$^{r}$.

25 Bernaudo, *L'Incendio*, 12.

26 Mormile, *Descrittione*, 72.

27 See Gianfrancesco, 'Vesuvio e Società'; Gianfrancesco, 'Narratives and representations', 172–9.

28 Guidoboni and Boschi, 'Vesuvius before the 1631 eruption', 417.

29 Nazzaro, 'Implicazioni di una Ermeneutica', 140.

30 De Martino, *Osservatione*, 13; Rosi, Principe and Vecci, 'The 1631 Vesuvius eruption', 153–61.

31 De Martino, *Osservatione*, 13; Giuliani, *Trattato del Monte Vesuvio*, 63; Bernaudo, *L'Incendio*, 4; Santorelli, *Discorsi*, 1.

32 Oliva, *Lettera*. Sig A2$^{v}$.

33 Oliva, *Lettera*, 7.

34 Braccini, *Relazione*, 10.

35 Braccini, *Dell'Incendio*, 39; Oliva, *Lettera*, Sig. A2$^{r}$.

36 Masino, *Distinta Relatione dell'Incendio*, Sig. C2$^{r}$.

37 Bibliothèque Méjanes, Aix-en-Provence (hereafter BMAEP), Ms 0212 (1030), Correspondance littéraire de Nicolas-Claude Fabri de Peiresc avec les savants de son temps, *Lettera di Dom Severo di Napoli Cartusino al Sig$^{re}$ De Peiresc*, fol. 200.

38 Braccini, *Dell'Incendio*, 39; Carafa, *In Opusculum de Novissima Vesuvuij Conflagratione*, 7.

39 BMAEP, Ms 0212 (1030), *Lettera di Dom Severo,* fol. 207. In early modern Naples a *palm* measured about 26.4 cm. A brief reference to Peiresc's interest in the 1631 eruption of Vesuvius is in Vermij, *Thinking on earthquakes*, 171–2. On Saverio Trafaglione see Toppi, *Biblioteca Napoletana*, 279; Morozzo, *Theatrum chronologicum*, 145.
40 'cotanto ardenti, che sembravano più tosto piombo, o stagno liquefatto', Braccini, *Dell'Incendio*, 40.
41 Benigni, *La Strage di Vesuvio*, Sig. A3$^{1-r}$.
42 Oliva, *Lettera*, Sig. A2$^{r-2}$.
43 Braccini, *Dell'Incendio*, 40; Carafa, *In Opusculum de Novissima Vesuvuij Conflagratione*, 36–43; Masino, *Distinta Relatione*, Sig. B2$^{v-2}$ - Sig. C1$^{r}$.
44 'Si mosse con gran impeto il mare', Masino, *Distinta Relatione*, Sig. C1$^{v}$.
45 Carafa, *In Opusculum de Novissima Vesuvuij Conflagratione*, 7.
46 Nazzaro, 'Implicazioni di una Ermeneutica', 141–3.
47 Masino, *Distinta Relatione*, Sig. C1$^{r}$.
48 Milesio, *Vera Relatione*, 5.
49 Masino, *Distinta Relatione*, Sig. B2$^{v-2}$.
50 Masino, *Distinta Relatione*, Sig. B2$^{v-2}$; Bove, *Nuove Osservationi*, 12–13.
51 De Martino, *Osservationi*, 50.
52 Giuliani, *Trattato del Monte Vesuvio*, 83–4.
53 Oliva, *Lettera*, Sig. A2$^{r-2}$.
54 Giuliani, *Trattato del Monte Vesuvio*, 85.
55 Oliva, *Lettera*, Sig. A2$^{v-2}$.
56 '[voragini] che si fossero aperte nell'istesso mare', BMAEP, Ms 0212 (1030), *Lettera di Dom Severo*, fol. 211.
57 'non potendo fermare li piedi sopra l'ancora cocenti ceneri', BMAEP, Ms 0212 (1030), *Lettera di Dom Severo*, fol. 213.
58 Braccini, *Dell'Incendio* (1632), 41.
59 Garsia, *I funesti Avvenimenti del Vesuvio*, Sig. A2$^{r}$; Recupito, *Avviso*, 143; Mormile, *Descrittione*, 69.
60 For an excellent analysis of early modern meteorology see Martin, *Renaissance Meteorology*.
61 Braccini, *Dell'Incendio*, 50 and 74.
62 'contra il costume del clima', BMAEP, Ms 0212 (1030), *Lettera di Dom Severo*, fol. 195; Amodio, *Breve Trattato del Terremoto*, 56.
63 'facendoci venire le vesti in faccia il vento', BMAEP, Ms 0212 (1030), *Lettera di Dom Severo*, fol. 196.
64 BMAEP, Ms 0212 (1030), *Lettera di Dom Severo*, fol. 200.
65 The literature on this topic is vast. See, for instance, Self, 'The effects and consequences'; Jones et al., 'An AOGCM simulation'; Kjeld C. Engvild, 'A review of the risks of sudden global cooling'; Behringer, *A Cultural History of Climate*, 18–19.
66 For an analysis of the medical debate on the plague in early modern Naples see Silvana D'Alessio, 'On the Neapolitan plague of 1656'.
67 'infettino ella [l'aria] dalla quale noi vivemo, & da questa mescolanza maligna nascono i morbi maligni, pestilentiali', De Martino, *Osservationi*, 11.
68 'palle odorifere [con] aceto rosato, acque de fiori de narancio [...] & altri odori polverizzati', De Martino, *Osservationi*, 11.
69 'Per rettificar il sangue, e nettar il stomaco [si beva] il succo di limone, agro di cedro [...] mescolandoci theriaca, bolarmeno, belzuarro', De Martino, *Osservationi*, 11.
70 'nell'estate, dal cui calore potrebbono facilmente corrompersi l'acque & i cadaveri, a da i cattivi, e fetidi vapori così dell'uno, como dell'altro infettarsi l'aria', Santorelli, *Discorsi*, 50.
71 'grosse et continue piogge [...] che allagò tutto il piano di Terra di Lavoro, da Nola insino alla Cerra', BMAEP, Ms 0212 (1030), *Lettera di Dom Severo*, fol. 191.
72 'questa città da tanti scoppi di lampeggianti tuoni, con spessa grandine, & agghiacciata neve', Masino, *Distinta Relatione*, Sig. E2$^{r}$.
73 Oliva, *Lettera*, Sig. A2$^{r-3}$; Braccini, *Dell'Incendio*, 37; BMAEP, Ms 0212 (1030), *Lettera di Dom Severo*, fol. 198.
74 Porrata Spinola, *Discorso sopra l'origine de' fuochi*, 3–4.
75 Padavino, *Novissima Relatione dell'Incendio*, Sig. A4$^{4-r}$.

76 Archivio di Stato di Napoli (hereafter ASN), Segreterie dei Viceré, stanza 159/n.1, viglietti originali, busta 34.
77 ASN, Segreterie dei Viceré, stanza 159/n.1, viglietti originali, busta 34.
78 Padavino, *Novissima Relatione*, Sig. A4$^{4-v}$; Braccini, *Dell'Incendio*, 38; Amodio, *Breve Trattato del Terremoto*, 59; Spinola, *Discorso*, 17.
79 Robinson, 'A relation of the raining of ashes', 377. References to Hungary and Corfu are in BMAEP, Ms 0212 (1030), *Lettera di Dom Severo,* fol. 216. A reference to the ashes in Istanbul is in Spinola, *Discorso*, 18.
80 BMAEP, Ms 0212 (1030), *Lettera di Dom Severo*, fols. 215–16.
81 Biblioteca Apostolica Vaticana (hereafter BAV), Ms. Vat.lat.8193, pt. 2, *Il Monte Vesuvio, doppo cinque mesi ch'era seguito dal suo incendio, il quale fu' a' 16. di Decembre 1631*, fol. 528$^{r}$. This account is dated Rome, 1 September 1637; it was written to inform a patron in Rome about what the author had experienced during his sojourn in Naples at the time of the eruption.
82 'non tanto dal diluvio delle ceneri, e dalla tempesta delle pietre infocate, che vomitò la montagna; quanto da i torrenti d'acqua, fuoco, e bitume', BAV Ms. Vat.lat.8193, pt. 2, *Il Monte Vesuvio, doppo cinque mesi ch'era seguito dalsuo incendio, il quale fu' a' 16. di Decembre 1631*, fol. 528$^{v}$.
83 'Non bastò al Monte far cruda stragge col fuoco [...] cacciò un rapido fiume d'acque bituminose, e sulfuree [che] aperse i sentieri; appianò le valli; svelse le selve; spiantò i boschi; diradicò gli edificij; e seco le capanne, le greggi, le biade, le piante, le selve, i tetti, i fondamenti [...] parte in tributo al mare', Bernaudo, *L'Incendio*, 11–12.
84 Braccini, *Dell'Incendio*, 25.
85 'che si stimò potere bastare a far fuoco per dieci anni in tutta la città', BMAEP, Ms 0212 (1030), *Lettera di Dom Severo*, fol. 211.
86 BMAEP, Ms 0212 (1030), *Lettera di Dom Severo di Napoli*, fol. 202.
87 Santorelli, *Discorsi*, 30.
88 BMAEP, Ms 0212 (1030), *Lettera di Dom Severo*, fol. 202.
89 Bernaudo, *L'Incendio*, 11; Braccini, *Dell'Incendio*, 1.
90 Padavino, *Novissima Relatione*, Sig. A4$^{3-v}$; Braccini, *Dell'Incendio*, 1.
91 'puzzor di solfo, e bitumene intollerabile all'odorato', De Martino, *Osservationi*, Sig. A3$^{r}$.
92 Padavino, *Novissima Relatione*, Sig. A4$^{3-v}$.
93 Bernaudo, *Del Monte Vesuvio*, 5; De Martino, *Osservationi*, 12.
94 Braccini, *Dell'Incendio*, 42.
95 Carafa, *In Opusculum de Novissima Vesuvuij Conflagratione*, 32.
96 BMAEP, Ms 0212 (1030), *Lettera di Dom Severo,* fols. 207.
97 De Martino, *Osservationi*, 26.
98 Santorelli, *Discorsi*, 44.
99 See for example Bernaudo, *L'Incendio*, 12; Padavino, *Novissima Relatione*, Sig. A4$^{3-v}$.
100 'Gli animali pure lasciati insepolti s'interrano, perche non corrompino co[n] il fetore l'aria', Padavino, *Novissima Relatione*, Sig. A4$^{3-v}$.
101 Bernaudo, *L'Incendio*, 12; Padavino, *Novissima Relatione*, fol. 529$^{v}$.
102 BAV, Ms. Vat.lat.8193, pt. 2, *Il Monte Vesuvio*, fol. 531$^{r}$.
103 BAV, Ms. Vat.lat.8193, pt. 2, *Il Monte Vesuvio*, fol. 531$^{v}$.
104 BAV, Ms. Vat.lat.8193, pt. 2, *Il Monte Vesuvio*, fol. 532$^{r}$.
105 See note 46 above; Rosi, Principe and Vecci, 'The 1631 Vesuvius eruption', 152–3.
106 'è quel fumo, che da Napoli, e d'altrove si vede', BAV, Ms. Vat.lat.8193, pt. 2, *Il Monte Vesuvio*, fol. 535$^{v}$.
107 BAV, Ms. Vat.lat.8193, pt. 2, *Il Monte Vesuvio*, fol. 536$^{v}$.
108 Biblioteca Ursino Recupero, Catania (hereafter BURC), Ms. E.20.$^{32}$ Giovanni Tommaso Giovino, *Lettione Academica dell'incendio e terremoto di Somma fatta nell'Accademia dell'Infuriati di Napoli, da Giov. Tomaso Giovino D.$^{r}$ di filosofia e medicina, accademico prima nell'Otiosi e poi nell'Infuriati, nell'anno 1632*, fol. 201$^{v}$. An entry on Giovino is in IAD.
109 'non si niega che tal'hora l'irato et giusto Dio à suo cenno per punire i misfatti [...] si serve di simili stromenti', BURC, Ms. E.20$^{32}$, G. T. Giovino, *Lettione,* fol.203$^{r}$.
110 As I have discussed elsewhere, one of the scholarly outcomes of the 1631 eruption was the reconstruction of the history of Vesuvius since antiquity. See Gianfrancesco, 'Vesuvio e Società', 77–9.

111 BURC, Ms. E.20$^{32}$, G. T. Giovino, *Lettione*, fols, 203$^{r-v}$. For an overview of the discourse on earthquakes from antiquity to the early modern period see Vermij, *Thinking on Earthquakes*, 21–34.

112 BURC, Ms. E.20$^{32}$, G. T. Giovino, *Lettione*, fol. 204$^{r}$. See also Masino, *Distinta Relatione*, Sig. A1$^{r-v}$.

113 De Martino, *Osservationi* (1632), Sig. A1$^{r}$.

114 'Scuotono, e fanno crollare la Terra, muovono i Monti, alla fine resi implacabili si fanno il varco, aprendo voragini', De Martino, *Osservationi* (1632), Sig. A2$^{r}$.

115 Santamaria, *Trattato Scientifico delle cause*. See also Santorelli, *Discorsi*, 9–14.

116 'e strinse li pori della terra, con la sua freddezza, onde à fatto restorono rinchiusi li spiriti secchi', Santamaria, *Trattato Scientifico delle cause*, 36.

117 'freddezza causa nelli pori della terra; onde li spiriti secchi più violenza patiscono e non potendosi ritenere in picciol luoco scotono le viscere della madre', Santamaria, *Trattato Scientifico delle cause*, 37.

118 'ardendo quelle materie bituminose, e sulfuree, e trova[n]do stretto il luogo, per donde doveva esalare, formava nell'uscire quello strepito, a guisa della bombarda', Braccini, *Relazione*, 10.

119 Capaccio, 'Incendio di Vesuvio. Dialogo', *Il Forastiero*, 60. Posidonius of Rhodes's analysis of earthquakes is mostly available in passages contained in Seneca's *Naturales Quaestiones*. See Vimercati (ed.), *Posidonio. Testimonianze e frammenti*, 171, 600–3. A brief discussion of Posidonius's categorisation of earthquakes is in Vermij, *Thinking on Earthquakes*, 28.

120 'Altro terremoto vien detto collesivo quasi che dui grandi e duri corpi fra di loro s'urtassero, suole ciò accadere quando da tal contrasto escono fuora sassi così perche facci ancora adesso [...] il Mo[n]te nostro Vesuvio', Amodio, *Breve Trattato del Terremoto*, 14.

121 Santorelli, *Discorsi*, 45–6. For a discussion of Seneca's ideas on seismology see Williams, 'Greco-Roman seismology'.

122 'e questi torrenti sono peggiori di tutti', Santamaria, *Trattato Scientifico*, 46.

123 BURC, Ms. E.20$^{32}$, G. T. Giovino, *Lettione*, fol. 209$^{r}$.

124 BURC, Ms. E.20$^{32}$, G. T. Giovino, *Lettione*, fol. 209$^{v}$.

125 'che gravidandosi per secoli, e secoli di tanta quantita di essalationi [...] facci empito grande, et cossì poi effetti si horrendi e maravigliosi', BURC, Ms. E.20$^{32}$, G. T. Giovino, *Lettione*, fols. 209$^{v}$–210$^{r}$.

126 'somiglianti alle vene et alle arterie che ha' l'huomo sparse per il suo corpo', BURC, Ms. E.20$^{32}$, G. T. Giovino, *Lettione*, fol. 208$^{v}$; de Pietri, *I Problemi Accademici del Signor Francesco De' Pietri*, 217–20. See also L. Gianfrancesco, 'Vesuvio e società', 79–80.

127 'nel mezzo fra due grandi spiragli d'essalationi, cioè Somma, e Pozzuoli [...] per dove agevomente la detta essalatione ritrovato la strada colà s'invola', BURC, Ms. E.20$^{32}$, Giovino, *Lettione*, fol. 207$^{v}$.

128 'Questo Monte nasconde dentro il seno grandissima copia di acque, di bitume, e di zolfo, la qual triplicata materia è caggione di fuochi chimati sempiterni [...] anche se non sono sempre visibili', Porrata Spinola, *Discorso*, 4, 15, 16.

129 Della Porta, *De Aëris Transmutationibus*, 95–8.

130 Giovino, *Lettione*, fol. 210$^{r}$. See also Giuliani, *Trattato del Monte Vesuvio*, 205–8.

131 Porzio, *De Conflagratione Agri Puteolani*.

132 Porzio, *De Conflagratione Agri Puteolani*, 3. On Porzio see Castelli, 'Simone Porzio', 119, 121–5.

133 Porzio, *De Conflagratione*, 6.

134 'Como dunque quì è andato mancando per molta robba, che si menò al molo: nell'Egitto per la molta terra dal Nilo portatavi', Santorelli, *Discorsi*, 43–4.

135 'onde l'acque parte per parte violentemente ancor ritornando al vacuo lasciaro il porto asciutto', Spinola, *Discorso*, 12–13.

136 'aprendo lo Spirito i pori della terra, e facendo spesse e grandi aperture, quella parte di mare che gli sovrastà viene inghiottita, più propriamente, che se n'entri per le caverne del monte, e che poi dallo Spirito vehemente,sian portate insieme col foco fora le voragini', Capaccio, 'Incendio di Vesuvio. Dialogo', in *Il Forastiero*, 61.

137 'I terremoti sogliono predirsi dagl'eclissi, et da i gran congiogimenti di Pianeti maggiori', BURC, Ms. E.20$^{32}$, G. T. Giovino, *Lettione*, fol. 204$^{r}$.

138 BURC, Ms. E.20$^{32}$, G. T. Giovino, *Lettione*, fol. 204$^{r}$.
139 'come anco i rumori delle [...] armi che in piu parti del mondo a' nostri danni si odono', BURC, Ms. E.20$^{32}$, G. T. Giovino, *Lettione,* fol. 204$^{v}$.
140 'Coniuntione di Saturno e di Giove fatta nella casa di Sagittario [...] essendo Sagittario il piu forte segno della triplicità ignea', BURC, Ms. E.20$^{32}$, G. T. Giovino, *Lettione,* fol. 205$^{r}$.
141 De Martino, *Giornali*, Sig. B1$^{r}$.
142 De Martino, *Giornali*, Sig. B1$^{r}$.
143 'La casa di Leone che predomina Italia [...] la casa del Toro predominatore della Campagnia [Campania] dovea soccedere il danno, cioè in Italia & in Campagna [Campania] felice', De Martino, *Giornali*, Sig. B1$^{r\text{-}v}$.
144 De Martino, *Giornali*, Sig. B1$^{r\text{-}v}$. Dom Severo also reported the appearance of a comet near Vesuvius in late November 1631, in BMAEP, Ms 0212 (1030), *Lettera di Dom Severo*, fol. 195.
145 Porrata Spinola, *Discorso*, 23.
146 Porrata Spinola, *Discorso*, 20–1. See Gianfrancesco, 'Vesuvio e società', 77–82.
147 Finella, *Incendio del Vesuvio*. See Gianfrancesco, 'Vesuvio e società', 81–2; Gianfrancesco, 'From propaganda to science', 19–20. An entry on Filippo Finella is in IAD.
148 Finella, *Incendio*, Sig. A4$^{2\text{-}v}$.
149 '& in Napoli particolarmente, che viene predominato dal segno d'Ariete casa di esso Marte, la cui calidità non sà altro apportare, che siccità, incendij, guerre, pestilenze, e morte', Finella, *Incendio*, Sig. A4$^{4\text{-}r}$.
150 'perche la Città di Napoli sta fondata sopra una certa Pietra, communemente chiamata Tufo, arida in se stessa, e senza sorte alcuna di pinguedine [...] che accender si potesse, di modo che per la Dio gratia resta questa Citta esente, e sicura da simile infortunio', Finella, *Incendio*, Sig. A4$^{4\text{-}r\text{-}v}$.
151 'hanno scritto molti, e ne sento ragionar publicamente, che questo incendio è cagionato dalle Stelle', Capaccio, 'Incendio di Vesuvio. Dialogo', in *Il Forastiero*, 39.
152 Capaccio, 'Incendio di Vesuvio. Dialogo', in *Il Forastiero*, 41.
153 'e havrebbono fatto più profitto all'anime loro, trà tante fatiche sparse al vento, e trà tante notturne osservationi, pensar come si havessero potuto congiungere con Dio', Capaccio, 'Incendio di Vesuvio. Dialogo', in *Il Forastiero*, 41.
154 Martin, *Renaissance Meteorology*, 4.
155 Cocco, *Watching Vesuvius*, 138–50.
156 Milesio, *Vera Relatione del Miserabile*, 5; Capradosso, *Il Lagrimevole Avvenimento dell'Incendio*. For an interesting analysis of the link between environmental history and the senses see Nash, 'The body and environmental history', 409–11.
157 'O fratello Antonio mio, non si può veder Napoli e non piangere dirottamente', Letter of Ascanio Capece to Antonio Capece, in Riccio, 'Nuovi documenti sull'incendio', 497.

# Bibliography

## Websites

The Italian Academies Database (IAD), www.bl.uk/catalogues/ItalianAcademies/Default.aspx

'Discompose' Disasters, Communication and Politics in Southwestern Europe – DISCOMPOSE – ERC StG2017 n° 759829 (unina.it).

## Archival sources

Archivio di Stato di Napoli (ASN), Segreterie dei Viceré, stanza 159/n.1, viglietti originali, busta 34.

Biblioteca Apostolica Vaticana (BAV), Ms. Vat.lat.8193.

Bibliothèque Méjanes, Aix-en-Provence (BMAEP), Ms 0212 (1030).

Biblioteca Ursino Recupero Catania (BURC), Ms. E.20.[32]

## Printed sources


Amodio, Giulio. *Breve Trattato del Terremoto. Scritto da D. Giulio Amodio Napolitano. In occasione dell'incendio successo nel Monte Vesuvio nel giorno 16. Di Dicembre 1631. Con una verissima relatione di quanto è successo da detto dì fino à 22. Di Gennaro 1632*. Naples: 1632.

Behringer, Wolfang. *A Cultural History of Climate*. Cambridge: Polity Press, 2010.

Benigni, Domenico. *La Strage di Vesuvio. Lettera Scritta all'Illustrissimo Signore Abbate Perretti dal suo Secretario*. Naples: 1632.

Bernaudo, Francesco. *L'Incendio del Monte Vesuvio*. Naples: 1632.

Bove, Vincenzo. *Nuove Osservationi Fatte Sopra gli Effetti dell'Incendio del Monte Vesuvio*. Naples: 1632.

Braccini, Giulio Cesare. *Relazione dell'Incendio fattosi nel Vesuvio alli 16. di Decembre 1631*. Naples: 1631.

Braccini, Giulio Cesare. *Dell'Incendio Fattosi nel Vesuvio a XVI. Di Dicembre M.DC.XXXI. E delle sue cause, ed effetti*. Naples: 1632.

Braudel, Fernand. *The Mediterranean and the Mediterranean World in the Age of Philip II*. New York; Evanston; San Francisco; London: Harper & Row, 1972.

Calaresu, Melissa and Helen Hills (eds). *New Approaches to Naples c.1500–c.1800. The power of place*. Farnham: Ashgate, 2013.

Capaccio, G. C. *Il Forastiero*. Naples: 1634.

Capaccio, G. C. *Descrizione Di Napoli ne' Principi del Secolo XVII*. Naples: F. Giannini, 1882.

Capradosso, ?. *Il Lagrimevole Avvenimento dell'Incendio del Monte Vesuvio Per la Città di Napoli, e Luoghi Adiacenti*. Naples: 1631.

Carafa, Gregorio. *In Opusculum de Novissima Vesuvuij Conflagratione, Epistola Isagogica*. Naples: 1632.

Casapullo, Rosa and Lorenza Gianfrancesco (eds). *Napoli e il Gigante. Il Vesuvio tra immagine, scrittura e memoria*. Soveria Mannelli: Rubbettino, 2014.

Cesaro, Francesco. *L'Opre Stupende e Maravigliosi Eccessi Dalla Natura Prodotti nel Monte Vesuvio della Città di Napoli*. Naples: 1632.

Cocco, Sean. *Watching Vesuvius. A history of science and culture in early modern Italy*. Chicago: Chicago University Press, 2013.

Cocco, Sean. 'Locating the natural sciences in early modern Naples'. In *A Companion to Early Modern Naples*, edited by Tommaso Astarita, 453–74. Leiden; Boston: Brill, 2013.

D'Alessio, Silvana. 'On the Neapolitan plague of 1656: Expedients and remedies'. In *Disaster Narratives*, edited by Domenico Cecere, Chiara De Caprio, Lorenza Gianfrancesco and Pasquale Palmieri, 187–204. Rome: Viella, 2018.

Della Porta, Giambattista. *De Aëris Transmutationibus*. Libri IV (1610), edited by Alfonso Paolella. Edizione Nazionale delle Opere di Giovan Battista della Porta, 14. Naples: Edizioni Scientifiche Italiane, 2000.

De Martino, Cesare. *Osservationi. Giornali del Successo nel Vesuvio dalli XVI. di Decembre M DC XXXI. Fino alli X. di Aprile M DC XXXII*. Naples: 1632.

De Pietri, Francesco. *Dell'Historia Napoletana*. Naples: 1634.

De Pietri, Francesco. *I Problemi Accademici del Signor Francesco De' Pietri. L'Impedito Accademico Otioso ove le più famose Quistioni proposte nell'Illustrissima Accademia de gli Otiosi di Napoli si spiegano*. Naples: 1642.

Engvild, Kjeld C. 'A review of the risks of sudden global cooling and its effects on agriculture', *Agricultural and Forest Meteorology* 115 (2003): 127–37.

Everson, J. 'The melting pot of science and belief: Studying Vesuvius in seventeenth-century Naples', *Renaissance Studies* 26 (2012): 691–727.

Finella, Filippo. *Incendio del Vesuvio del Lanelfi*. Naples: 1632.

Garsia, Gio. Andrea. *I funesti Avvenimenti del Vesuvio. Principiati Martedì 16 di Decembre MDCXXXI*. Naples: 1632.

Gianfrancesco, Lorenza. 'From propaganda to science: Looking at the world of academies in early seventeenth-century Naples', *California Italian Studies* 3 (2012): 1–31.

Gianfrancesco, Lorenza. 'Vesuvio e società: Informazione, propaganda e dibattito intellettuale a Napoli nel primo Seicento'. In *Napoli e il Gigante. Il Vesuvio tra immagine, scrittura e memoria*, edited by Rosa Casapullo and Lorenza Gianfrancesco, 65–77. Soveria Mannelli: Rubbettino, 2014.

Gianfrancesco, Lorenza. 'Narratives and representations of a disaster in early seventeenth-century Naples'. In *Disaster Narratives*, edited by Domencio Cecere, Chiara De Caprio, Lorenza Gianfrancesco and Pasquale Palmieri, 172–9. Rome: Viella, 2018.

Giuliani, Gianbernardino. *Trattato del Monte Vesuvio e de' Suoi Incendi*. Naples: 1632.

Guidoboni, Emanuela and Enzo Boschi, 'Vesuvius before the 1631 eruption', *Eos, Transactions, American Geophysical Union* 40 (2006): 417–23.

Jones, Gareth S., Jonathan M. Gregory, Peter A. Stott, Simon F. B. Tett and Robert B. Thorpe. 'An AOGCM simulation of the climate response to a volcanic super-eruption', *Climate Dynamics* 25 (2005): 725–38.

Martin, Craig. *Renaissance Meteorology. Pomponazzi to Descartes*. Baltimore: Johns Hopkins University Press, 2011.

Masino, Michel'Angelo. *Distinta Relatione dell'Incendio del Sevo Vesuvio. Alli 16. di Decembre 1631 Successo*. Naples: 1632.

Milesio, Giacomo. *Vera Relatione del Miserabile, e Memorabile Caso. Successo nella Falda della Nominatissima Montagna di Somma, Altrimente Detta Mons Vesuvij*. Naples: 1631.

Mormile, Giuseppe. *Descrittione Dell'Amenissimo Distretto della Citta di Napoli et Dell'Antichita della Citta di Pozzuolo*. Naples: 1617.

Morozzo, Carlo Giuseppe. *Theatrum chronologicum sacri cartusiensis ordinis lectori exhibens*. Turin: 1681.

Nash, L. 'The body and environmental history in the anthropocene'. In *The Routledge Companion to the Environmental Humanities*, edited by Ursula Heise, Jon Christensen and Michelle Niemann, 403–13. London and New York: Routledge, 2017.

Nazzaro, Antonio. 'Some considerations on the state of Vesuvius in the Middle Ages and the precursors of the 1631 eruption', *Annali di Geofisica* 4 (1998): 555–65.

Nazzaro, Antonio. 'Implicazioni di una Ermeneutica delle Fonti Vesuviane sull'Eruzione del 1631: Forma del Vulcano e Risposta al Rischio'. In *Napoli e il Gigante. Il Vesuvio tra immagine, scrittura e memoria*, edited by Rosa Casapullo and Lorenza Gianfrancesco, 137–77. Soveria Mannelli: Rubbettino, 2014.

Oliva, Maria Nicolo. *Lettera. Nella quale dà vera, e minuta relatione delli Segni, Terremoti, & Incendij del Monte Vessuvio, cominciando dalli dieci del mese di Decembre 1631. Infino alli cinque di Gerraro 1632*. Naples; Rome: 1632.

Padavino, Marco Antonio. *Novissima Relatione dell'Incendio Successo Nel Monte di Somma A 16. Decembre 1631. Con un' Aviso di Quello è successo nell'istesso dì nella Città di Cattaro, nelle parti d'Albania*. Venice and Naples: 1632.

Pollera, Giovan Domenico. *Relatione dell'Incendio del Monte di Somma*. Naples: 1632.

Porrata Spinola, Giovanni Francesco. *Discorso sopra l'origine de' fuochi gettati dal Monte Vesevo, ceneri piovute, et altri successi, & pronostico d'effetti maggiori*. Lecce: 1632.

Porzio, Simone. *De Conflagratione Agri Puteolani. Epistola*. Florence: 1551.

Recupito, Giulio Cesare. *Avviso Dell'Incendio del Vesuvio*. Naples: 1635.

Riccio, Luigi. 'Nuovi documenti sull'incendio vesuviano dell'anno 1631 e Bibliografia di quella eruzione', *Archivio Storico per le Province Napoletane* 14 (1889): 489–555.

Roberts, Lewis. *The Merchants Mappe of Commerce. Wherein, the universal manner and matter of trade is compendiously handled*. London: 1638.

Robinson, Henry. 'A relation of the raining of ashes, in the Archipelago, upon the eruption of Mount Vesuvius, some years ago', *The Royal Society, Philosophical Transactions* 1 (1665).

Rosi, Mauro, Claudia Principe and Raffaella Vecci. 'The 1631 Vesuvius eruption. A reconstruction based on historical and stratigraphical data', *Journal of Volcanology and Geothermal Research* 58 (1993): 153–61.

Santamaria, Agnello. *Trattato Scientifico delle cause, che concorsero al fuoco, & terremoto del Monte Vesuvio vicino Napoli. Utilissimo à Theologi, Filosofi, Astrologi, et ad ogni studio*. Naples: 1632.

Santorelli, Antonio. *Discorsi, Accidenti, e Pronostici Dell'Incendio del Monte di Somma dell'anno 1631*. Naples: 1632.

Schottus, Franciscus (Franz Schott). *Itinerarii Italiae Rerumq*[ue]. *Romanorum Libri Tres*. Antwerp: 1625.

Self, Steve. 'The effects and consequences of very large explosive volcanic eruptions', *Philosophical Transactions Royal Society* 364 (2006): 2073–97.

Tortora, Alfonso. *L'eruzione Vesuviana del 1631. Una Storia d'età Moderna*. Rome: Carocci, 2014.

Tortora, Alfonso. 'Il Vesuvio ed il suo territorio. La grande eruzione del 1631', *Rendiconto Dell'Accademia Delle Scienze Fisiche E Matematiche* Serie IV 84 (2017): 119–29.
Tortora, Alfonso, Domenico Cassano and Sean Cocco (eds). *L'Europa moderna e L'Antico Vesuvio*. Battipaglia: Laveglia & Carlone, 2017.
Toppi, Niccolò. *Biblioteca Napoletana*. Naples: 1678.
Vermij, Rienk. *Thinking on Earthquakes in Early Modern Europe. Firm beliefs in shaky grounds*. Abingdon; New York: Routledge, 2021.
Vimercati, Emmanuele (ed.). *Posidonio. Testimonianze e frammenti*. Milan: Bompiani, 2004.
Williams, Gareth D. 'Greco-Roman seismology and Seneca on earthquakes in "Natural Questions 6"', *The Journal of Roman Studies* 96 (2006): 124–46.

5

# Some effects of the plague on medicine in Naples: the choices and dialogue among the *novatores*

Silvana D'Alessio

## Introduction: texts and accounts during the plague

The plague spread in Naples from January 1656, killing more than half of the population. Its rapid circulation was mainly caused by the imprudent behaviour of the viceroy and his ministers and the long delay in tackling the epidemic.[1] After spreading the fable that the disease was caused by plague spreaders, the viceroy invited some doctors to meet and discuss the most appropriate measures to take against the disease. The surgeons Marco Aurelio Severino and Felice Martorella performed autopsies on the corpses of two infected people. Soon after they published a document, *Consultatio medicorum praevia sectione cadaverum pro praeservatione et curatione pestis* (2 June 1656).[2]

At this point, movement of infected people had not been prohibited; doctors could therefore only speak of remedies in the medical sense.[3] Preventative measures included fires 'in houses', with scented berries and herbs, 'acqua teriacale' (theriac), mithridatium, a compound of sulphur, rue, garlic, carnations, saffron and walnuts, powdered Armenian bole (clay)[4] and clay from Asia Minor ('terra sigillata').[5] There were also several antidotes, such as bezoar (a hard mass that forms in the intestines of animals and was believed to be an antidote since it remained undigested)[6] and sulphur (which, together with several plants, was supposed to be kept in the mouth). Other recommendations included suffumigations, sponges to be held under the nose and spheres with holes into which were placed fragments of theriac, vinegar, rue and oils, such as that concocted by Andrea Mattioli. This was thought to possess the power to ward off poison because it contained scorpion venom.[7]

Other favoured treatments included the antiplague elixir of Oswald Croll (a compound of sulphur, theriac, berries and oils),[8] the bezoardic vinegar of the spagyric Ernst of Bavaria, the diascordium of Fracastoro (which consisted of the teucrium scordium plant, cinnamon, tormentil, gentian and other ingredients, not least opium)[9] and a certain powder of Giovanni Battista l'Eremita, the recipe for which (tormentil, sandalwood, dittany, deer horn and Armenian bole) had been used in a previous epidemic in Naples.[10] The strongest remedies were those designed to purge bodies either by inducing vomiting, through the use of antimony or vinegar of squills, or through the practice of bloodletting (provided that certain rules were followed). It was also advisable to make the patient sweat, using ivy, flower of sulphur, theriac and powders prepared by the same friar, 'eremitano'. As we can see, several remedies served more than one purpose.

The doctors who wrote the *Consultatio* did not comment on the properties of the remedies that they recommended. In most cases these were ancient remedies that were considered generally effective. Nor was this the occasion to reason about the causes of the disease; other doctors would do this after the epidemic had ended.

The most ancient treatments were theriac, mithridatium, Armenian bole and *terra sigillata*. The first two date back to the first century BC. The recipe for theriac prepared by Bartolomeo Maranta was well known in Naples; he had been able to trace it back to the original version. It contained dozens of ingredients, including pieces ('trocisci') of scilla (a poisonous plant), vipers, pepper, opium, red roses and many other 'simples'.[11] In the preface to his volume, Maranta compares theriac and mithridatium to two ancient Greek heroes, Hercules and Theseus: one 'killed the hydra, the lion, the wild boar, Diomedes' mares and many other ferocious monsters' ('amazzò l'Hidra, il Leone, il Cinghiale, i Cavalli di Diomede, Cacco, e molti altri fieri mostri'), while the other 'in a constant effort to imitate him' visited 'various places and always freed their people from the oppression and cruelty of tyrants' ('sforzandosi sempre di imitarlo, discorse vari paesi liberando sempre le genti dalle oppressioni, e crudeltà dei tiranni'). Mithridatium, a remedy for the bites of wild animals, had been devised by Mithridates VI of Pontus. Andromachus, physician to the emperor Nero, added viper to the recipe, terming the new antidote 'theriac' (ϑήρειον). In *De Theriaca ad Pisonem*, the author attributed to theriac both a preventive function, since it 'corrected' the 'corrupted air', and healing properties against the plague.[12] According to the author of *De Theriaca*, Hippocrates stopped the spread of the plague that had arrived in Greece from Ethiopia by lighting fires in which he burnt not only wood and flowers but also ‘greasy and odorous ointments’ (‘unguenti grassissimi e odoratissimi’). This account – well known among physicians and philosophers of that period – led many to believe that theriac was both a means of purifying air corrupted by the plague and an antidote to that disease.[13]

In several cases the reason for which a remedy was considered useful was given. For example, when discussing quinine, discovered in Ecuador,[14] the apothecary Giuseppe Donzelli explained that it could combat the ‘bad humours’ residing in the stomach that were the source (‘fomite’) of malignant fevers.[15]

The section on buboes (swollen and inflamed lymph nodes) that appears in the *Consultatio* was also tied to the Galenic framework. The authors suggested intervening with emollients if the buboes were about to burst, leaving nature to take its course; only if they did not mature on their own were leeches and ‘vescicatori’ (stinging substances) to be used, followed by the application of ointments. In some circumstances buboes could be removed with ‘the cold iron’.

The many remedies recommended against the plague were meant to reassure, as far as possible, a frightened population. As Ludovico Antonio Muratori would later write, these were supposed to be the weapons in the fight against such a powerful disease.[16] The remedies also provided an opportunity for physicians to demonstrate the importance of their knowledge, and obviously to earn some money. The need for these weapons in other contexts in which plague spread, as well as the great reputation for authority that Neapolitan physicians enjoyed, explains the success of the *Consultatio* beyond the Kingdom of Naples. Sebastiano Bado, a physician at the Ospedale Pammatone in Genoa, wrote in a short treatise published in 1656 that he approved of the remedies proposed by the ‘Neapolitan’ physicians, even though they recommended serving the theriac hot. Bado continued:

> I approve it, because if there is a medicine that is effective against the plague, that is theriac, which operates like fire and consequently like other hot medicaments. Galen mentions Hippocrates who stopped the plague with fire, with the virtue of theriac which, like fire, purges and cleans all pestilential seeds and filth.[17]

For Bado the plague did not derive from putrefaction, but rather from an ‘occult quality’ that was countered by ancient remedies, which he listed in order of importance: theriac, mithridatium, Armenian bole and finally the compound of ‘dried figs, rue, walnut and a little salt’ that was mentioned in the *Consultatio*. Bado was also pleased to remind readers that Neapolitan physicians had recommended the use of quinine ('china'), a medicament whose value in the treatment of malignant fevers he had already extolled. The important thing, he wrote, was to purge the body 'of imbalanced humours in order to make it transpirable' ('dagl'humori soverchi, e farlo transpirabile').[18]

In *Il pestifugo Esculapio* Giovann'Antonio Bumaldi (Ovidio Montalbani), a very versatile author from Bologna,[19] wrote about the wonders of 'Teriaca' (always capitalised). The physician claimed certain victory:

> We have won when from the pharmaceutical flasks the theriacal viper (*Echnidi triacale*), that is, the Apolline serpent, emerges to defend us, and whose teeth are mighty and capable of devouring the serpent.[20]

Montalbani called theriac the 'Aesculapius', the 'Alexian Hercules' praised by Galen in his *Theriaca ad Pisonem*.[21] Adopting the bellicose language already used in several Hippocratic texts, he described it as the 'shield' that would repel 'the hostile blows and will make [the disease] weak and harmless'.[22]

As previously noted, the plague offered a great opportunity for the sale of all types of remedies and secrets. The anonymous author of the account edited by de Blasiis comprehensively described how physicians and members of the clergy took advantage of the plague, noting 'apothecaries are so cunning that they would even squeeze pellitory-of-the-wall to make an antidote'. Many sold sacred images, claiming that 'they too would help to fight the disease'.[23] Carlo Morexano, a physician from Messina who remained in Naples during the epidemic, also noted:

> Among the many charlatan doctors, there was one who until that time had sold used furnishings. He went around claiming to possess a secret to cure the plague, which earned him the trust of the credulous populace, and he obtained the right to sell it. His secret remedy was nothing more than a roughly made mixture of rose preserve with some grains of antimony, which he made people drink with six ounces of barley water.[24]

The physician earned 2,000 *scudi* in only two days, but his patients and later the physician himself died: 'his medicament could not heal him'. The anonymous account describes how not just the compound of walnuts, dried figs and salt, but also walnuts themselves were in great demand, to the extent that they reached record high prices. In some cases, the names of the drugs were linked to miracles and successes from the past that encouraged their use, for example the 'radice della Carlina' (thistle) – supposedly named because an angel had shown it to Charlemagne during the epidemic that struck his army in 781.[25]

## Testing the facts

The epidemic caused great strains in the relations between physicians and political authorities in Naples. At the end of May, the protomedico Francesco Liotta, who wanted to close the quarters where the disease had taken hold, was replaced by Francesco Mosca,[26] 'a man of less learning' but well regarded 'for his long beard and old age'.[27] Once he had been nominated protomedico, Mosca published a notice outlining that, since many people were taking all the remedies they wanted, they were no longer able to take those they needed. The only remedy they were allowed to have from then on was a so-called 'vomitatorio'. As the announcement, never previously cited, explains:

> Those who will be infected are immediately required to inform the doctors and the deputies of the *Ottine* (areas of the city of Naples) who will provide the *vomitatorio* that we will order. The apothecaries will distribute it to all *Ottine* with instructions on how to apply it and with the grace of Our Lord, his Blessed Mother and our Patron Saints. We hope for good results and heath. The *vomitatorio* will be distributed to the poor for free, with a prescription signed by the doctor operating within each *Ottina* as it has been ordered by the Most Illustrious Deputies that the expenses [of such an operation] be covered by this most trusted city, although those who can afford to will pay one *carlino* per *drama* (*c*.4 grams). It is expressively advised that this *vomitatorio* should not be taken by those in good health, because they would be damaged rather than preserved [from the plague]. (Naples, Egidio Longo, 8 June)[28]

It is unclear to which remedy the new protomedico is referring in his announcement. Leonardo Di Capua, however, discussed this emetic in his *Parere sull'incertezza della medicina*:

> Let this noble city say that during the time of the past outbreak of plague many people died with agonising abdominal pain caused by

> that most poisonous quicksilver (*ariento vivo*), which they called angelic powder, which was then suggested by the *Protomedico* for the collective recovery of the sick, and which was contained in an edict that was disseminated in print. And reasonably, some were uncertain whether more men died of the powerful force of that remedy than they did of the plague.[29]

The drug recommended by the protomedico is here described as quicksilver ('argento vivo'), and it did indeed cause more deaths than the plague itself. These observations can be found in the Seventh *Ragionamento*. Here Di Capua extolled chemical medicine, hailing it as the greatest of the arts at the doctor's disposal – one that enabled him both to penetrate the secrets of nature and to give the right remedies to patients. Di Capua also emphasised the errors that were committed by incompetent doctors, including administering incorrect dosages, using inappropriate types of recipients and lacking the right experience – as was the case, for example, with certain Galenists who did not have the skills to use chemistry. Furthermore, doctors used apothecaries rather than preparing medicines themselves.

Another mistake was to use many ingredients in one compound. Leonardo Di Capua expressed views common to at least some of the academicians who – after the tragedy of the plague – met first at the home of the philosopher and mathematician Tommaso Cornelio and subsequently in the palace of the Marquis Concublet (during the so-called 'second period' from 1663 onwards).[30] Following in a long tradition, Di Capua wrote that the 'uncertainty of medicine' does not invalidate medical practice. Rather such uncertainty was a limit that was meant to push scholars to reflect on past errors in order to avoid them in the future, by drawing on the experience and teachings of certain philosophers and men of letters who had had the courage to distance themselves from ancient authorities. One such scholar was Erasistratus (330–250 BC), who – among other things – disliked the use of many ingredients in the preparation of a compound[31] such as teriaca. As we shall see, Erasistratus also intervenes in a dialogue by Cornelius on the plague.

The great tragedy of the plague had undoubtedly been an unprecedented test case; if certain medicines had caused death, it was necessary to reveal this. Even if the space devoted to the effects of the 'vomitive' (quicksilver) used during the plague of 1656 in the *Parere* was limited, it was nevertheless a fitting, tragic example of the danger caused by foolish and imprudent conduct among doctors.[32]

During the epidemic, people became more aware of the sources of the plague as well as the therapies that could be used to counter it. Some ancient remedies were abandoned because they were difficult to find. Others became the subject of debate among physicians. One such debate, which occurred between Neapolitans and Salernitans, can be found among the papers of the Pontifical Secretariat. The text, written at the start of August 1656, was entitled *Modo di medicare la peste doppo più consulte tra medici di Napoli e di Salerno, esperimentato nella medesima città di Salerno per mezo del quale ne sono sanati e sanano un'infinità.* (*Means of treating the plague following discussions among the doctors of Naples and Salerno …*) The advice given by the account was to use emollients, not cupping glasses or iron; the bubo had to be allowed to mature, then be cleaned and treated with grease. It continued:

> If the disease manifests with buboes in the armpits and groins, or anywhere else, nothing must be done but rub them with almond oil, camomile and butter, six or eight times a day. After this, prepare a decoction of mallow and violet and once they have dissolved, squeeze the water and mix this decoction with pork lard and chicken fat. Then cover the bubo with pork lard, on which you will apply the said decoction so as to make the bubo mature. If cut, such bubo can be medicated like an ordinary sore. If the bubo does not mature, apply a vesicant in both sides of the groin and under the armpits, making sure that buboes are not cut in cold blood or treated by using cupping glasses, as this would cause death.[33]

In the accounts of physicians who had remained in Naples or had treated infected people, we see that their understanding of the causes of disease varied. However, the humoral conception of disease – which led physicians to believe that remedies intended to expel harmful humours were useful – remained. As we will see, although some physicians, such as Gatta, aimed to explain the affliction by invoking infected 'corpuscles' or 'atoms'; they considered implementing 'political' remedies – such as quarantine – to be the priority.

## Printed accounts

After the failure of Mosca's efforts, the viceroy appointed as protomedico Carlo Pignataro, chair of Theoretical Medicine at the University of Naples.[34] Towards the end of the epidemic, Pignataro wrote a letter to

Michele Giustiniani, later published in his *Historia del contagio di Avellino* (Rome, 1662).[35] This letter offered a brief and 'cautious' historical account of the epidemic from the point of view of a notoriously Galenist protomedico. The epidemic began, he wrote, in May (so much later than when the first cases became known);[36] it immediately affected children and women, a view consistent with the Galenic belief that their natures made them more susceptible to diseases: 'strage puerorum primo deinde mulierum saevire coepit lues, postmodum debaccata in omnes est'.[37]

The letter contains no mention of the illness's origin, but it seems clear that it was transmitted via the air. A passage clarifies Pignataro's conviction regarding the 'density' and even the fatal nature of the air: birds had been witnessed breathing the air, he wrote, then dropping down to the ground dead.[38] Pignataro then quoted a significant verse from *On the nature of things* by Lucretius, 'mussabat tacito medicina timore' (VI, 1179) ('stammered medicine out of fear'). He used this quotation both to underline the limits of medicine and to introduce the notion that medical remedies had done more damage than good.[39]

Following this approach, Pignataro highlighted the failure of chemical medicine in general. According to him, emetics and diaphoretics had proven effective in only a very few cases;[40] the milder remedies adopted later had proved more efficacious. Swallowing snow brought great relief to those with a fever; its consumption increased, even among the lower classes compared with the previous year, even though the population had diminished.[41] In addition to administering the snow, it had proved beneficial to clean the air with a fire sustained with bitumen.[42] Pignataro's view was not shared by physicians such as Gatta and Morexano, however. Not long after he wrote this letter, they began to discuss plague-carrying atoms and to stress that contagion was the carrier of the illness, which led to a more empowering, less fatalistic outlook for civic leaders.

## Faithfulness to Galen and atomism

Geronimo Gatta, wanting to prepare a work that could help future populations who found themselves in a similar situation, courageously wrote a completely infection-based explanation of the disease.[43] In this work he declared that the plague came from 'corpicelli' or 'atoms' that were transmitted, as Girolamo Fracastoro had indicated, either through exhalations or by coming into contact with things touched by infected people.[44] Gatta's work relied upon the aphorisms of the Italian physiologist Santorio Santorio. In sharing one of Santorio's maxims without reservation, Gatta rejected both the Galenic notion that the subject was responsible for his illness[45] and the idea that the disease could develop from the environment, for example in stagnant water, marshland or rotting food. The air was only responsible for causing disease in the sense that it was a vector for spreading contagion from person to person.

According to Gatta, most medical writings were of little use. Instead the three remedies highlighted by Ingrassia and Giovann'Antonio Foglia (one of Gatta's teachers), 'fire, gold and gallows' ('fuoco, oro e forche'), were crucial.[46] The fire ('fuoco') purged the goods of the merchants and purified the air of 'atoms' ('corpicelli') produced by the infected; meanwhile burning herbs and sulphur helped the lungs because sulphur – as had been correctly identified – was 'a close friend of the lung and an enemy of the pestilential seed' ('amicissimo del polmone e inimico del Pestilente seme').[47] The gold ('oro') was needed not only to pay the guards, to stop them accepting offers from people who wanted to violate the bans and to cover all outgoings during a period of isolation (from food to medicine); it was also useful as a substance itself, as a diaphoretic or as part of the 'philosophers' stone' – a compound that Gatta, whom we could call one of the 'chemical' physicians, held to be effective.[48] The gallows were there to punish anyone who violated the bans and to discourage others from doing the same.

With regard to actual medical treatments, Gatta was devoted to the teachings of the ancients – but not to the extent that he ignored their internal contradictions or inconsistencies, which reflected his desire to seek the truth and make a genuinely helpful contribution. A good example is his examination of bloodletting, one of the remedies recommended by Galen. Various physicians indicated the works in which Galen had proposed its use. The short work *De cucurbitulis et scarificationibus*,[49] for instance, provided an account of how Galen had beaten the plague by letting blood from his foot two days after he had fallen ill.[50] This differed from his commentary on the Hippocratic *Aphorisms*,[51] in which he stated that humours that were not 'cooked'[52] could not be shifted. Here Galen appeared to contradict himself, Gatta noted, because the humours could not have been 'ready' after only two days. Perhaps he had not been infected with the plague after all, the physician concluded, but the crucial aspect was not to empty the veins of the clots that, in his opinion, formed in those infected with the plague. More importantly, bloodletting was not to be carried out on patients who were already very weak:

> with bloodletting performed even on the extreme parts of a weak body the pulse becomes increasingly weak.
> (con un semplice evacuativo di sangue etiandio nelle estreme parti del corpo incontinente si vede il polso mancar per strada)[53]

Gatta shared Santorio's (and Galen's)[54] pessimistic view of drugs and medicine in general. To this end he quoted the adage that the poor, who could not afford drugs, were more fortunate than the rich (aphorism no. 139 from Santorio's *De statica medicina*).

Gatta then contradicted himself slightly by writing that it was good to remove poisonous humours (a belief shared by several physicians based in Naples). The illness progressed quickly, so it was important to act early, removing anything that was alien to nature (a Galenic concept).[55] For this reason Gatta believed that purging was beneficial, provided that it was administered in line with certain criteria, namely the famous questions: 'quid, quantum, quando, quomodo, ubi' (why, how much, when, how, where)? On this subject he cited numerous Paracelsian physicians who, in his view, had created effective compounds.[56] It would appear that at the time there was widespread appreciation of those remedies, and in the period immediately after the plague some physicians felt free to express it.[57]

In Gatta's treatise there were detailed instructions on how to prepare 'anti-venom' ('scaccia-veleno') drugs. One diaphoretic consisted of 'filings of Venus' (copper), fragments of Mercury and 'mercurial salt' ('sale mercuriale'), to which needed to be added a large amount of ethyl alcohol ('spirito') and 'oil of sulphur' ('olio di zolfo').[58] Gatta claimed to have seen at first-hand how successful it was after having administered it with a little theriac:

> Those who took it as a treatment have all been saved, such as my two sons, a servant and other strangers, except for two individuals out of all those to whom I administered it.
> (quanti l'hanno preso per cura tutti si son salvati, come son doi miei figli maschi una serva, e altri estranei, eccetto due persone di quanti n'ho dato).[59]

Gatta was also aware of the contemporary controversy regarding the dangers of certain metals and metalloids. After having recommended the use of antimony as an emetic, he added that, contrary to what ‘idiots and common people’ (‘ideoti e volgari’) believed, it could actually be given to babies only a few months old – a view that reappeared later in Donzelli’s *Teatro farmaceutico*.[60] In Gatta’s treatise there were as many traces of readings and conversations on the atomistic nature of things (which can also be found in contemporary works by non-medical authors such as Camillo Tutini and Niccolò Pasquale)[61] as there were ancient and modern ‘remedies’ based on Hippocratic and Galenic texts.

The medical advice Gatta gave in his treatise bore some similarities to the contents of Carlo Morexano’s short booklet *Il torchio delle osservationi della peste di Napoli nell’anno M.DC.LVI* (Naples, 1659). Morexano, a philosopher and physician from Messina, dedicated his booklet to the treasurers of the Santa Maria della Pietà hospital in that city, where he was soon to work.[62] He recorded that he was at Severinus’s side while he treated the plague victims in the last days of his life.[63]

Unlike Gatta, Morexano did consider theories that linked the plague to astral conjunctions, though he ultimately rejected them. His opinion agreed with that expressed by Jean Baptise van Helmont’s *Tumulus pestis* (an author greatly revered by the Investiganti): the stars are not the cause of traumatic events, but they may signify that they are going to happen.[64] From other texts, Morexano derived some of his most effective arguments against the hypothesis that human events, including outbreaks of disease, could be conditioned by the movements of the stars.[65] He maintained that if the stars conditioned events on the Earth, for instance, then they would do so entirely, because the Earth is no more than a mere ‘dot’ with respect to them; therefore ‘All that the Earth contained would remain subject to superior influences’.[66]

Evidently influenced by those who had supported the rumour of plague-spreaders, that is, the protomedico Pignataro as much as by Paracelsian physicians themselves (towards whom he had a marked attachment), Morexano did not, rule out the idea that the plague could have been spread deliberately for political purposes. The explanation that he favoured was that of contagion: disease diffused itself by means of corpuscles, or ‘atoms’, that were easily spread in the first days by means of processions and hunts for the plague-spreaders.[67] Past experience, then, allowed him to conclude that interactions among the healthy may also be responsible for the spread of disease.[68] From this evidence, we can conclude that the phenomenon of healthy carriers was even then well understood.

Like Gatta, Morexano noted the failure of various remedies. Drinking one's own urine, a practice recommended by Galen and endorsed only a few decades earlier by Ficino and Ingrassia, was evidently inefficacious.[69] Also ineffective were small bags of arsenic, to be placed over the heart or under the armpit, or amulets that induced blisters. At times he condemned bloodletting, maintaining that it was debilitating and could lead to death.[70] However, he did recommend various remedies including those derived from vipers, similar to those administered by Hippocrates,[71] various diaphoretics and emetics already mentioned in the *Consultatio*: Croll's antimony oxysulphide ('croco di metalli'), Leonardo Fioravanti's philosopher's stone, Donzelli's pink syrup (made with roses and honey) and the bezoar stone. Having freed the body from corrupted humours, purgatives prepared the body to receive the antidote ('alessifarmaco').

For the most part, Morexano had consulted the same texts as Gatta. His relationship with the Neapolitan elite was clear, illustrated by the fact that he referred not only to Marco Aurelio Severino as his 'teacher', but also to Donzelli and Pietro Castelli (the last deeply influenced by Paracelsus).[72] Also in this case, one can say that one of the lasting effects of plague was a new-found appreciation for spagyric remedies.

## The audacity of Tommaso Cornelio

During the plague the philosopher and medic Tommaso Cornelio, close friend of Severino and professor of mathematics, escaped to the town of Vico. He later travelled to villa Pazzigno, located near Naples. Here he was with Vincenzo Protospataro, when his pupil Lucantonio Porzio arrived (an event that was recounted by Mosca in his biography of Porzio).[73] The two men were pleased to meet one another. After a short time Cornelio asked him whether he still possessed notes from his lectures, to which Porzio replied that he did, and that was willing to enter Naples in order to go to his villa and search for them. Equipped with a heath certificate ('bolletta'), he entered the city and finally reached his residence, now ransacked by thieves. Here Porzio found the lecture notes that Cornelio had requested.

When the plague finally ended, Cornelio returned to Naples. He had previously brought with him a number of important new scientific works including, for example, some of the writings of Bacon, Galileo, Gilbert, Sennert, van Helmont and Gassendi. It is also well known that he had spoken on eclipses at the Oziosi academy in May 1652.[74] When once more resident in Naples, Cornelio gathered Gennaro and Francesco D'Andrea, Leonardo Di Capua and the young Marzio Carafa in his home. There he read to them the dialogues that he had written, contained in *Progymnasmata physica* (1663).[75] One of the most celebrated of these dialogues was 'Severino's letter from the afterlife' (*Marcus Aurelius Severinus Crathigena Timaeo Locrensi Municipi Suo*), written by Cornelio as a homage to his close friend whom the plague had torn from his friends and city.[76] This letter, to which Garin and Torrini have justly drawn attention, not only reveals tensions within the medical circles and the ideals current within the Investiganti academy, but also offers a critical analysis of some classical and modern theories of the plague. I argue here that it followed a model established in Traiano Boccalini's satirical literary work *Ragguagli*, in which some philosophers and doctors discuss the diseases that had afflicted the century ('il Secolo'). Everyone feels free to indicate the causes and remedies that he wants, but nobody is convincing; their speeches reveal themselves to be too long and far removed from the real necessities of the patient (the century itself).[77]

In Cornelio's 'Severino's letter from the afterlife', the character of Severino recounted that in the afterlife he had met various other deceased physicians and philosophers. These included Mario Schipano, 'a talented physician' who had challenged 'the doctrines of Aristotle',[78] as well as Bernardino Telesio, Tommaso Campanella and the physician Agostino Doni. Severino also recalled conversations with the engineer and Lyncean academician Nicola Antonio Stigliola. Every passage in this dialogue is rich in meaning, but the most important part for us comes when Severino speaks about what some philosophers had said about the plague.

The discussion began with a most bitter reflection on the current state of medicine:

> And then he began to denounce the disgraceful actions of doctors, and to attack more vehemently those who, neglecting the contemplation of nature and the pursuit of noble arts, find satisfaction in having a long beard, an antiquated attitude, as well as a certain ridiculous attire to convey gravity.
> (Ad haec detestari Medicorum flagitia coepit, atque in eos veehementer invehi, qui naturae contemplatione, bonisque artibus posthabitis, satis habent promissam barbam pascere, et obsoletiori habitu, atque vestitu perridiculam quamdam gravitatem affectare).

Medicine was soiled by petty men who were appointed to practise medicine, and appropriated the treatises, long beards and togas of antiquity.[80]

At this point, there was an intervention from a man with a pronounced forehead, grim eyes and a shaggy beard: the physician Carlo Pignataro. He began to speak in a hideous voice, saying that it was necessary to return medicine to its ancient state, to stifle the physicians' ability to innovate, to take 'chemical studies from the hands of students, to remove the books of Paracelsus, Van Helmont, Harvey and other *novatores* and drive all those who wished to study medicine to the lessons of Galen alone'.[81] Speaking next about the plague, Pignataro affirmed that it was necessary to determine whether the plague was a disease that depended on the imbalance of the humours or 'a disease of all substance'. If it was of the first type of disease, then one could administer a contrary medicament or antidote; if it was of the second type, then an antidote ('alexipharmacon') was necessary: two vague and similar possibilities.

Pignataro's 'vapid loquacity' reduced the bystanders to helpless laughter[82] until Mercury called Hippocrates to speak, as the founding father of medicine. Evidently there had been an expectation that he could put the conversation on the right track. The medical art, he said, had become debased when it became venal; purged of its 'illiberal' traits, corruption and greed, medicine could once more be 'most worthy' ('dignissima'). With regard to the plague, he specified that he had never used medicine; if he had managed to free Attica from this disease, he had achieved this feat through prudent political interventions rather than by pharmaceutical means.[83] This was a fundamental clarification, thanks to which the reader was led to understand that it was absurd and dishonest to attribute to Hippocrates the choice of one or another remedy against the plague. The oration was received with great applause and Momus – 'chastiser of Gods and men', whose attribute is free speech – 'praises the wisest ancient'. After a short period several ancient physicians made interventions, finishing with Galen, who had been eagerly anticipated.

Galen spoke at length – a defect for which he was often rebuked in the 'letter'. Above all he spoke of himself, recalling his studies, his many works, the principles and axioms that he had derived from nature. In particular, he emphasised that all things derived from the four elements and that their balance governed health, while imbalance led to disease and death. He also demonstrated the medical theory of opposites ('contraria contrariis curantur'), which declared that one could cure through the use of contrary principles: heat could be treated with cold, wetness with dryness, and so forth.[84] If there were principles that could not be attributed to the four elements, these were defined as 'occult qualities'. Galen also explained that the therapies were suggested to him by the gods, and that on many occasions he had opened a vein spontaneously, without any rational reason for having done so other than having been commanded to do so in a dream ('ita quidem iubentibus somniis').

Van Helmont then asked in a quiet voice when he would stop talking in this garrulous manner ('nugator') and added that it was not a trivial thing to have instituted a bloodletting sect that had killed several people.[85] Momo exalted Van Helmont, praising him for speaking freely, but he also made it known to him that the plague was not yet buried, as he had believed would happen. Momo was here alluding to Van Helmont's *Tumulus pestis* (*The Plague-Grave*) that, according to its author, had brought the problem of that disease to an end.[86] Here Van Helmont justified himself, saying that the fates had neither allowed him to discover the composition of the 'Alkahest' (the principle that would return bodies to health)[87] nor allowed him to experiment with his drug derived from toads. However, his declarations were evidently not persuasive.

Paracelsus was next to address the council. Attacking Galen, he declared 'never in the world has there existed a plague that has inflicted such a devastating massacre on the human race than has your medicine'.[88] He then gave his opinion on the origins of the plague: it was the arsenic-laden air, secreted by the 'Yleid', that struck the human race by means of the 'basilisk of the firmament'. Paracelsus's fable continued thus: the plague forms under the influence of Mercury ('quandocunque videlicet Mercurius ex venefica vi a Solis evestro sibi delata exaltatur, ac lethale virus concipit ex quicuplici ente divino'). He added that the excrements of the stars were the envoys of the plague, and that from their form one could predict future outbreaks of disease.[89] After the enormous tragedy that had just occurred, the Paracelsian theory seemed absurd.[90]

The Lutheran doctor Erastus then stood up and declared that such an abstruse discourse was offensive to all physicians; Momo instead defended Paracelsus, explaining that he had found not only a new form of treatment, but also a new 'way of philosophising'.[91] The ancients, furthermore, could not deliver such a scolding to anyone, seeing as they had sent so many patients to their deaths. At this point Mercury called upon Stigliola to speak; he had been depicted as a good man (goodness was evidently a guarantee of intellectual honesty and authentic philanthropy, necessary virtues if one desired to care for humans). Stigliola offered an assessment of the views of the ancients and moderns, and in doing so demonstrated how the problem of the plague could be correctly understood only by abandoning fantastical theories.[92]

Why did Cornelio enable Stigliola to give voice to the opinions that were evidently his own? Here it is important to highlight, first and foremost, his ties with the Neapolitan Linceans. Stigliola had been a member of the academy: a scholar of architecture, but also of scientific questions. He had shared with Galileo an interest in the telescope and its potential to reveal the mysteries of the heavens.[93] In Naples he had made his mark not only through his knowledge and critical spirit, but also through his active promotion of culture by means of his press, the celebrated 'Stamperia Stigliola' in Porta Nolana, and his school that had been frequented by Severino himself. To these activities there had been added a trait of Stigliola's personality that emerges both from this 'letter' by Cornelio and the printer's note in his volume on the telescope, *Il telescopio over ispecillo celeste*, of 1627. Having declared Stigliola to be one of the excellent literati ('eccellenti Letterati') of this time, the author of that note added that he should be considered as such because

> with diligent labour he delighted more in being rather than merely appearing to be learned; he has always lacked that pride that so often accompanies those who carry the knowledge of the sciences and a wisdom that exceeds the common condition of humanity.[94]

Stigliola was therefore humble – a fundamental characteristic that allowed him to speak in a sensible manner, one close to the truth about the plague. He began by making introductory remarks on the fact that the precepts of the ancients were mostly unsubstantiated and noted that the problem was that physicians were overly attached to their preconceptions (an issue that Boccalini had encountered too). Rational physicians were more skilled at speaking than treating; they affirmed that one could know the 'reason' ('ratio') of things, but that they could not know the beginning of things ('rerum initia'). Consequently, in practice they produced 'cures that were risky and dangerous to health'.[95] Among other things, they ignored the force and nature of the blood; many still believed that the blood was produced by the liver and dispersed through the body by means of the veins, trusting more in reasoning than in observation.[96] They claimed that the empirics were not better than them, although they relied upon experience.[97]

> Their ignorance was compensated for with arrogance and a way of self-representation that was meant to evoke respect: short hair, long beard, a 'grave and severe' demeanour and refined speech. The truth was that the causes of the maladies – let alone those of the plague – had not yet been understood.

For this reason, Stigliola exclaimed disconsolately that he was surprised to learn that several volumes had been written about the plague when no one understood its nature.[98] Cornelio, an enemy of unnecessary verbosity, shared this view and criticised the large production of writings on a little-known subject. It was therefore more honest to admit ignorance and focus on the little knowledge gained by experience, rather than producing pointless writings. No medical treatment had been found to cure the plague: use/knowledge ('usus') and death toll ('ratio mortali') appeared to have failed. The plague appeared to be similar to malignant fevers, although highly contagious. The effects were purplish and livid pustules, as seen in autopsies, which Stigliola (Cornelio) evidently exhibited as a way to reach well-founded knowledge.

The plague's causes were the breath and the exhalations emitted by the sick body which mixed with the air.[99] To prove that corrupted air did not have a role in the transmission of the disease, Cornelio added that the plague seemed to spread more effectively in clean rather than bad (fetid) air. This, he argued, was possibly because the infective particles ('atomi pestiferi') could move more freely.[100] There were no effective remedies; in fact, Stigliola claimed that he preferred vipers to dangerous medicines.[101] The most important measure to take was, therefore, isolation in order to avoid those who may have been infected and were spreading the lethal 'virus'.

In the concluding analysis of this letter, it is noteworthy that chemical remedies were not considered to be safe, despite the admiration of Cornelio (and the doctors close to him) for chemistry.[102] Evidently, those remedies had proved to be dangerous or ineffective. Aiming to provide the contemporary reader – and indeed future generations – with a piece of work that was based on certainty, it was not deemed honest and prudent to disseminate remedies whose effectiveness had not been proved. This was a drastic decision, which reduced knowledge about the plague to very few principles. This stance also followed the approach of Santorio Santorio, whose lessons on the plague had been reduced to a few aphorisms.[103] Cornelio's intention here was to disseminate the little knowledge on the plague he possessed, which was based on experience and on a corpuscularian vision of the disease.

Finally, it is worth considering Ludovico Muratori's interest in Cornelio's dialogue. Although in his work *Il Governo della Peste* he showed an understanding of Cornelio's view (as expressed in the dialogue), Muratori nonetheless chose to provide infected patients with some answers. He stated that

> In his marvellous Dialogue composed in the style of [the writer] Lucian, the renowned physician Tommaso Cornelio advises to leave the plague-stricken patients in the hands of Nature rather than in the hands of doctors because, as Cornelio states, they often use harmful remedies which cause damage that is even worse than the disease itself.[104]

In other words, Muratori's understanding of Cornelio's choice did not prevent him from suggesting various remedies intended to fight, like weapons, a much-feared disease.

## Porzio and Di Capua

In his biography of Porzio, Mosca writes that during the epidemic Porzio moved from his native town of Positano to attend university in Naples. Among those lecturing there (which included Pignataro), Porzio chose to attend Cornelio's lectures because 'he (Cornelio) did not accept the probable as true and often made the doctrines of others appear to be false'.[105] After the outbreak of plague, Porzio attended the first gatherings of the Investiganti academy in Naples that were held in Cornelio's house.[106] Following the closure of the Investiganti, Porzio moved to Rome in 1670.[107] There he met, among others, Marquis Filippo Nerli and his uncle Cardinal Francesco Nerli, who had been friends with Galileo and Torricelli. Thanks to Nerli he was named chair of medicine at the Sapienza University in Rome.

Significantly, Mosca writes that during his first lecture Porzio debated the 'uncertainty of medicine'; explaining that fever was not 'heat', as commonly believed. He also lectured on the circulation of the blood: a topic not yet being debated in Rome.[108] The majority of those in attendance did not receive Porzio's debate, which was also critical of 'the doctors', favourably.[109] Influenced by the teachings of Cornelio, Porzio's unfavourable position led some doctors to feel threatened in their roles. In 1683 Porzio moved to Venice and then to Vienna, where he wrote a booklet entitled *De Militis in Castris sanitate tuenda* to concretely help the soldiers of the imperial army engaged in Buda.[110] It is in this booklet that he specifically mentioned the plague of 1656. Porzio emphasised that fighting the plague was first a political endeavour and then a medical one – as proved in Florence in 1631 and more recently in Venice, where the plague had not caused carnage.

Porzio went on to illustrate the correct measures that Grand Duke Ferdinand II took in Florence, and which successfully stopped the spread of the pestilence in 1631. He did so by

> closing the streets, deploying the army in the city, and tasking some nobles to control that no one would violate the orders. The Grand Duke himself and his collaborators frequently inspected the city at night and during the day.[111]

Moreover, once those infected within the city had been located, the Grand Duke ensured that they were isolated from the rest of the population. Ferdinand II also ordered that corpses be removed, so that they would be out of sight and not contaminate the streets. In other words, the Grand Duke in Florence did what had not been done in Naples. Porzio courageously added that

> In Naples there was no diligence [in managing the pestilence], although some monasteries remained immune while the plague spread in the city. By contrast, in Rome not only most of the monasteries but also some families and households escaped the plague.[112]

Employing some vivid metaphors, Porzio explained how the plague disappeared. He likened the plague to fire that burned only if it was kept fed with wood, just as yeast that was used to make bread needed to be fed.[113] If a pinch of yeast were not mixed with water and flour it could no longer ferment and thus activate a dough, so too 'ashes are incapable of igniting a fire'.[114]

Preventing 'communication' and removing the infected bodies through which it survived was thus the only way to extinguish the plague, according to Santorio. Moreover, as Gatta and Cornelio had already written, the causes of the plague were the 'exhalations' and the touch of the sick, which infected even clothes and objects – hence the necessity to expose them to the air for a long time. Porzio concluded that

> if the Viceroy Count of Castrillo had acquired this knowledge and put in into practice, he could have saved hundreds of thousands of men, no less than Ferdinand II of Florence did.[115]

Another eminent figure who Porzio admired was Domenico Contarini – a member of the Venetian patriciate who from 1680 served for almost five years as the city's ambassador in Vienna.[116] Alluding to his prudence, Porzio writes that Contarini, like Ferdinand II, would have successfully kept the plague far from the borders of a state ('extra fines Regnorum'). Turning to Contarini himself, and possibly alluding to Viceroy Castrillo's failure to act promptly against the plague in Naples, Porzio observed that 'you [Contarini] do not approve the conduct of those princes who only see with foreign eyes and only listen with foreign ears'.[117]

This passage is extremely important as it suggests that Viceroy Castrillo's conduct during the plague in Naples had aroused disappointment and frustration among his subjects. Having come from a 'foreign' land to govern a kingdom under Spanish rule, Castrillo retrospectively appeared different from the Grand Duke and the Doge, both the leaders of independent states.[118] Despite the viceroy's attempts to remedy the situation, he had displayed a superficial attitude and failed to pay attention to the warnings of judicious scholars. The damage and death toll caused by the plague had thus been a consequence of the authorities' incompetence.

Away from Naples, Porzio felt the freedom to express an opinion shared by some other Investiganti who had experienced the tragedy of 1656, as he had. Regarding the treatments against the plague and the remedies discussed earlier in this chapter, Porzio recommended only sulphur, which Cornelio had suggested he take during the epidemic.[119] Moreover, Porzio discussed the plague in another booklet entitled *Quid pestis Galeno sit*.[120] Here he challenged Galen's idea that 'innate heat' and 'extreme dampness' could lead to the epidemic, but conceded that that new knowledge on the plague was not complete.[121] Following Cornelio, Porzio explained that the causes of some phenomena remained unexplained. He noted, for instance, that some people did not become infected, despite being close to the sick; others inexplicably recovered from the disease or benefited from garlic and roses.[122]

Some points were clear, however. The plague struck almost everyone, irrespective of sex, age or social status. One simple and yet effective way in which the disease spread was the presence of 'infected air in the vicinity of the sick', although the highly contagious 'seeds of the plague' ('pestis semina') were also present in wool cloths.[123] It was therefore necessary that doctors and magistrates ban contact with plague sufferers. Doctors should in addition 'warn the sick person's family, as they may also infect others'.[124]

It is noteworthy that in his *Parere* Leonardo Di Capua follows Cornelio and Porzio, especially in relation to the theme of the 'uncertainty of medicine'. In the short passages concerned with the plague, Di Capua states the impossibility of defeating the disease. He firmly denied, as had been often stated, that Hippocrates had defeated the plague, and instead urged others publicly to take this stance. Medicine had been ineffective against the plague, as Pignataro had admitted and Lucretius had written ('mussabat tacito medicina timore').[125]

Giuseppe Donzelli, a renowned physician and apothecary, distanced himself from the line of thinking shared by Cornelio, Porzio and Di Capua. Donzelli's *Teatro farmaceutico* in fact contains several remedies against the plague. Donzelli's trust in the resources of contemporary pharmacopeia was likely the result of his knowledge and role as an apothecary. Despite leaving Naples, he had the opportunity to verify the effects of some medical remedies against the plague. For instance, he states that the so-called 'common sublimated' (possibly sublimated mercury) was 'ineffective' during the 1656 plague in Naples.[126]

By contrast, remedies such as antimony oxysulphide ('croco di metalli'), resinous yellow amber ('succino'),[127] flowers of antimony of Croll, sulphur, Calomel ('mercurio dolce'), oils and compounds such as the 'confettione liberante' seemed efficacious. This was a concoction made with 'dictamnus albus', clay ('terra sigillata'), coral which caused sweating and thus was believed to both defend and heal from the plague,[128] Armenian bole[129] and syrup of 'teocrium scordium', believed to contrast rotten pustules and balance 'corrupted humours'.[130] Other ingredients included scorpion oil (known as 'olio del gran duca di Toscana'); aloe, myrrh, 'the good and ancient theriac', the mithridate,[131] precious stones and tinctures – not to mention deer horn, believed to fight putrefaction and pestilential fever.[132]

## Conclusion

In conclusion, the 1656 plague had a strong impact, especially on the *novatores* doctors. Following that tragic event, two groups emerged among the doctors that we can define as *novatores* (to distinguish them from orthodox Galenists). One group that included Gatta, Morexano and Donzelli was in favour of some medicines to fight the pestilence; the other was highly critical of all remedies for the plague. As discussed in this chapter, the debate that emerged around Cornelio shows the reasons for which none of the Investiganti academicians wrote a treatise on the plague. According to the doctors and philosophers who followed Cornelio, factual knowledge of the plague was very limited, with the result that one could write only very little on this topic. More importantly, it was crucial that a leader ('principe') would act promptly and wisely.

It is thus not surprising to note that those leaders who, in different contexts, were believed to have acted justly did receive praise as positive examples to follow – in contrast to the cloying rhetoric lavished on those who had let the pestilence spread. Nonetheless, the group of Investiganti academicians who came to deny the validity of Galenic medicine and the efficacy of almost any remedy achieved little. Despite this, they were gloriously responsible for a coherent stance regarding the importance of state authorities in treating plague, as it showed the most powerful weapons to be in their own hands.

## Notes

1 For a general panoramic, see De Renzi, 'Napoli nell'anno 1656'; Fusco, '*Peste, demografia*' and Fusco, 'La grande epidemia'; for the beginning in January, see D'Alessio, 'Su alcune lettere', 106.
2 De Renzi, *Napoli nell'anno 1656*, 188–92.
3 D'Alessio, 'On the plague in Naples, 1656'.
4 Mattioli, *Il dioscoride*, 479.
5 MacGregor, 'Medicinal *terra sigillata*', 113–36. This compound was believed to neutralise certain poisons.
6 Donzelli, *Teatro farmaceutico*, 70; Do Sameiro Barroso, 'Bezoar stones'.
7 Mattioli, *Il dioscoride*, 202.
8 Muratori, *Del governo della peste*, 242.
9 Teucrium scordium is a plant that smells like garlic; Mattioli, *Il dioscoride*, 465; Donzelli, *Teatro farmaceutico*, 202.
10 The powder was to be given to adults with some theriac, water or wine; the compound was supposed to make them sweat. See Napolitano, *Trattato utilissimo*, 13 onwards. The friar had been governor of the lazaretto in 1576; on Neapolitan convents as places of chemical research and experimentation, see Gianfrancesco, 'Books, gold and elixir'.
11 Maranta wrote that he produced the volume after encouragement by Ferrante Imperato: Maranta, *Delle theriaca*. See also Marra, 'La vipera'.
12 Galen (attributed to), *L'antidotario*, 126; 'For plague, like a kind of beast, does not just kill a few, but spreads over entire cities [...] men cannot escape the evil, but draw the air into themselves like a poison through their mouths'. He then mentions Hippocrates' use of fires to stop the plague, *On Theriac to Piso, Attributed to Galen*, 149; Flemming, 'Galen and the plague', 219–44.
13 Maranta, *Della theriaca*, 164.
14 'China china' has pain-relieving and anti-fever properties; on Pietro Castelli – an author known to have been close to Severino – who wrote *Responsio chimica de effervescentia*, Messan, 1654, with the final part on china, see Dollo, *Modelli scientifici*, 14.

15 This concept was explained by Donzelli and demonstrates the continued presence of the Galenic framework in which the illness takes root if the subject is not in good condition. Donzelli, *Teatro farmaceutico*, 236.

16 Muratori, *Dal governo*, 252: 'può esser utile il conoscere, ed aver pronte molte armi diverse per tentare di far fronte a sì gagliardo e sì strano nemico'.

17 'Lo provo, perché se vi è nessuna medicina, che vagli alla peste, quest'è la triaca, operante a guisa di fuoco, e per consequenza gli altri medicamenti caldi: perloche Galeno accorda il fatto d'Ippocrate, che sgombrò la peste col fuoco, con la virtù della triaca, che pure a guisa di fuoco purga, e netta ogni seme pestifero, e immondezza', Bado, *Saggio*.

18 Bado, *Saggio*, 31.

19 Montalbani, *Il pestifugo Esculapio*, 7. The recipe was made up of viper, 'thebaic opium' (the juice of mature poppies, a substance known to the Egyptians), opobalsalm, myrrh and various plants, from scordium to 'quinquefolia' (creeper) and 'calamentha' (mint). Ovidio Montalbani (1601–1671) was a scholar of rhetoric, philosophy, medicine, mathematics and astronomy. On him see Marchi, 'Ovidio Montalbani'.

20 'habbiamo vinto, quando dalle caverne de vasi farmaceutici esca per nostra difesa l'Echnidi theriacal viper Triacale, cioè il serpente d'Apolline, i cui denti vitali sono ben possenti a divorare le serpentine', Montalbani, *Il pestifugo Esculapio*, 7.

21 Montalbani, *Il pestifugo Esculapio*, 7.

22 'Montalbani, *Il pestifugo Esculapio*, 8.

23 Anonymous, 'Relazione della pestilenza', 353.

24 'Tra il gran numero di tali medici selvatici vi fu uno, che sino a quel tempo haveva professato di vendere suppellettili usate. Andava questo millantando per tutto di possedere un secreto per curare la Peste; onde hebbe tanto seguito da credula plebe, che ottenne facoltà di venderlo; non era altro il suo mal secreto, che una mistura di conserva di rose con alcuni grani d'antimonio, preparato rusticamente, facendovi bere appena sei oncie d'acqua d'orzo', in Morexano, *Il torchio*, 28. The secret was so-called because it should be kept secret to the physicians: see Gentilcore, *Malattia*, 90.

25 Morexano, *Il torchio*, 29. The name 'carlina' comes from the word 'cardina', or 'little thistle' ('piccolo cardio'). It was only later that the Charlemagne connection took hold. Mattioli, whose works were well known to Neapolitans, wrote: 'Chiamasi il bianco Camaleone volgarmente in Toscana, come quasi per tutto il resto d'Italia, Carlina; imperocché si crede il volgo (come si dice) che dall'Angelo fosse ella dimostrata a Carlo Magno'; Mattioli, *Discorsi*, 398.

26 According to Giuseppe Mosca, Liotta was relieved of his duties because he believed that the illness circulating in Naples was the plague: Mosca, *Vita di Lucantonio Porzio*, 6. Sebastiano Bartoli, a member of the accademia degli Investiganti, described Liotta thus: 'quondam Regni Archiater [*Protomedico*], vir apprime dignus rem Spagyricam theorice, ac practice quoquomodo pertractat'; Bartoli, *Artis medicae dogmatum communiter receptorum*, 142. See also Gatta, *Di una gravissima peste*, 99.

27 Anonymous, *Relazione della pestilenza*, 349.

28 'S'esortano tutti quelli che di tal morbo saranno afflitti che subito che comparisce debbiano manifestarlo, e publicarlo alli Medici, e Deputati del Ottine che se li darà il vomitatorio a tal effetto maturamente ordinato da noi, e si distribuirà per tutte le Ottine, dalli Spetiali di medicina con l'applicatione del qual rimedio con la gratia di N.S. e della sua Santissima Madre, e Santi Protettori si spera ogni buon esito, e salute, qual vomitatorio a poveri si dispensa gratis con la ricetta firmata dal Medico del Ottina poiche s'è dat'ordine dal Illustrissimi Signori Deputati che si faccia a spese della Fedelissima città e alle persone commode uno carlino la dramma. Avertendo espressamente che questo vomitatorio non si pigli da persone di bona salute, li quali volessero pigliarlo per preservarli, che li sarà di danno, e non di utile', Napoli, Egidio Longo, 8 di giugno.

29 'Dica questa nobilissima città quanti, e quanti nel tempo della passata pestilenza con dolori acerbissimi di viscere n'avesse fatti morire quel velenosissimo ariento vivo precipitato, ch'angelica polvere chiamavano, proposto allor dal Protomedico di que' tempi a comun salvamento degli ammalati, e con pubblico editto divolgato colle stampe. E ragionevolmente per avventura dubitonne alcuno, se più huomini allora per la potentissima violenza di quel medicamento, o per la medesima pestilenza mancassero', in Di Capua, *Parere*, 530. The powder was made of hellebore and antimony and could kill. Mosca, *Vita*, 6.

30 Fish, 'The Academy', 526–37.
31 Di Capua quotes from Plutarch on that point: Di Capua, *Parere*, 340.
32 It is in the same 'ragionamento' that he speaks about antimony, condemning the Galenist doctors: Carlino, *La via dell'incertezza*, 328–34.
33 *Modo di medicare la peste doppo più consulte tra medici di Napoli e di Salerno*, Segr. di Stato, Naples vol. 55, cc.281–82$^{v}$. It is published in Tanturri, *Terapie contro la peste*, 181–94.
34 The protomedico had consultive powers, gli 'uffici di sanità' have real powers. See Gentilcore, 'Il regio protomedicato'. His figure is less important than the Deputazione della sanità that was elected during the plague. On the contrasts among Deputazione and the government see Fusco, 'Il regno di Napoli'.
35 Pignataro, 'Illustrissimo'.
36 'cum mense Maio saeva lues in florentissimam hanc Urbem sese immisit. ... Graves hinc subortae altercationes ...', Pignataro, 'Illustrissimo', 198.
37 Pignataro, 'Illustrissimo', 201.
38 'ex quibus putridis iam atque corruptis parum abfuit, quin ex toto inficeretur aer, cuius militiam aves persensere lethalem, dum inter volandum mortuae concidebant', Pignataro, 'Illustrissimo', 202.
39 Pignataro, 'Illustrissimo', 202.
40 'Pugnatum vero est inter initia fortibus, atribusque medicamentis, quocirca magno fuere in usu vehementiora vomitoria chimico e fonte petita validissima dejectoria ex communi medicina desumpta, efficacia sudorifica, sectio venarum pedis [...] quam paucissimis profuerunt', Pignataro, 'Illustrissimo', 202–3.
41 Pignataro, 'Illustrissimo', 205.
42 'nihil pro ipso aere expurgando operosius igne potest inveniri praesertim si ex bitumine, similibusque rebus, quemadmodum peractum a nobis et suscitetur', Pignataro, 'Illustrissimo', 206.
43 Gatta, *Di una gravissima peste*, 97.
44 The atoms come from infected people: on Fracastoro's view see Nutton, 'The reception of Fracastoro'; Alessandro Pastore and Enrico Peruzzi, eds, 'Girolamo Fracastoro'; on Gatta's view more details are in D'Alessio, 'L'aria innocente'.
45 This is aphorism 129 from the section of aphorisms in Santorio's *De statica medicina*, published in 1634. On this see Nutton and D'Alessio, 'Santorio Santori on plague', 232.
46 Gatta, *Di una gravissima*, 97. On Ingrassia's strategy see Cancila, 'Salute pubblica'.
47 Gatta, *Di una gravissima*, 99.
48 Gatta, *Di una gravissima*, 221. The medical community was divided on the use of gold. It certainly did not bring any health benefits. See Renzo Console, '*Pharmaceutical use of gold in the 16th and 17th centuries*', in *A History of Geology and Medicine*, 171–92.
49 Gatta, *Di una gravissima*, 113–14; Galen, *De cucurbitulis*, 7.
50 Gatta, *Di una gravissima*, 119.
51 The commentary on the 22nd aphorism, first series, of the Hippocratic *Aphorisms*; Gatta, *Di una gravissima*, 121.
52 Vegetti, 'Il pensiero ippocratico', 48.
53 Gatta, *Di una gravissima*, 125.
54 Gatta cites the treatise *De optima secta ad Transibulum*: 'nullum remedium impune administrari potest in corpore quin non noceat'.
55 Gatta refers to *De Sanitate tuenda*, l. IV: Id., *Di una gravissima*, 131.
56 On the use of purges Gatta cites various Paracelsians. See Donato, 'La peste'.
57 Perfetti, 'L'alchimia a Napoli'; Verardi, 'La scienza e i segreti'; The Investiganti Academy defended chemistry, the private teaching of which had been prohibited, see Torrini, '*L'accademia*', 866.
58 Gatta shares with other authers, for example Kircher, some of the principles developed by Paracelsus. See Partini, *Athanasius Kircher*.
59 Partini, *Athanasius Kircher*, 217.
60 'non sol non offende come alcuni Ideoti, e volgari dicono, ma purga, e leggiermente, e gagliardamente come vorrà chi sa prepararlo', *Di una gravissima*, 222; Donzelli, *Teatro farmaceutico*, 36.
61 On Atomism in Naples, see Garin, *Dal Rinascimento*, 95; Festa and Gatto (eds), *Atomismo*, 321–39; Labellarte, *Atomismo*.

62 C. Morexano, *Il torchio*.
63 D'Alessio, 'Un allievo'.
64 'Coeli non ad causam, sed ad signa futurorum; ac dumtaxat intemporum [...]', Van Helmont, *Tumulus pestis*, 30.
65 Morexano, *Il torchio*, 9. Among others, the quotes from Girolamo Cardano's *De Subtilitate* (1550) and Julius Caesar Scaliger's *Exotericarum exercitationum liber quintus decimus de subtilitate ad Hieronymum Cardanum* (1557).
66 Cardano, for example, had noted that in 'Calecut (Città d'India) e per tutto il suo distretto non sia mai stato Peste [...]', Morexano, *Il torchio*, 9.
67 Morexano, *Il torchio*, 19–20.
68 'perché è accaduto, che una persona attualmente sana ha infettato, chi seco ha pratticato', Morexano, *Il torchio*, 33.
69 Morexano, *Il torchio*, 30.
70 'ne trovo io che Galeno curasse mai la peste con altro, che con gli antidoti, ne meno si trovarà, che nel lib. *De differentiis febrium*, faccia motto di cavar sangue a gli appestati, anzi nel libro *De succorum varietate atque vitio* si vede, che condanna apertamente la sagnia', Morexano, *Il torchio*, 35.
71 'Di questo elettuario io ne ho veduto mirabile effetto', Morexano, *Il torchio*, 37.
72 Morexano, *Il torchio*, 32 and 43.
73 Mosca, *Vita*, 6–9.
74 Torrini, 'L'accademia', 858–9; Serrapica, 'Per una teoria dell'incertezza', 36–9, with more bibliography.
75 Mosca, *Vita*, 10.
76 Cornelio, *Progymnasmata*, 186. 'Timeo' is Cornelio.
77 Boccalini, 'Generale Riforma', 258–85.
78 Among the first were Della Porta, Telesio, Quattromani, Campanella, Schipani, Stigliola and Severino. See Origlia, *Istoria*, 191.
79 Cornelio, *Progymnasmata*, 190; Stelliola, *Encyclopedia pythagorea*; Ricci, 'Nicola Antonio Stigliola', 66–74.
80 Cornelio, *Progymnasmata*, 174. Trusianus (in the *Progymnasmata*) is the typical exemplum of these doctors. Torrini, *L'accademia*, 864.
81 'Nimirum compescendum esse dixit ingeniorum aestum, praecidendamque Medicis libertatem novandi, excutienda item de manibus discentium Chemiae studia, submovendos Paracelsi, Helmontii, Harveij, et aliorum Novatorum libros ad unius tantum Galeni lectionem adigendos omnes', Cornelio, *Progymnasmata*, 204.
82 'at omnia protinus magno astantium risu, atque cachinnis, ut par erat, accepta fuere', Cornelio, *Progymnasmata*, 204.
83 'Nam quod memoriae proditum es Atticam olim mea cura ex imminentis pestilentiae impetu ereptam servatamque fuisse, ad praecepta potius civilis prudentiae, quam medicae professionis est referendum', Cornelio, *Progymnasmata*, 177.
84 Cornelio, *Progymnasmata*, 208.
85 On the theme, Porzio wrote *Erasistratus*, 183.
86 Cornelio, *Progymnasmata*, 209.
87 On the Alkahest see Hedesan, 'Jan Baptist Van Helmont', 91–100.
88 'nullam unquam in orbe terrarum extisisse pestilitatem quae mortalium generi cladem maiora inflixisset, quam ipsius medicinam', Cornelio, *Progymnasmata*.
89 'Plerumque vero excrementa stellarum pestis esse praenuncia, et ab eorundem forma, atque figura, future morbi genus posse praedici', Cornelio, *Progymnasmata*, 210.
90 On the plague as excrement of the stars see Paracelsus, *De pestilitate*, 386.
91 Cornelio, *Progymnasmata*, 211.
92 Cornelio, *Progymnasmata*, 211.
93 Gabrieli, 'Intorno a Nicola Antonio Stelliola'; Manzi, 'Un grande nolano'.
94 'con particolar industria dilettandosi più di essere, che di parer dotto, è stato sempre mai lontano da quella alterezza, che per ordinario porta seco la cognitione che s'ha delle scienze e il saper oltre la commune condizione de gli huomini', printer's note to the reader ('Lo Stampatore al Lettore') in Stigliola, *Il Telescopio*: I–IV (pages not numbered).
95 'ex quo fit ut ancipites periculosasque curationes morbis admovere cogantur', Cornelio, *Progymnasmata*, 212.

96 Cornelio, *Progymnasmata*, 212.
97 Cornelio, *Progymnasmata*, 213.
98 'Miror autem cur tot, et tanta quotidie de peste volumina exarentur, cum nemo non fateatur incomprehendam esse illius naturam, et omnino ab humana intelligentia remotam. Scilicet insitum est nonnullis quanto minus sciunt quae scribunt, tanto plura conscribere', Cornelio, *Progymnasmata*, 213.
99 'Hinc vero liquet caussa contagij; quippe halitus expirationesque a morbido corpore exhalantes circumfuso aeri permiscentur, eique lethale virus communicant', Cornelio, *Progymnasmata*, 214.
100 Cornelio, *Progymnasmata*, 214.
101 Cornelio, *Progymnasmata*, 215.
102 See Torrini, 'Uno scritto'.
103 On this theme and the influence of Santorio, see Zurlini, 'The uncertainty'.
104 Cornelio, *Progymnasmata*, 288.
105 Mosca, *Vita di Lucantonio Porzio*, 4.
106 Fish, 'The Academy', 550.
107 Mosca, *Vita di Lucantonio Porzio*, 19.
108 Mosca, *Vita di Lucantonio Porzio*, 21. Porzio was 'Borelli's faithful spokesman in the Roman enviroment': Torrini, *Dopo Galileo*, 171; Dini, 'Lucantonio Porzio', 8.
109 Mosca, *Vita di Lucantonio Porzio*, 21–2.
110 Mosca, *Vita di Lucantonio Porzio*, 43.
111 *Del governo della peste*, 385.
112 'Neapoli vero etsi nulla adhibita hujusmodi diligentia fuerit (erat autem tunc Regni Prorex comes de Castrillo) Monalium tamen Monasteria quaedam, cum tota Urbs latissime peste vastaretur, a peste fuere immunia. Ac Romae non tantum pleraque Monasteria, sed etiam nonnulla familiae, domusque Civium nihil omnino a peste passae sunt', Porzio, *De Militis*, 5.
113 Bartoli, in *Il lago di Agnano*, used a metaphor to say that it was necessary to intervene soon: 'Accurate policy must not neglect to extinguish even the smallest sparks, so as not to have to repair a large fire' ('La Politica accurata non deve trascurare di estinguere scintille anco minutissime, per non obligarse a riparare un gran fuoco'). Other maxims say that it is not wise to use the 'deceit' to rule, because the Popolo could vendicate themselves.
114 'Est veluti fermentum, ex quo sit panis, cuius si tantillum accipias, quod novae massae farinae ex aquis compactae admisceas, haec omnia fermenti naturam acquirent; ita ut hac ratione fermentum futuros semper habeamus in usus. [...]', Porzio, *De Militis*, 8.
115 'Ac si Comes de Castrillo de his scientiam habuisset vel forte cum habuerit, si ea uti licuisset, non minus ac Ferdinandus III Florentiam, centena millia hominum Neapoli servare potuisset', Porzio, *De Militis*, 9.
116 Benzoni, 'Contarini, Domenico'.
117 'Neque illorum Principum vitae institutio tibi placet, qui alienis tantum oculis vident, alienis tantum audiunt auribus', Porzio, *De Militis*, 11. Mosca, *Vita di Lucantonio*, 48.
118 See Bertoloni Meli, 'The Neoterics'.
119 'Cujus rei sufficientem habeo ex eo usque tempore experientiam, in quo pestis saevissima Neapolim graviter infestavit', Porzio, *De Militis*, 65.
120 Porzio, *Quid pestis*, 136.
121 'Ac ut de homine potissimum loquar quaecunque corporea sunt in homine continuo in fluxu sunt, et reparatione indigent ... Calor quicunque utcunque acceptus; humiditas quaecunque, utcunque accepta continuo in fluxu est: ut stupiditas is in homine jam adulto calidum quaerere insitum sive nativum; stupiditas sit humiditatem aliquam quaerere nativam, sive radicalem', Porzio, *Quid pestis*, 143.
122 'ut obscura, et occulta ratio propter quam allium, et rosa paucis quibusdam pro veneno fuerint', Porzio, *Quid pestis*, 145.
123 Porzio, *Quid pestis*, 136–47.
124 In this last passage Porzio alludes to a book on the plague that he had allegedly written when he was in Conversano. Porzio, *Quid pestis*, 147.
125 Di Capua, *Parere*, rag. IV, 254.
126 Donzelli, *Teatro farmaceutico*, 48.
127 Donzelli, *Teatro farmaceutico*, 154.
128 Donzelli, *Teatro farmaceutico*, 172.

129 Donzelli, *Teatro farmaceutico*, 121.
130 Donzelli, *Teatro farmaceutico*, 315.
131 Donzelli, *Teatro farmaceutico*, 498.
132 Donzelli, *Teatro farmaceutico*, 114.

## Bibliography


Aceti, Antonio. *Un genio cosentino negletto* (Cosenza, 1833). Reprinted in *Omaggio a Tommaso Cornelio*, edited by G. Mocchi, vol. 1, 187–556. Rubbettino: Soveria Mannelli, 2005.

Amenta, Niccolò. *Vita di Lionardo di Capoa*. Venice, 1710.

Anonymous. 'Relazione della pestilenza accaduta in Napoli l'anno 1656', edited by G. de Blasiis, *Archivio storico per le Province Napoletane* 1 (1876): 323–57.

Anonymous [attributed to Sebastiano Bartoli]. *Il lago d'Agnano utile et innocente con l'infusione de' lini e senza quella dannosissimo alla cittadinanza di Napoli, et a' massari della Campagna felice*, s.t., 1664.

Bado, Sebastiano. *Cortex Peruviae redivivus, profligator febrium*. Genoa: ex typographya Benedicti Guaschi, 1656.

Bado, Sebastiano. *Saggio d'alcuni rimedii preservativi dalla peste dedicato all'Illustriss. Signori Protettori dello spedale di Pammatone Francesco Viale, Gio. Battista Centurione, Gieronimo Fiesco, da un de Medici dello stesso spedale*. Genoa: Per Benedetto Guasco, 1656.

Benzoni, Gino. 'Contarini, Domenico', *Dizionario Biografico degli Italiani*, vol. 28 Rome: Istituto della Enciclopedia Italiana, 1983.

Bertoloni Meli, Domenico. 'The Neoterics and political power in Spanish Italy: Giovanni Alfonso Borelli and his circle', *History of Science* 34 (1996): 57–89.

Boccalini, Traiano. '*Generale Riforma dell'Universo da i sette Savij della Grecia, e da altri Letterati pubblicata di ordine di Apollo*.' In *Ragguagli di Parnaso*, edited by L. Firpo, vol. I: 258–85. Bari: Laterza, 1948.

Calvi, Giulia. 'L'oro, il fuoco, le forche: la peste napoletana del 1656', *Archivio storico italiano* 139 (1981): 405–58.

Cancila, Rossella. 'Salute pubblica e governo dell'emergenza: la peste del 1575 a Palermo', *Mediterranea-ricerche storiche* 37 (2016): 231–72.

Capaldi, Donatella. *Momo. Il demone cinico tra mito, filosofia e letteratura*. Naples: Liguori, 2011.

Carlino, Andrea. *La via dell'incertezza: una lettura scettica del* Parere sull'incertezza della medicina *di Lionardo di Capua*. In *Storia sociale e politica. Omaggio a Rosario Villari*, edited by A. Merola, G. Muto, E. Valeri and M. A. Visceglia, 323–51. Milan: Franco Angeli, 2007.

Conforti, Maria. 'Surgery, medicine and natural philosophy in the library of Marco Aurelio Severino (1580–1656)', *Bruniana & Campanelliana* 10/2 (2004): 283–98.

Conforti, Maria. *Medicina sotto il vulcano. Corpi e salute a Napoli in età moderna*, Milan: Editrice Bibliografica, 2021.

Cornelio, Tommaso. *Progymnasmata physica*. Venice: Sumptibus Bartholomaei Nicola Moreschi, 1683.

Console, Renzo. 'Pharmaceutical use of gold from antiquity to the seventeenth century'. In *A History of Geology and Medicine*, edited by Christopher J. Duffin, Richard T. J. Moody and Christopher Gardner-Thorpe, 171–92. London: Geological Society, 2013.

D'Alessio, Silvana. 'On the plague in Naples, 1656: Expedients and remedies'. In *Disaster Narratives in Early Modern Naples: Politics, communication and culture*, edited by Chiara De Caprio, Domenico Cecere, Lorenza Gianfrancesco and Pasquale Palmieri, 187–204. Rome: Viella, 2018.

D'Alessio, Silvana. 'L'aria innocente. Geronimo Gatta e le sue fonti', *Mediterranea–ricerche storiche*, 15 (2018): 587–612.

D'Alessio, Silvana. 'Un allievo di Marco Aurelio Severino sulla peste di Napoli', *Medicina Historica* 4 (2020): 43–5.

D'Alessio, Silvana. 'Su alcune lettere del viceré e di un suo ministro durante la peste (Napoli, 1656)'. In *A novant 'anni dalla missione Egidi a Simancas. Studi mediterranei*, edited by M. Verga and P. Spinato, 1–47. Cagliari, Milan and Rome: ISEM-CNR, 2021. D'Andrea, Francesco. 'Riflessioni sopra la seconda scrittura circa la materia degli atomi'. In *D'Andrea*

*atomista. L'Apologia e altri inediti nella polemica filosofica della Napoli di fine seicento*, edited by Antonio Borrelli. Naples: Liguori, 1995.
De Renzi, Salvatore. *Napoli nell'anno 1656*. Naples: Domenico De Pascale, 1867.
Di Capua, Leonardo. *Parere del signor L. di Capoa. Divisato in otto Ragionamenti*. Naples: Per Antonio Bulifon, 1689.
Dini, Alessandro. *Filosofia della natura, medicina, religione. Lucantonio Porzio (1639–1724)*. Milan: Franco Angeli, 1985.
Do Sameiro Barroso, Maria. 'Bezoar stones, magic, science and art from the late Middle Ages to the end of the 17th century'. In *A History of Geology and Medicine*, edited by Christopher J. Duffin, Richard T. J. Moody and Christopher Gardner-Thorpe, 193–208. London: Geological Society, 2013.
Donato, Maria Pia. 'La peste dopo la peste. Economia di un discorso romano (1656–1720)', *Roma moderna e contemporanea* 1 (2006): 159–74.
Donzelli, Giuseppe. *Teatro farmaceutico dogmatico, e spagirico del dottor Giuseppe Donzelli napoletano, barone di Digliola*. Venice: Appresso Gasparo Storti, 1681.
Festa, Egidio and Romano Gatto (eds). *Atomismo e continuo nel XVII secolo: atti del Convegno internazionale Atomisme et continuum au XVIIe siècle*. Naples: Vivarium, 2000.
Fusco, Idamaria. *Peste, demografia e fiscalità nel regno di Napoli del XVII secolo*. Milan: Franco Angeli, 2007.
Fusco, Idamaria. 'Il Regno di Napoli nelle emergenze sanitarie del XVII secolo. Istituzioni, politiche e controllo dello spazio marittimo e terrestre', *Storia urbana* 38 (2015), 55–74.
Findlen, Paula. 'Microscopic musing: Athanasius Kircher and the Roman plague of 1656–57', Harvard Library Bulletin, https://harvardlibrarybulletin.org/microscopic-musings-athanasius-kircher-and-roman-plague 16 June 2023.
Fish, Max H. 'The academy of investigators'. In *Science, Medicine and History*, edited by E. Ashworth, 521–63. London: Oxford University Press, 1953.
Flemming, Rebecca. 'Galen and the plague'. In *Galen's Treatise Περί Αλυπίας (De indolentia) in context. A tale of resilience*, edited by Caroline Petitt, 219–44. Leiden; Boston: Brill, 2019.
Gabrieli, Gabriele. 'Intorno a Nicola Antonio Stelliola filosofo e linceo napoletano', *Giornale critico della Filosofia Italiana* 10 (1929): 469–85.
Galen. *De cucurbitulis, scarificatione, Hirundinibus, Derivatione, & Revulsione sermo ex septimo & octavo Medicinalium collectionum ad Iulianum Imperatorem libro*: Venice: Apud Iuntas, 1576.
Galen. *L'antidotario di Claudio Galeno Pergameno, interpretato da Michelangelo Angelico vicentino*. Vicenza: Domenico Amadio, 1613. Garin, Eugenio. *Dal Rinascimento all'Illuminismo: studi e ricerche*. Pisa: Nistri-Lischi, 1970.
Gatta, Geronimo. *Di una gravissima peste che nella passata primavera, e estate dell'anno 1656 depopulò la città di Napoli [...]*. Naples: Luc'Antonio di Fusco, 1659.
Gentilcore, David. 'Il Regio Protomedicato nella Napoli Spagnola', *Dynamis: Acta Hispanica ad Medicinae Scientiarumque Historiam Illustrandam* 16 (1996): 219–36.
Gentilcore, David. 'Tempi sì calamitosi: Epidemic disease and public health'. In *A Companion to Early Modern Naples*, edited by Tommaso Astarita, 281–306. Leiden; Boston: Brill, 2013.
Gentilcore, David. *Malattia e guarigione*. Lecce: Besamuci, 2020.
Gianfrancesco, Lorenza. 'Books, gold and elixir: Alchemy and religious orders in early modern Naples', *Ambix* 65 (2018): 250–74.
Hedesan, Georgiana D. *An Alchemical Quest for Universal Knowledge: The 'Christian Philosophy' of Jan Baptist Van Helmont (1579–1644)*. New York: Routledge, 2016.
Hedesan, Georgiana D. 'Jan Baptist Van Helmont and the medical-alchemical perspectives of oison'. In *Toxicology in the Middle Ages and Renaissance*, edited by Philip Wexler, 91–100. London: Elsevier/Academic Press, 2017.
Jouanna, Jacques. 'Air, miasma and contagion in the time of Hippocrates and the survival of miasmas in post-Hippocratic medicine (Rufus of Ephesus, Galen and Palladius)'. In *Greek medicine from Hippocrates to Galen*, edited by Philip van der Eijk, 119–36. London; Leiden: Brill, 2012.
Labellarte, Alberto. *Atomismo e corpuscolarismo nella Napoli di fine Seicento*. Rome: Armando editore, 2020.
Leigh, Robert (ed.). *On Theriac to Piso, Attributed to Galen*. Leiden; Boston: Brill, 2016.
MacGregor, Arthur. 'Medicinal *terra sigillata*: A historical, geographical and typological review'. In *A History of Geology and Medicine*, edited by Christopher J. Duffin, Richard T. J. Moody and Christopher Gardner-Thorpe, 113–36. London: Geological Society, 2013.

Manzi, Pietro. 'Un grande nolano obliato: Nicola Antonio Stigliola', *Archivio storico per le province napoletane* 2 (1973): 131–6.
Maranta, Bartolomeo. *Della theriaca et del mithridato libri due*. Venice: Marc'Antonio Olmo, 1572.
Marra, Massimo. 'La vipera e l'oppio: la teriaca di Andromaco a Napoli tra XVI e XVII secolo', *Airesis*, online. Accessed 12 February 2023. https:/aispes.net/biblioteca.
Mattioli, Pietro Andrea. *Il dioscoride dell'eccellente dottor medico M.P. Andrea Matthioli da Siena [...], con la giunta del sesto libro dei rimedi di tutti i veleni [...]*. Mantua: Appresso Iacomo Rossinello, 1649.
Mattioli, Pietro Andrea. *Discorsi di M. Pietro Andrea Mattioli Sanese, medico cesareo, ne' Sei libri di Pedacio Dioscoride Anazarbeo Della materia medicinale*. Venice: Presso Niccolò Pezzana, 1744.
Montalbani, Ovidio. *Il pestifugo Esculapio cioè regole più sicure per iscampare da ogni contagioso pericolo; considerationi di Gio. Antonio Bumaldi bolognese*. Bologna: Per Giacomo Monti, 1656.
Morabito, Wanda. 'Dediche, elegie ed epigrammi nei Progymnasmata physica di Tomaso Cornelio', *Calabria sconosciuta* (1984): 63–5.
Morexano, Carlo. *Il torchio delle osservationi della peste di Napoli nell'anno M.DC.LVI*. Naples: Sebastiano di Alecci, 1659.
Mosca, Giuseppe. *Vita di Lucantonio Porzio*. Naples: Presso Gennaro Migliaccio, 1765.
Muratori, Ludovico Antonio. *Del governo della peste e delle maniere di guardarsene*. Modena: Per Bartolomeo Soliani Stamp. Duc., 1710.
Napolitano, Giovanni Battista. *Trattato utilissimo, et necessario a ciascuno per conservarsi sano nel tempo della peste, e per liberare sicuramente chi da quella fosse già tocco, e infettato*. Venice: Gio. Comenzini, 1576.
Nutton, Vivian, 'The reception of Fracastoro's theory of contagion. The seed that fell among thorns?', *Osiris* 6 (1990): 196–234.
Nutton, Vivian and D'Alessio, Silvana. 'Santorio Santori on plague: Ideas and experience between Venice and Naples'. In *Santorio Santori and the emergence of quantified medicine, 1614–1790*, edited by Jonathan Barry and Fabrizio Bigotti, 217–38. London: Palgrave Macmillan, 2022.
Origlia, Giovanni Giuseppe. *Istoria dello Studio di Napoli*, vol. II. Naples: A. Forni, 1754.
Paracelsus, Theophrastus. 'De pestilitate' (1535). In Paracelsus, *Opera omnia*, vol. I. Geneva: Ioan. A. and S. de Tournes, 1658.
Pastore, Alessandro and Peruzzzi, Enrico (eds). *Girolamo Fracastoro fra Medicina, Filosofia e Scienze della natura. Atti del convegno internazionale di studi in occasione del 450Åã anniversario della morte Verona-Padova, 9–11 ottobre 2003*. Florence: Olschki, 2006.
Perfetti, Amelia. 'L'alchimia a Napoli tra cinquecento e seicento: Leonardo Fioravanti e Giovan Battista Della Porta'. In *Napoli Viceregno spagnolo. Una capitale della cultura alle origini dell'Europa moderna (sec. XVI–XVII)*, edited by M. Bosse and A. Stoll, vol. 1, 311–28. Napoli: Vivarium, 2001.
Partini, A. M. *Athanasius Kircher e l'Alchimia. Testi scelti e commentati*, translated by Pasquale Faccia. Rome: Edizioni Mediterranee, 2015.
Pignataro, Carlo. 'Illustrissimo Domino Abbati Michaeli Iustiniano'. In *Historia del contagio di Avellino*, edited by Michele Giustiniani, 197–208. Rome: Ignatio de Lazari, 1662.
Porzio, Lucantonio. 'Erasistratus, sive de sanguine missione.' Venice: Pezzana, 1683.
Porzio, Lucantonio. 'Quid pestis Galeno sit'. In *Opuscula, et Fragmenta varia*, edited by L. A. Portii, 136–47. Naples: Antonio Bulifon,1701.
Porzio, Lucantonio. 'De Militis in Castris, sanitate tuenda'. In *Opera omnia, medica, philosophica et mathematica* [...], edited by F. Porzio, vol. II, 12–102. Naples: Felice Carlo Mosca, 1736.
Rascaglia, Mariolina. 'I libri proibiti di un "philosopho libero"'. In *Sebastiano Bartoli e la cultura termale del suo tempo*, edited by R. Maria Zaccaria, 95–112. Florence: Olschki, 2012.
Ricci, Saverio. 'Colantonio Stigliola, "mathematico", "theologo" e "incingiero"'. In *Leonardo e il Rinascimento nei Codici napoletani. Influenze e modelli per l'architettura e l'ingegneria*, edited by A. Buccaro and Mariolina Rascaglia. 283–91. Naples: CIRICE-CB Edizioni, 2020.
Serrapica, Salvatore. *Per una teoria dell'incertezza tra filosofia e medicina. Studio su Leonardo Di Capua (1617–1695)*. Naples: Liguori, 2003.
Serrapica, Salvatore. '"Malo nodo malus cuneus": La diffusione di Van Helmont nella Napoli "investigante"'. In *Sebastiano Bartoli e la cultura termale del suo tempo*, edited by Rafaella Zaccaria, 45–63. Florence: Olschki, 2012.
Slater, John, Maríaluz López-Terrada and José Pardo-Tomás (eds). *Medical Cultures of the Early Modern Spanish Empire*. London: Routledge, 2014.

Stigliola, Nicola Antonio. *Il Telescopio over Ispecillo Celeste*. Naples: 1627.
Tanturri, Alberto. 'Terapie contro la peste', *Ricerche di Storia Sociale e religiosa*, XXXIII, 65 (2004): 181–94.
Torrini, Maurizio. 'Uno scritto di Leonardo Di Capua in difesa dell'arte chimica', *Bollettino del Centro di Studi Vichiani* 4 (1974): 126–39.
Torrini, Maurizio. *Dopo Galileo: Una polemica scientifica, 1684–1711*. Florence: Olschki, 1979.
Torrini, Maurizio. 'L'Accademia degli Investiganti. Napoli 1663–1670', *Quaderni storici* 16 (1983): 845–83.
Trabucco, Oreste. 'La corrispondenza tra Pietro Castelli e Marco Aurelio Severino'. In *Filosofia e scienze nella Sicilia dei secoli XVI e XVII*, edited by Corrado Dollo, vol. I, 109–31. Catania: Centro studi per la Storia della Filosofia in Sicilia, 1995.
Trabucco, Oreste. 'Anamorfosi di un medico "eretico" in *Sebastiano Bartoli*'. In *Sebastiano Bartoli e la cultura termale del suo tempo*, edited by R. M. Zaccaria, 65–94. Florence: Leo Olschki, 2012.
Trabucco, Oreste. 'Delle cagioni delle febbri maligne di G. A. Borelli. Una lettura contestuale', *Giornale critico della Filosofia Italiana* 20 (2000): 236–80.
Zurlini, Fabiola. *The Uncertainty of Medicine: Readings and reactions to Santorio between tradition and reformation (1615–1721)*. In *Santorio Santori and the Emergence of Quantified Medicine*, edited by Jonathan Barry and Fabrizio Bigotti, 103–17. Palgrave Studies in Medieval and Early Modern Medicine series. London: Palgrave Macmillan, 2022.

# Part III

# Naples and the early modern world

6

# *Kunst- und Naturalien-Kammern* of seventeenth-century Naples: Johann Daniel Major's view from Germany

James Clifton

## Introduction

The Neapolitan apothecary and collector Ferrante Imperato (1525?–1615?) sounded like many a modern scholar labouring in an isolated backwater when he complained to a correspondent in 1590 of little local interest:

> Please be kind to me because here in Naples I seem to be at the end of the world. I don't know who to talk to in this profession since we have such a shortage of men that delight in this profession of ours.[1]

Yet Imperato was well integrated in the humanistic republic of letters, with correspondents throughout Europe, and his collection enjoyed a continent-wide fame that long survived the death of its founder. Thomas Bartholin (1616–1680), professor of medicine at the University of Copenhagen, described his visit to the Imperato museum in a letter, written in Naples on 26 March 1644, to his compatriot in Padua Johan Rode, also a physician: he noted that the museum was now in the hands of Ferrante's grandson, while a great-grandchild, 'knowing beyond his years, shows details to foreigners'.[2] Republishing in Venice Imperato's great book *Dell'Historia Naturale*, more than seven decades after its first appearance in 1599, G. B. La Noù averred in a dedication to Johann Friedrich, Duke of Braunschweig-Lüneberg, that the work was 'buried under the ashes of forgetfulness' ('sepolta sotto le ceneri della dimenticanza'), yet its republication was a testament to its continuing appeal even beyond the borders of Italy. The Imperato museum had always been a social space – welcoming both local scholars and foreign visitors – as well as a space for collecting and experimentation; both are suggested by the famous woodcut frontispiece to *Dell'Historia Naturale*, reproduced as an engraving and reversed in the second edition seen here (Fig. 6.1). Here, in a kind of hybrid space, shelves and cabinets store books and specimens for consultation, while further specimens are arranged along the walls and ceiling for display. A young man, perhaps Ferrante's son Francesco (*c*.1570–*c*.1630) uses a long pointer to draw the attention of three well-dressed visitors to particular objects, much as his own grandson may have done decades later with Bartholin.[3]

Imperato's fame abroad is confirmed in a German treatise on Neapolitan collections, Johann Daniel Major's *Vorstellung etlicher Kunst- und Naturalien-Kammern in Italien zu Neapolis und Alt-Rom* (*Presentation of Several Art- and Naturalia-Chambers in Italy at Naples and Ancient Rome*). In the 1670s Major (1634–1693) published several brief treatises on what he called *Kunst- und Naturalien-Kammern* ('Art and Naturalia Chambers') around the world.[4] Of the lengthy series of such treatises he planned, Major produced only a few; these were gathered and

**Figure 6.1** Frontispiece in Ferrante Imperato, *Historia Natvrale … nella qvale ordinatamente si tratta della diuersa condition di Minere, Pietre pretiose, & altre curiosità. Con varie Historie di Piante, & Animali, sin'hora non date in luce*, 2nd ed. Venice: Presso Combi & La Noù, 1672. © Sarah Campbell Blaffer Foundation, Houston.

republished after his death in Michael Bernhard Valentini's monumental compendium, *Museum Museorum* (1704).[5] From 1665 professor of medicine at the new Christian Albrecht University in Kiel, where he founded the university botanical garden, Major was a prolific writer, translator and editor on diverse subjects. He inclined towards medicine and natural philosophy, however, often with a focus on objects available to him from curiosity merchants and collections, including his own – which was directed towards experimentation and open to students and, eventually, the public.[6]

Major's remarks on Neapolitan collectors and collections, which span 23 unnumbered pages in the original publication, paint an incomplete picture of the culture of natural philosophy in *seicento* Naples, to say the least. However, they do provide an indication of its international reputation. Although Major advocated travel for the naturalist and had made it as far as Padua himself (where he completed a doctorate of medicine in 1660), he did not visit Naples; for his account he drew on various verbal sources, mostly printed, which he duly cites.[7] To determine which collections to address in his planned series of treatises, Major circulated a provisional list, seeking addenda, and *Index Alphabeticus*.[8] His request implies that he is seeking objects in currently existing collections – he is hoping to put together a *Catalogum Rerum Naturalium Selectiorum* that 'are to be found' ('zu finden sind') in collections both personal and institutional.[9] Yet in the *Vorstellung etlicher Kunst- und Naturalien-Kammern in Italien zu Neapolis und Alt-Rom* itself he gives no consideration as to whether the collections he addresses are still intact and accessible or not. He also discusses ancient collections that have long since disappeared.

Major also stipulated that he was more interested in 'naturalia' than 'artificialia', that is, the natural material rather than any art that had enriched them – which largely, but not entirely, holds true of the *Vorstellung*.[10] In the *Index Alphabeticus* six early modern collectors are listed under the heading Neapolis, though the Neapolitan Fabio Colonna appears in the list under Roma. Major does devote a chapter to Colonna in the *Vorstellung etlicher Kunst- und Naturalien-Kammern in Italien zu Neapolis und Alt-Rom* and indicates in the first paragraph why the naturalist, scion of an august Roman family, but born and resident in Naples, belongs among the Neapolitan collectors.[11] The only addition to the list in the *Vorstellung* is a brief chapter on Giovan Vincenzo Della Porta, which, as we shall see, is particularly uninformative.

The *Vorstellung etlicher Kunst- und Naturalien-Kammern in Italien zu Neapolis und Alt-Rom* comprises a title page, an introduction dedicated to the physicians Sebastian Scheffer and Johann Wilhelm Schmidt and 15 chapters. In the first chapter, as if taking a mental voyage from his first two treatises – one on the West and East Indies and one on Africa – Major steers 'das Schiff unserer Schrifftlichen Abhandelung, nach Constantinopel, Malta und Messina' ('the ship of our written treatise toward Constantinople, Malta and Messina').[12] Here Major speaks generally of Italy, touching on its cartographic shape, the fruitfulness of its land, the bountifulness of its agriculture, minerals and stone, the beauty of its buildings, sculpture and gardens; he also considers its importance, especially in the current century, for the renaissance of the sciences and liberal arts through academies, libraries, museums and galleries, along with their collections of books, paintings, antiquities, medallions and coins, mathematical instruments, and choice of natural and art objects.[13] Major's second chapter narrows the focus to Campania and Naples, which he praises for magnificence, fertility and refreshing climate, valued since antiquity and recognised by poets such as Virgil, Horace, Statius and Jacopo Sannazaro. In the third chapter, one of the longest in the treatise, Major turns to specific collections, beginning with that of the ancient Roman general and statesman, Licinius Lucullus, at his villa on the Bay of Naples. Major then invites the reader to enter the city itself with him:

> Let us therefore leave the pleasurable landscape of the Neapolitan environs and stroll into the city itself, and, through indefatigable writing, wrest from the power of oblivion what foremost curious and gallant, upper- and middle-class people, as possessors of noble rarities, expensive houses and palaces in this city were found, or are still to be found in our times.[14]

He devotes nine chapters to early moderns: the Spanish viceroy(s), prince Tiberio Carafa, Fabio Colonna, Vincenzo Cioffi, Mario Schipani, Ferrante Imperato and his son Francesco, Donato D'Eremita and Giovan Vincenzo Della Porta. With the final three chapters he returns to antiquity, but strays from the Neapolitan locus, addressing Caesar Augustus's rarities, whether Cicero was concerned with rarities and Lucullus's aviaries at Frascati.[15]

An examination of the *Vorstellung etlicher Kunst- und Naturalien-Kammern in Italien zu Neapolis und Alt-Rom* provides some sense of the transalpine reputation of Neapolitan collectors, especially of *naturalia*, and of the pursuit of natural philosophy in Naples more generally in the late seventeenth century. Although our point of departure is a single text from a single writer, with consequent idiosyncrasies, it seems fair to conclude that knowledge in northern Europe of Neapolitan scientific endeavours was partial at best.[16] I shall consider, in order, Major's nine chapters dedicated to early modern collectors and collections.

## The gallery of the viceroy at Naples (Das IV. Capitel. Von der Gallerie des Vice-Ré zu Neapel)

Major's brief description of the 'gallery of the viceroy' in the Palazzo Reale, 'filled with powerful artworks and expensive things', is, as Major acknowledges, drawn from the first-hand account in the *Mercurius Italicus* of the Swabian Johann Heinrich von Pflaumer (1584–1671), who studied in Siena, where he took a law degree in 1607, and travelled extensively in Italy during his student years.[17] The account of his travels was first published in 1625, though Major seems to have used the second edition of 1650.[18] Major lists the objects Pflaumer found in the Palazzo Reale: all kinds of arms and armour, old coins, a musical clock, marble tables inset with 'mosaic' and 'statues of precious stone'.[19] Major editorialises only minimally to explain that what Pflaumer meant by 'statues of precious stone' were more likely to be 'small simulacra, idols or seals (as much as small signs)' ('Simulacra parva, Idola, oder Sigilla [so viel, als kleine Signa])'. Pflaumer does not name the viceroy in residence during his visit, nor does he indicate whether the objects belonged to the viceroy personally or to the palace. Major does not mention the palace garden, which Pflaumer notes briefly.

## Prince Tiberio Caraffa's palace (Das V. Capitel. Von dem Palast des Fürsten Tiberii Caraffae)

At fewer than 100 words, Major's chapter on the palace of Tiberio Carafa, prince of Bisignano (1580–1647), is his briefest. Quoting Pflaumer's Latin directly, Major writes that the palace is worth seeing for its abundance of ancient marble statues, as well as a model of the Colosseum in bronze.[20] Major also notes that Pflaumer is in accord with the mention of the collection in the *Itinerarium Italiae Nov-Antiquae* by Martin Zeiler (1589–1661), but does not name the sculpture singled out there (a Scipio Africanus).[21] Major also reports that in *De Unicornu observationes Nouae*, Thomas Bartholin praises both the palace and the prince's curiosity in the knowledge of natural things ('Curiosität in natürlicher Dinge Wissenschafft').[22] In neither this nor the preceding chapter does Major specify objects of *naturalia*, which may in part account for their brevity.

## Fabio Colonna's curiosity (Das VI. Capitel. Von Herren Fabii Columnae Curiosität)

Major was personally familiar with most of the published work of Fabio Colonna (1567–1640) and drew on it directly, rather than relying primarily on brief notices from secondary sources.[23] Major was aware that Colonna was deeply embedded in naturalist networks – especially involving the Lincei academy, of which he was a member – in both Rome and Naples (where a branch was inaugurated in 1612).[24] He quotes Johann Faber's assessment of Colonna as a sedulous investigator of natural things, learned in law, mathematics (especially optics), botany and zoology, which appeared in the massive (and massively important) volume published by the Linceans, the *Rerum medicarum novae Hispaniae thesaurus*, Colonna's contributions to which Major admired.[25]

In the same year that Major published the *Vorstellung etlicher Kunst- und Naturalien-Kammern in Italien zu Neapolis und Alt-Rom*, he also published an extensively annotated edition of Colonna's *Purpura* of 1616, which treats the molluscs that produce a purple dye and in which Major provides a *vita* of Colonna.[26] In the *Vorstellung*, Major reviews Colonna's publications, beginning with the *Phytobasanos* of 1592 – which, as Major describes it, treats the strengths and medicinal uses of various herbs known to Theophrastus, Dioscorides, Pliny and Galen, as well as some little-known fish and aquatic mammals. He names Colonna's *Minus cognitarum Stirpium ac etiam rariorum nostro coelo orientium Stirpium Εκφρασιζ [Ekphrasis] ... Item de Aquatilibus aliisque animalibus quibusdam libellus*, published in its final form in 1616, as an elaboration of the *Phytobasanos*, and ultimately turns to the *Purpura*. Major draws attention to both Colonna's reputation for careful observation of specimens and the illustrations in the *Phytobasanos*, drawn and etched by Colonna himself.[27] In praising *Purpura*, Major also notes the many life-sized illustrations in the book, which were also etched, though Major replaced them with less expensive woodcuts in his edition (Fig. 6.2). He explains not reproducing them in the *Vorstellung* as an effort to save time, paper and expense.[28]

Cap. I.

quopiam Recentiorum visi, quamvis illius Historia satis ab *Aristotele* (z.) & *Plinio* (a.) descripta inveniatur: quam hic inserere & repetere, multum molestumq; esset. Illas tantùm notas recitabimus, quas in testaceo à Nobis proposito, observare licuit. (z.) *lib. 4. de N. Animal. c. 4. & lib. 5. c. 15.* (a.) *lib. 9. cap. 36. & inde.*

§. 27. Testa hæc parva est, Regiæ Nucis, adhuc carnoso tegmine obducta, magnitudine, (quanquam ejusdem generis inanes Testas quadruplo majores habeamus, atque maximas asperiores) à vulgari *Purpurâ Rondeletii*, quæ Neapoli Sconciglio appellatur, in his differens; Cortice quippe rugosiore, colore fusco, aut cinericeo, veluti Limo obducto, quandoque flavicante, & Fasciis distincta oblique purpurascen-

C 2

**Figure 6.2** *Murex trunculus* in Fabio Colonna, *Opusculum de purpura*, edited by Johann Daniel Major, 13. Kiel: Joachim Reumann, 1675. © Sarah Campbell Blaffer Foundation, Houston.

Major has no specific information about Colonna's collection as such, but extrapolates from his careful observation, well documented by reputation and obvious in his texts, averring that he must have kept many original specimens ('Cörper') in a special 'logiment' for his own 'useful amusement' and that of others.[29] He names a number of examples of shells from the *Aquatilium et terrestrium aliquot animalium observationes* section of *Purpura*, noting that some are particularly common in rocky areas around Naples;[30] another shell, the '*Murex Auritus Marmoreus maximus Exoticus* … oder grosse außländische Stachel-Schnecke' ('large foreign spiny snail'), was given to Colonna by a friend, a builder of fountains, in the city.[31] Major concludes that Colonna described and depicted such things as a reminder and must have kept the originals in his cabinet.[32] However, Colonna greatly relied on, and recommended, the much larger Imperato collection for his studies.[33] Major himself points to the instances in the *Purpura* in which Colonna refers to objects in Imperato's collection, and praises both the apothecary and his museum.[34]

## Vincenzo Cioffi's museum (Das VII. Capitel. Von dem Muséo, Herrn Vincentii Cioffi)

Major opens his next chapter with a disclaimer. He is not in fact certain that this Neapolitan nobleman or prominent man by the name of Vincenzo Cioffi had a significant curiosity cabinet, since his information comes from a single source, *De lucernis antiquorum reconditis* (1652) by Fortunio Liceti – whose single-mindedness, Major complains, provides no information beyond his own purposes. But we might assume, Major avers, that among the ancient lamps that were of sole interest to Liceti, there must have been 'many other rare things, which had manifested nature as well as art'.[35] Major quotes Liceti's introductory sentence directly:

> An illustrious man, Vincenzo Cioffi, a Neapolitan, had a museum arrayed with remains of antiquity, and from that he gave me several images of ancient lamps.[36]

He notes that Liceti subsequently refers to this museum as a *Gazophylacium* or treasury ('Schatz-Behältnüß'), but gets no closer to its contents aside from lamps.[37] Liceti's discussion of Cioffi's lamp collection is in fact extensive and includes many illustrations,[38] but Major closes the chapter by dutifully listing a few further references to Cioffi himself and his museum – which in Major's mind, however, provide no additional information. As with two previous brief chapters, there is no mention of specific 'naturalia'.

## Mario Schipani's cabinet (Das VIII. Capitel. Von Marii Scipani Cabinett)

Mario Schipani (d. after 1654) is a somewhat shadowy figure whose place in Neapolitan intellectual circles has been pieced together.[39] A physician and natural philosopher, he was a friend and patron of Pietro Della Valle, whom he sent on his famous, decade-long travels to Turkey, Persia and India from the 1610s to the 1620s. Della Valle dedicated to Schipani the resultant *Viaggi*, which comprises 54 letters 'to the erudite Mario Schipano, for many years his friend, among the dearest' ('All'erudito, e fra' più cari, di molti anni suo Amico Mario Schipano') in Naples.[40] Della Valle politely apologises for the fact that his desire to give Schipani accounts of the strange things he has seen must exceed the latter's curiosity to hear them recounted at such length, even tediously.[41] But Schipani's curiosity appears not to have been exhausted; he fed it further through his friendship with Fabio Colonna and his loose association with the Linceans, as well as ties to Tommaso Campanella, Federico Cesi, Galileo Galilei, Paolo Antonio Foscarini and Giambattista Della Porta.[42]

Major begins his chapter on Schipani with some general remarks, asserting that there is no one better suited to oversee a collection of *naturalia* than a physician (like himself, we may infer),[43] but he singles out the Roman physician, anatomist and botanist Pietro Castelli (1574–1662), who was responsible for the botanical gardens in Messina from 1635 and to whom Major had devoted a chapter in an earlier treatise.[44] Before veering into a tangential discussion of exotic guineafowl at the end of his chapter on Schipani, Major, drawing once again on Thomas Bartholin as well as the *Gammarologia* of Philipp Jakob Sachs von Lewenheimb (1627–1672), provides information about Schipani's collection. Major quotes Bartholin's *De Unicornu observationes Novae* again, calling Schipani a *Hippocrates Campaniae redivivus* (Hippocrates of Campania, reborn).[45] Paraphrasing a letter from Bartholin to the Danish physician Johan Rode in Padua (p.163), written from Naples on 26 March 1644, Major calls Schipani a great man, a venerable elder and good friend of Bartholin's father Casper, with a choice library; he also credits him with medical innovations.[46] He also plucks the reference to Schipani's 'museum, in which [there are] various types of nature' ('Musea, in quibus varia Naturae genera') from Philipp Jacob Sachs von Lewenheimb's *Gammarologia*. Major acknowledges that Sachs provides no details,[47] but reports that Bartholin saw in Schipani's cabinet, among other things, an African guineafowl (presumably taxidermied); this was probably a so-called 'helmeted guineafowl' (now *Numida meleagris*) with a bony projection on its head, given the context of Bartholin's book on 'unicorns'.[48] In a relatively lengthy excursus, Major adduces various references to, and nomenclature and illustrations of, the bird by Ulisse Aldrovandi, Georg Marggraf, Johannes Jonstonus, Ole Worm and Adam Olearius, although the illustrations he cites do not depict the helmeted genus (see for example Fig. 6.3).[49]

**Figure 6.3** *Gallina Guineae* in Ole Worm, *Museum Wormianum, seu Historia rerum rariorum, tam Naturalium, quam Artificialium, tam Domesticarum, quam Exoticarum, quae Hafniae Danorum in aedibus Authoris servantur*, 297. Amsterdam: Lodewijk and Daniel Elzevir, 1655. © Sarah Campbell Blaffer Foundation, Houston.

## Ferrante Imperato's famous naturalia chamber (Das IX. Capitel. Von der berühmten Naturalien-Kammer, Ferrandis Imperati)

In two chapters Major gives the most attention by far – and rightly so – to the vast collection of the apothecary Ferrante Imperato and his son Francesco, yet he was unable to gain access to the publications of either, as he regretfully acknowledges.[50] These works are, of course, Ferrante's *Dell'Historia Naturale* of 1599 – the famous frontispiece of which, here from the second edition of 1672, was published three years before Major's *Vorstellung etlicher Kunst- und Naturalien-Kammern in Italien zu Neapolis und Alt-Rom* – which hints at the scope of the collection (see Fig. 6.1) and Francesco's *Discorsi intorno a diverse cose naturali* of 1628.[51] Major relied instead on a significant number of published secondary sources for his comments – Pflaumer, Zeiler, Colonna, Giovanni Battista Ferrari, Athanasius Kircher, Ulisse Aldrovandi, Johann Vesling, Thomas Bartholin, Sachs von Lewenheimb, Heinrich Volgnad and Simon Schultz – as well as a letter of 4 November 1674 from the Hamburg curiosity collector David Schellhammer. Based on these reports, he is able to declare that no one in Naples before or after Ferrante Imperato, whom he describes as one of the foremost (and very learned) 'aromatorii' or 'materialists' of that city, had the curiosity and fortunate diligence to bring together so many kinds of beautiful natural rarities under one roof.[52] He quotes Bartholin's letter to Rode, in which he says that he saw in Imperato's possession a letter to him from Ulisse Aldrovandi declaring that even the crumbs from Imperato's 'table of a rich man' ('mensa divitis') would be worthy of Aldrovandi's own collection.[53]

Major cites in passing Zeiler's *Itinerarium Italiae Nov-Antiquae*, noting that he refers to Imperato in German as a *Gewürzhändler* (spice merchant) and indicates a few of the types of natural wonders in his house.[54] Zeiler turned his reader's attention to Pflaumer's *Mercurius Italicus* and so, too, does Major – recapitulating his list closely, but with occasional abbreviating, editorialising or loosely translating, in three paragraphs. Among the objects Pflaumer and Major name are several shells, a piece of cloth made from the 'sea silk' of molluscs,[55] silk from India, a dwarf, a Brazilian armadillo, 'Chirastes' (Major admits that he does not know what this is), a chameleon (an Asian colour-changing animal, Major explains), a kind of tailed ape with a human face,[56] a pelican, a crocodile, a tooth and tail from a hippopotamus (a piece of hide of which Major had received from Schellhammer for his own collection),[57] a seal (*Meer-kalb*), a sea turtle, a salamander, a remora, many kinds of birds (among them an albatross, the 'avis diomedea' on which Friedrich Lachmund had recently written a treatise dedicated to Major),[58] mandrake root and a magnet from Elba (a so-called white magnet, mentioned by Boethius, a small one of which Major himself had 'among my things').

Emphasising Pflaumer's first-hand knowledge of Imperato's collection ('von Ihm observirt worden wären'), Major continues to describe the exhibits. They include an Ethiopian magnet, different kinds of paper with various scripts (Japanese, Egyptian, Longobard), Indian ink, 'grown silver', an incombustible wick and a crystal encompassing water (both described by Girolamo Cardano),[59] a piece of petrified wood from which one can strike fire and a shiny stone knife used by Jews for circumcision and Indians for felling trees. In the final paragraph of this chapter, still following Pflaumer, he adds a Ceratites ('Bocks-Horn-Stein', 'Cornu Ammonis') and several petrified objects: a wood tablet, a crab, sponge, fungus, net and a nut, which, Pflaumer suggests and Major reports, Imperato may have taken from the nearby Silarus (Sele) river.

## More on Ferrante Imperato and his son Francesco (Das X. Capitel. Ein mehrers von Demselben; und seinem Sohn, Francisco Imperato)

In the second chapter on the Imperato collection, Major turns to sources other than Pflaumer, beginning with Fabio Colonna and the text of his that Major knows best, the *Purpura*, in which Colonna mentions Imperato's museum several times. Colonna says he observed shell specimens, both local and foreign, in the museum of 'our most learned Imperato', which he refers to, in various passages, as 'a sufficiently copious treasury of all the things of Nature', a 'very large museum' and 'a very rich repository of nature'.[60] Colonna calls Imperato himself 'that very rich provider of natural things, our Imperato'.[61]

Major then reports similar experiences and encomia from numerous authors. In his *De florum cultura* of 1633, the Jesuit horticulturalist Giovanni Battista Ferrari, visiting from Rome, calls Imperato (in Major's abbreviated paraphrase)

> a man most experienced in botanicals, or herbs, and his museum, a museum equipped with strange and utterly wondrous materials, natural as well as medicinal things, in which he had seen various live herbs or books in which dried originals of herbs had been glued.[62]

Similarly, a 'much more famous Jesuit', Athanasius Kircher, called Imperato a famous antiquarian and described, in a chapter on images appearing in stones, a specimen with an image of wheat that he had seen in the museum in Naples.[63] Major reports that in his *Musaeum metallicum* Ulisse Aldrovandi recalls having seen a petrified jawbone of an elephant on a visit to Imperato's museum,[64] and that Bartholin saw there a piece of fossilised ivory or horn as thick as an arm.[65] After naming a couple of other objects, drawn from secondary and tertiary sources,[66] Major summarises, exalting Imperato as

> a brave, curious, learned and gallant man, and his rarities-house was none other than a compendiously collected house of the entire wonderful nature, and it remained so afterwards, and doubtlessly was enlarged daily with beautiful additions by his son Francesco.

He then cannot resist adding yet one more object, this one reported in the recent personal letter from Schellhammer: an aetites (eaglestone), the size of a head.[67]

## Donato d'Eremita's rarities (Das XI. Capitel. Von den Raritäten Donati, deß Einsidlers)

In the subsequent chapter, Major turns his attention to another apothecary, but one of a very different type: Donato D'Eremita. Within a couple of years of his arrival in Naples from Tuscany in 1609, this Dominican friar was appointed prefect of a new apothecary shop at the convent of Santa Caterina a Formello, and there he remained until his death in 1630.[68] D'Eremita's networks included Giambattista Della Porta, Ferrante Imperato, Nicola Antonio Stigliola and Fabio Colonna in Naples, the Lincean Johann Faber in Rome and Pietro Castelli in Rome and Messina. In his brief chapter, Major relies on a variety of sources for his information about the friar and his collection of 'all sorts of beautiful rarities' ('allerhand schöne Raritäten'). One of these was the letter that Major received from David Schellhammer, cited at the end of the previous chapter, but with no details. Another, which Major quotes directly, is Bartholin's letter to Rode from 1644, more than a decade after D'Eremita's death. This letter notes D'Eremita's two books (claiming they were in reality written by Pietro Castelli) and the museum at Santa Caterina, in which various things are exhibited, including the skeletons of diverse animals.[69] The two books that Major mentions, but shows no sign of having consulted, are the *Dell'Elixir Vitae libri quattro* (1624), which was dedicated to the Tuscan grand duke Ferdinando II, and the posthumously published *Antidotario* (1639).[70] Major could have found in the illustrations to *Dell'Elixir Vitae*, reused in the *Antidotario*, not only a variety of laboratory equipment but also a group of figures, D'Eremita and Giovan Battista Della Porta among them, in a stylised representation of the convent laboratory (Fig. 6.4).[71] Meanwhile in the *Antidotario*, a laudatory poem by Marc'Antonio Perillo, member of the Neapolitan Accademia degli Incauti, addressed to D'Eremita's 'Museo', offers a few hints about the contents of the collection: Arab and Greek medical texts, flowers, herbs, saplings, metals and gems.[72] Major might also have noted the several approving references to D'Eremita in *Rerum medicarum novae Hispaniae thesaurus*, which he knew.[73]

Major also quotes from another correspondent, Danish naturalist Ole Borch (1626–1690). Borch's letter dates from 22 August 1674, but must be a report on observations made while travelling in Italy a decade earlier. In it Borch affirms the persistence of the collection at Santa Caterina a Formello decades after D'Eremita's death, as well as offering a few further indications of its contents, including a two-headed child, a two-headed calf and a child with four feet.[74]

**Figure 6.4** *Nobilissimo antidoto dell'Elyxir Vitae* in Donato D'Eremita, *Dell'Elixir Vitae libri qvattro*, plate 1. Naples: Secondino Roncagliolo, 1624. © Getty Research Institute, Los Angeles.

## Giovan Vincenzo Della Porta (Das XII. Capitel. Von Joh. Vincentio Portâ)

Major's chapter on Giovan Vincenzo Della Porta, the final chapter on Neapolitan collections before he moves on towards Rome, is so brief and so surprising that it is worth quoting in its entirety:

> But who Johannes Vincentius Porta in Naples was, or still is, and whether one counts him among the friends and descendants of the famous Johannes Baptista Porta or not, and how far or narrow his curiosity extend, I do not trust myself to report.
>
> However, because I found even just the name in the catalogue of the *Pinacotheken*, or art- and naturalia-chambers, of Italy, sent to me in October of last year by the excellent physician and polyhistor at Augsburg, Herr Doktor Georg Hieronymus Welsch, I cannot let

> his report entirely go by; perhaps in the near future some special notice could be found.[75]

Major's extremely limited information seems to come solely from a list of noteworthy Italian collections sent to him by the Augsburg physician Georg Hieronymus Welsch (Velschius, 1624–1677). Welsch had studied in Padua and had himself acquired a collection of medallions, antiquities and portraits of famous people.[76] Although Major calls Giambattista Della Porta (1535–1615) 'famous' in this chapter – he was indeed the most famous Neapolitan natural philosopher throughout Europe, as well as the most prolific writer – he does not discuss him and his collection, nor does he mention him elsewhere in the treatise. Most of the cited sources with which Major was directly acquainted do not mention Della Porta. But in the Lincean *Rerum medicarum novae Hispaniae thesaurus*, which he did know, he could have found Della Porta listed in the index and mentioned several times. Here he is described as 'a most learned man and most skilful in searching the mysteries of nature' ('doctissimus vir et in naturae arcanis perquisendis solertissimus') and 'that man whose writings already gained him the fame of immortality' ('vir ille, cui sua iam immortalitatis famam scripta conciliarunt'),[77] with references to some of his works (*Magia Naturalis*, *Taumatologiae*, *De Distillatione*). Likewise, in the *Musaeum metallicum* (which Major cites in his second Imperato chapter), Aldrovandi names Della Porta dozens of times, mostly with reference to the *Magia Naturalis*, calling him 'a most diligent investigator of natural things' ('diligentissimus rerum naturalium inuestigator').[78]

Major does not appear to know that Giovan Vincenzo Della Porta is Giovan Battista's older brother – not only a collector and antiquarian, but also a correspondent of Abraham Ortelius, *inter alia*.[79] Together, the brothers had a museum that was well known, at least in Italy, though not often referred to in print. In an unpublished comment, Aldrovandi praised Giovan Vincenzo for 'having through his own efforts created a most noble Museum to which scholars come from the farthest corners of Europe, drawn by its fame'.[80] However, Major was evidently aware of such a reputation in only the vaguest of terms.

## Conclusion

There had long been connections among intellectuals between Germany (or northern Europe more broadly) and Italy,[81] but Johann

Daniel Major's *Vorstellung etlicher Kunst- und Naturalien-Kammern in Italien zu Neapolis und Alt-Rom* demonstrates how tenuous this link was in the second half of the seventeenth century. To be sure, as Vera Keller has pointed out, 'Major's second- or third-hand accounts of global collections were hardly impressive examples of research',[82] and he could presumably have gleaned more information even with his resources at hand. His interest or knowledge did not extend to the flourishing academic culture in the Neapolitan primo seicento in which several of his subjects were embedded and which extended outward to encompass others, collectors included.[83] His accounts were also largely out of date: all the subjects of Major's treatise were dead by the time of his writing, and most of the books and personal accounts he cites dated from decades earlier. Major was enmeshed in a robust network of physicians and collectors of *naturalia* in northern Germany and its environs; there was – or, perhaps better, had been – an equally robust network of naturalists in Naples (with connections to Rome and elsewhere in Italy). However, the intersection of these two networks, separated by decades and the length of the European continent, was slight. In the late seventeenth century, many challenges to the transmission of knowledge in the republic of letters had not yet been overcome.

## Notes

1 'de gratia compatite con me perché qua in Napole mi par star nell'ultimo del mondo, ne so in questa professione con chi raggionare tanto havemo carestia de huomini che si dilettano de questa nostra professione', Olmi, '"Molti amici in varij luoghi"', 13; Findlen, *Possessing Nature*, 129.
2 Bartholin, *Epistolarum*, 201; quoted by Major, *Vorstellung*, n.p. (*c*.10, §.10).
3 For a reading of the frontispiece, see Findlen, *Possessing Nature*, 117.
4 On Major, see Reinke, *Der älteste Botanische Garten Kiels*, 8–19; Reinbacher, *Leben, Arbeit und Umwelt des Arztes*; Keller, 'Professionalizing Doubt'; Keller, 'Johann Daniel Major (1634–1693) and the experimental museum'; Keller, 'Vernacular knowledge, learned medicine and social technologies in the Leopoldina', as well as Keller's forthcoming monograph *Curating the Enlightenment: Johann Daniel Major and the experimental century* (Cambridge University Press).
5 Valentini, *Museum Museorum*, vol. 1. Major's treaties are paginated separately.
6 For Major's extensive bibliography, see Moller, *Cimbria Literata*, 2: 504–21. In his treatise on Neapolitan collections, he occasionally refers to his own. For example, in describing a shell that Fabio Colonna had seen in Ferrante Imperato's collection, he notes there was one as well 'auch in meinen Scriniolis' (Major, *Vorstellung*, n.p. (c. 10, §. 4)). For the purpose of his own collection, see Keller, 'Johann Daniel Major (1634–1693) and the experimental museum'. For the botanical garden, see Reinke, *Der älteste Botanische Garten Kiels*. For his collection, see Major, *Musei Cimbrici*.
7 At the beginning of his second chapter on Ferrante Imperato, Major, *Vorstellung*, n.p. (c. 10. §. 1), acknowledges his sources in general: 'Wir fahren weiter fort; und wollen erwegen, was andre glaubwürdige Scribenten mehr, von seiner, deß Imperati, so wol Person, als Raritäten-Kammer, für nachricht geben'.

8 Major, *Index Alphabeticus*; Keller, 'Johann Daniel Major (1634–1693) and the experimental museum', 463.

9 'die intention ist einen vollständigen Catalogum Rerum Naturalium Selectiorum, die so wol in Kayser- König- Cur- und Fürstlichen und anderer Fürnehmen Standes-Personen, als Republiquen, Klöstern, Kirchen, Academien, Bibliothecen, und Privatorum Häusern, zu finden sind', Major, *Index Alphabeticus*, n.p.

10 'Die Universal-reflexion mehr auf Naturalia (mögen preciosa oder minus preciosa seyn, gilt deßfalls gleich) als Artificialia, ut talia, zu nehmen. Von welchen Artificialibus demnach zu seiner Zeit nicht so sehr beobachtet werden wird, wie künstlich, als aus was vor Materie sie gemachet seyn', Major, *Index Alphabeticus*, n.p.

11 Major, *Vorstellung*, n.p. (c. 6, §. 1).

12 Major, *Vorstellung*, n.p. (c. 1, §. 1).

13 'gantz Italien in viel schöne gelehrte Genossenschafften, so daselst Academien genennet werden, vertheilt, also ausser denselben nicht minder zu Ihrer eigenen, als vieler Frembden und durchreisenden Gemüthß-Ergötzlichkeit, viel herrlicher Bibliotheken, Musea, und Gallerieen mit Büchern, Gemählden, Antiquitäten, Medaglien und Müntzen, Mathematischen Instrumenten, außerlesenen Naturäl- und Kunst-Sachen, aufs curioseste accommodirt, und in diesem Seculo vorauß, Lob-würdigst annoch zu samlen und halten pfleget', Major, *Vorstellung*, n.p. (c. 1, §. 7).

14 'So lasset uns demnach aus der wiewol Lustigen Landschafft des Neapolitanischen gebiethes, in die Stadt selbst hinein spatzieren, und, was zu unsren zeiten vorauß sich für Curiös- und galante, hoch- und mittlern Standes-Leuthe, als Besitzer fürnehmer Raritäten, in erwehnter Stadt köstlichen Häusern und Palaesten befunden, oder noch sich finden lassen, durch unverdrossne Schrifft, der macht der Vergessenheit entreissen', Major, *Vorstellung*, n.p. (c. 4, §. 1). On Major and the type of the 'gallant' scholar, see Keller, 'Johann Daniel Major (1634–1693) and the experimental museum', 460–1.

15 Major, *Vorstellung*, n.p.: 'Das XIII. Capitel. Von Alt-Rom, und etlichen Raritäten Käysers *Augusti*'; 'Das XIV. Capitel. Ob *Cicero* auf Raritäten was gehalten?'; 'Das XV. Capitel. Von den Vogel-Häusern *Luculli*, zu *Frascati*, ausserhalb Rom'.

16 On foreigners' visits to Naples and the view of Naples from abroad more generally, see De Seta, 'I viaggiatori stranieri' and Rak, 'L'immagine di Napoli'.

17 On Pflaumer, see Fischer, *Die literarische Tätigkeit des Johann Heinrich von Pflaumern*.

18 Pflaumer, *Mercurius Italicus* 2: 37, defines the gallery as what the Italians call a 'rarissimi artificij pretijq[ue] rerum receptaculum'. The Palazzo Reale, designed by Domenico Fontana and begun in 1600, would have been fairly recently constructed when Pflaumer visited; see De Cavi, 'Il palazzo reale di Napoli'.

19 'Allerhand Arthen von Armatur und Rüstung; viel alte Müntzen; eine Musicalische Uhr, oder Glock-Spiel; etliche Marmolsteinerne Tische, auf Mosaisch- oder Musivische Manier köstlich eingelegt; ja Statuae von Edelgestein', Major, *Vorstellung*, n.p. (c. 4, §. 3). Cf. Pflaumer, *Mercurius Italicus* 2: 37: 'Spectabis horologium rotularum motu variam harmoniam edens; mensas marmoreas eximiâ arte tessellatas; Statuas è pretiosis lapidibus fabrè fictas, numismata & arma omnis generis aliq[ue]'.

20 Major, *Vorstellung*, n.p. (c. 5, §. 1). Cf. Pflaumer, *Mercurius Italicus* 2: 59.

21 'Ferners ist deß Herzogs von Gravina prächtiger Palast wie auch deß Caraffae, wegen der alten marmolsteinern Statuen; sonderlich deß Scipionis Africani, wie Henznerus schreiber zu sehen: Item der Ursiner deß Fürsten von Bisignano, unnd anderer von denen unnd sonderlich deß Adriani de Guilielmo Hauß Schraderus fol. 223. b. & fol. 247. seqq. zu lesen', Zeiler, *Itinerarium Italiae Nov-Antiquae*, 165, cols 1–2. Carafa's sculpture collecting receives passing notice in Dodero, *Ancient Marbles in Naples*, 237.

22 Major, *Vorstellung*, n.p. (c. 5, §. 2). Cf. Bartholin, *De Vnicornv observationes Nouae*, 49: 'Genus hoc spectaculi splendida *D. Tiberij Caraffae* Bisidiani Principis, nobilitate generis & Rerum naturalium amore comme[n]datissimi, aula Neapoli exhibit'.

23 E.g. Bartholin, *Epistolarum*, 206.

24 For the Linceans and Naples, see Olmi, 'La colonia lincea di Napoli' and Speciale, 'La ricerca botanica dei Lincei a Napoli'.

25 Major, *Vorstellung*, n.p. (c. 6, §. 8). Cf. Recchi et al., *Rerum medicarum novae Hispaniae thesaurus*, 550: 'Dom. Fabius Columna Lynceus, sedulus maxime rerum Naturalium Scrutator; qui non modo in Juris scientia multum pollet: sed in Mathesi (Optica praecipue) ac Plantarum

Animaliumque Cognitione, Neapoli nunc ab omnibus, ceu Oraculum consulitur'. Major, *Vorstellung*, n.p. (c. 6, §. 7) commented on Colonna's contributions to the volume: 'Endlich finde ich in denen, Anno 1651. zu Rom gedruckten, Mexicanischen Medicinal-Beschreibungen Nardi Antonii Recchi, einige Annotationes und lehr-reiche Zusätze, die mehr-gedachter Herr Columna von vielerley mehrern Erfahrenheiten darzu gethan, und auch deßfals nicht ermüden mögen, seine in dem Natur-Studio ungemeine Curiosität, mehr und mehr an Tag zu geben'.

26 Colonna, *Opusculum de purpura*; the *vita* is the first *annotatio* (following and paginated separately from Colonna's text).

27 'Und zwar solches alles aus eigener Erfahrung und unverdrossenem Fleiß, so gar, daß Er auch kein bedencken getragen, Pinsel und Stech-Eisen, nebst der Feder, selbst in die Hand zunehmen, und unterschiedene feine Figuren, in Kupfer-stich zu bringen', Major, *Vorstellung*, n.p. (c. 6, §. 3). The question of Colonna's authorship of the etchings is vexed, see Tognoni, 'Nature described'. The ability to draw was part of Major's concept of the ideal, hands-on naturalist; Keller, 'Johann Daniel Major (1634–1693) and the experimental museum', 462.

28 'also hat Er, nebst dieser, noch viel andre, einigen Purpur-Safft gebende, Muscheln und Schnecken mehr, in verwahrung gehabt, und derer Abrisse zugleich beygefüget; die ich aber, Zeit, Papier, und Unkosten zu sparen, alhier nicht wiederholden mag', Major, *Vorstellung*, n.p. (c. 6, §. 6).

29 'daß Er so viel artige Cörper muß haben selbst in Originali gehabt, und zu sein- und anderer nützlicher Belustigung, in einem absonderlichen logiment aufgehoben', Major, *Vorstellung*, n.p. (c. 6, §. 4).

30 Major, *Vorstellung*, n.p. (c. 6, §. 5). For example, the '*Buccinum Lapideum laeve*, oder steinern glattes Düt-Horn; (eine Arth von Schnecken)'; Cf. Colonna, *Purpura*, lv: 'Neapoli frequens in scopulis'.

31 Cf. Colonna, *Aquatilium et terrestrium aliquot animalium observationes* in *Purpura*, lxii: 'Neapoli ab amico Fontium constructore dono habuimus quindecim ab hinc annis'.

32 'und dergleichen Dinge mehr zum Angencken hat theils beschrieben, theils Copeylich zu gleich vorgestellt, und selbst-ständig in seinem Cabinett aufgehoben', Major, *Vorstellung*, n.p. (c. 7, §. 5).

33 Findlen, *Possessing Nature*, 226, 229, 233.

34 See note 35 below.

35 'viel andre rare Sachen mehr, die so wol die Natur, als Kunst hervor gebracht gehabt', Major, *Vorstellung*, n.p. (c. 7, §. 2).

36 'Clar[us] Vir, Vincentius Cioffius, Neapolitanus, instructum Cimeliis Antiquitatis Museum habuit: é quo transmisit ad me plusculas icones antiquiorum Lucernarum', Major, *Vorstellung*, n.p. (c. 7, §. 3); Licetus, *De Lvcernis antiqvorvm reconditis*, col. 806.

37 Licetus, *De Lvcernis antiqvorvm reconditis*, col. 814.

38 Licetus, *De Lvcernis antiqvorvm reconditis*, cols. 806–49, 934–50, 954–5. For brief references to Cioffi's collection, see also Radcliffe, 'Ricciana', 423; Vaiani, 'Alle origini della ricerca sulle lucerne antiche', 22, 28.

39 Gabrieli, 'Il "Liceo" di Napoli'; Ben-Zaken, *Cross-Cultural Scientific Exchanges*, 48, 55–6, 65–75. See also Speciale, 'La ricerca botanica dei Lincei a Napoli', 74, 75 and Clifton, 'Mattia Preti's *Madonna of Constantinople*', 337.

40 Della Valle, *Viaggi*, title page.

41 'La voglia, in me, di dar ragguaglio a V. S. di queste cose strane, hà ecceduto, come penso, la curiosità, che può essere in lei, di sentirle raccontar tanto alla distesa, per non dir tediosamente', Della Valle, *Viaggi*, 140.

42 Carlo Celano (1617–1693), *Notizie del bello*, 1714, extolled Schipani's modesty and capabilities in 'ogni scienza', noting also that in his home he retained 'una erudita libreria, e fra greci ed arabi'; quoted in Clifton, 'Mattia Preti's *Madonna of Constantinople*', 337.

43 Noted by Keller, 'Vernacular knowledge, learned medicine, and social technologies in the Leopoldina', 145.

44 Major, 'Vorstellung etlicher Kunst- und Naturalien-Kammern in Africa und an Gräntzen Europa', 60.

45 Major, *Vorstellung*, n.p. (c. 8, §. 3); cf. Bartholin, *De Vnicornv observationes Nouae*, 49.

46 Major, *Vorstellung*, n.p. (c. 8, §. 3–4); cf. Bartholin, *Epistolarum medicinalium*, 202.

47 Sachs von Lewenheimb, *Gammarologia*, 52; Major, *Vorstellung*, n.p. (c. 8, §. 4).
48 Bartholin, *De Vnicornv observationes Nouae*, 49: 'quum veras Meleagrides apud *Marium Schipanum* ... videre licuerit'; Major, *Vorstellung*, n.p. (c. 8, §. 4).
49 Aldrovandi, *Ornithologiae tomvs alter*, 337–8 (*Gallina guinea*); Marggraf, *Historiae Rervm Natvralivm Brasiliae*, 192 (*Gallina Africana*); Jonstonus, *Historiae Naturalis de Auibus*, plate 57 (*Gallina Africana*); Worm, *Museum Wormianum*, 297 (*Gallina guineae*); Olearius, *Gottorffische Kunst-Kammer*, pl. XV, fig. 3 (which, Major notes, is not in accord with the other depictions). Major also cites Claudius Salmasius's commentary on Suetonius's *vita* of Caligula.
50 'Gedachten Francisci Natural-Discurs, wie auch Ferrandis Imperati Historiam Nat. habe ich biß dato nicht habhafft werden können. Deßwegen muß mich nochmals mit einer kleinen allegation aus andern behelffen', Major, *Vorstellung*, n.p. (c. 10, §. 12).
51 Ferrante Imperato, *Dell'Historia Natvrale*; Ferrante Imperato, *Historia Natvrale*; Francesco Imperato, *Discorsi intorno d diverse cose naturali*. Major does not mention Francesco Imperato's *De Fossilibus opusculum* (1610), though he would have been attracted to the assertion on the title page that the illustrations were 'faithfully drawn from life'.
52 'Unter privat-Personen aber zu Neapolis hat es keiner an Curiosität und glücklichem Fleiß, allerhand schöne Natural-Raritäten in eine Behausung zu bringen, dem Ferrandes Imperatus zuvor- oder nach-gethan: welcher ist gewesen einer (und zwar gelehrter) von den fürnehmsten Aromatariis oder Materialisten selbigen Orths', Major, *Vorstellung*, n.p. (c. 9, §. 1); Major paraphrases, in part, Pflaumer, *Mercurius Italicus* 2: 59.
53 Major, *Vorstellung*, n.p. (c. 10, §. 10); Bartholin, *Epistolarum medicinalium*, 201; the implicit reference is to Luke 16: 19–21.
54 Zeiler, *Itinerarium Italiae Nov-Antiquae*, 165, col. 2.
55 This *byssus* was sometimes purported to be mermaid hair; see Keller, 'Vernacular knowledge, learned medicine and social technologies in the Leopoldina', 139–40.
56 Major, *Vorstellung*, n.p. (c. 9, §. 2): 'Eine arth von geschwäntzten Affen und mit Menschen-Gesicht'; this from 'Animal simie magnitudine, forma humana, caudatum, oculis nigris & lucidis' (Pflaumer, *Mercurius Italicus* 2: 60).
57 On Schellhammer, see Keller, 'Professionalising doubt'.
58 Lachmund, *De Ave Diomedea dissertatio*; the title page indicates that the bird specimen was in Lachmund's own 'museum'. Also listed here is a bird of paradise, which Pflaumer asserts is footless, but Major points out that this common opinion has been proven false; Major, *Vorstellung*, n.p. (c. 9, §. 2).
59 Cardano, *De Svbtilitate libri XXI*, 63, 362 (Book 1, Book 7); Cardano, *The* De Subtilitate *of Girolamo Cardano*, 26–7, 393.
60 Colonna, *Opusculum de purpura*, 33 (c. 12, §. 5): 'apud doctissimum Imperatum nostrum in suo Museo, rerum omnium Naturae satis copioso Thesauro'; 31 (c. 11, §. 4): 'apud Imperati Museum amplissimum'; 36 (c. 15, §. 2): 'ex ditissimo naturae promptuario Imperati'. Cf. Major, *Vorstellung*, n.p. (c. 10, §. 2): 'nennt Er ... den Imperatum einen hochgelehrten Mann, und gedachtes Museum einen gnungsam-erfülten Schatz aller natürlichen Dinge'; 'ein Amplissimum oder weitläufftiges Museum'; 'einen reichsten Vorrath (Promtuarium) der Natur'.
61 Colonna, *Opusculum de purpura*, 38 (c. 16, §. 5): 'ditissimus ille naturalium rerum promus noster Imperatus'; cf. Major, *Vorstellung*, n.p. (c. 10, §. 3): 'einen reichesten Promum, oder hervor-geber natürlicher Dinge'.
62 'einen in Botanicis, oder der Kräuter, erfahrensten Mann, und sein Museum, ein mit fremdem und gantz wundersamen Vorrath, so wol Natural- als Medicinalischer Dinge, außgerüstetes Museum; in welchem Er unterschiedene Herbaria viva oder Bücher, darinnen die trocknen Kräuter in Originali eingeleimt, gesehen habe', Major, *Vorstellung*, n.p. (c. 10, §. 6); cf. Ferrari, *De florum cultura*, 437–8. Major, *Vorstellung*, n.p. (c. 10, §. 9) reports that Johann Vesling, *De plantis Aegiptiis observationes et notae*, 34, praised Imperato in similar terms.
63 Major, *Vorstellung*, n.p. (c. 10, §. 7); cf. Kircher, *Mundus subterraneus* 2: 39–40.
64 Major, *Vorstellung*, n.p. (c. 10, §. 8); cf. Aldrovandi, *Musaeum metallicum*, 825 ('maxilla Elephanti petrificata'). Major here interpolates an unusual source (which I have not been able to identify), a description in Italian 'in einem geschriebenen kurtzen Reise-Tractätlein, weiß nicht wessen Autoris': 'Il Studio di Ferrando Imperato, pieno di Rarità, cioè Cocodrilli, Serpenti, Osse di Giganti, Fonghi impetriti, Vitelle di Mare, Pellicani, Remolo, picciolo pesce,

che fà fermare i Vascelli in Mare, e diverse altre Cose curiose'. He feels compelled to explain the use of the term 'studio' here. Major employed a great variety of terms for collections and the places where they were stored, exhibited and used. He examined these in two relatively lengthy chapters of the *Unvorgreiffliches Bedencken von Kunst- und Naturalien-Kammern ins gemein*, n.p. (c. 4–5). These chapters include a particularly long term in Greek invented by his friend, Philipp Jakob Sachs von Lewenheimb, and which, Major playfully asserts, is ten ells long, if not half the diameter of the Earth, and could not be pronounced before two soups had grown cold (c. 4. §. 5). On the nomenclature of early modern collections, see Findlen, 'The museum'; Findlen, *Possessing Nature*, 48–50.

65 Major, *Vorstellung*, n.p. (c. 10, §. 10); Bartholin, *De Vnicornv observationes Nouae*, 277. Bartholin (but not Major) reports that the specimen was indicated by both Ferrante Imperato, *Dell'Historia Natvrale*, 690–1 (as 'Unicorno fossile'), and Francesco Imperato, *Discorsi intorno a diverse cose natvrali*, 3.

66 Major, *Vorstellung*, n.p. (c. 10, §. 12) refers to articles in the *Miscellanea* of the Academia Naturae Curiosorum in Halle. Neither author had seen the objects in person: Volgnad, 'De globis vitulinis', 400, cites Imperato, *Dell'Historia Natvrale*, while Schultz, 'De Ovo gallinaceo serpentifero', 308, cites Aldrovandi's description of an Imperato object. On the journal, founded by Sachs von Lewenheimb in 1670, and the Academia, of which Major was a member from 1664, see Keller, 'Professionalising doubt'; Keller, "Vernacular knowledge, learned medicine, and social technologies in the Leopoldina'; Kramer, 'The *Curiosi* as *Collectores*'.

67 'ein wackerer, Curiöser, gelehrt- und galant Mann, und sein Raritäten-Hauß anders nicht, als ein compendiös-eingezogenes Hauß der gantzen wunderbarlichen Natur gewesen, auch nachgehends geblieben, und mit Täglichen schönen Zusätzen ohn allen Zweiffel von vorhin-gedachtem seinem Sohn Francisco, vermehret worden ist', Major, *Vorstellung*, n.p. (c. 10, §. 13).

68 On D'Eremita, see Gianfrancesco, 'From propaganda to science', 16–17; Gianfrancesco, 'Books, gold, and elixir'.

69 'Exstat quoque, schreibt Er, Muséum F. Donati Eremitae, Pharmacapolae olim S. Catharinae Monachorum, qui de arte Pharmaceuticâ librum edidit, & de Elixire vitae; quorum tamen operum autor est P. Castellus, qui apud eum diverterat .... In hujus Muséo varia exponuntur Sceleta variorum animalium', Major, *Vorstellung*, n.p. (c. 11, §. 1); cf. Bartholin, *Epistolarum medicinalium*, 202. Major also cites the '3. oder 4. Worthen' from Sachs von Lewenheimb, *Gammarologia*, 52, but notes that they came from Bartholin.

70 Bartholin's charge that Pietro Castelli was responsible for writing D'Eremita's books is unsubstantiated. See Clericuzio, 'Chemical medicine and Paracelsianism in Italy', 70, n. 52.

71 On this plate, see Gianfrancesco, 'Books, gold, and elixir', 263–4.

72 'Al Mvseo del P. F. Donato D'Eremita Dell'Ord. de Pred. nel Monasterio d S. Chatarina a Formello', in D'Eremita, *Antidotario*, n.p.: 'Ciò che medica penna araba, e greca / In carte impresse di Natura, e d'Arte / Quì Donato riserba ... / Et se'i fiori, l'herbette, & gli arboscelli, / I metalli, le gioie, & le miniere / Hebber nel lor terren ninfa amorosa'.

73 For example, Recchi et al., *Rerum medicarum novae Hispaniae thesaurus*, 787, 866 and 890.

74 'Neapoli hodie Pharmacopolium S. Catharinae adjunctum habet Museum rariorum Naturalium, non poenitenda diligentia collectum. Ubi visus mihi est Infans biceps, Vitulus biceps. Visus & Infans quatuor pedum, ab uno pectore descendentium', Major, *Vorstellung*, n.p. (c. 11, §. 3). Shortly after the publication of Major's treatise, another brief description of the Santa Caterina a Formello collection was published in Antonio Farina, *Compendio delle Cose piu Curiose di Napoli, e di Pozzuoli, con alcune notitie del Regno*, 36: 'noto solo una bellissima Galleria, ove sono molte curiose antichità, e la Spetiaria copiosa altresì di curiosità, frà le quali veggonsi molti Mostri naturali, e due radici di Mandragora maschio, e femina, degne d'esser vedute' ('I note only a very beautiful gallery, where there are many curious antiquities, and an apothecary shop full of curiosities, among which one can see many natural monstrosities, and two mandrake roots, one male and one female, which are worth seeing'). Quoted and translated in part by Gianfrancesco, 'Books, gold and elixir', 255.

75 'Wer aber Joh. Vincentius Porta zu Neapolis gewesen, oder noch sey, und ob man denselben zu den Freunden und Nachkommen deß berühmten Joh. Baptistae Portae zu rechnen habe, oder nicht, und wie weit oder eng sich seine Curiosität erstrecktet, getraue ich mir nicht zu

melden. Jedoch, weil ich, wiewol nur den blossen Nahmen, in deme von dem fürtrefflichen Medico und Polyhistore zu Augspurg, Hn. D. Georg. Hieron. Velschio, am Monath Octobr. vorigen Jahrs mir-übersendeten Catalogo, dero Ihm bekanten Pinacotheken, oder Kunst- und Naturalien-Gemächer Italiens, angetroffen; so mag ich auch dessen Meldung nicht gäntzlich vorbey lassen gehen; vieleicht möchte sich nechst-künfftig was specialere Nachricht finden', Major, *Vorstellung*, n.p. (c. 12, §. 1–2).

76 'Deßgleichen meritiret auch des Hrn. D. Velschens Cabinet besehen zu werden, welches sonderlich in Medaillen und Antiquitäten, desgleichen in Bildnissen berühmter Leute vielen andern weit zuvorgehet', Jencquel, *Museographia*, 24.

77 Recchi et al., *Rerum medicarum novae Hispaniae thesaurus*, 75, 496. In a funny story, told by Federico Cesi in Naples, about a request to Della Porta for medicine for a woman in labour, Della Porta rose from his bed in the morning and went into the museum ('mane e lecto in Musaeum se recipit Porta'), but no details on the museum are offered.

78 Aldrovandi, *Musaeum metallicum*, 555.

79 Jencquel, *Museographia*, 205 seems to have taken his list of *Raritäten-Kammern* directly from Major and knows nothing more of Giovan Vincenzo Della Porta: 'zu welchen [Raritäten] auch Johann. Vincentii Portae Curiositäten gehören. Wovon man aber keine besondere Particularia zu sagen weiß'. For Giovan Vincenzo's correspondence with Ortelius, through whom he obtained Dürer prints, see Meganck, *Erudite Eyes*, 15, 160, 198 and 202.

80 Quoted and translated by Findlen, *Possessing Nature*, 113. The partial contents of the Della Porta museum are known through an inventory published by Fulco, 'Per il "museo" dei fratelli Della Porta'.

81 Olmi, '"Molti amici in varij luoghi"', 7–8.

82 Keller, 'Johann Daniel Major (1634–1693) and the experimental museum', 463.

83 On the academic culture in Naples, see especially Gianfrancesco, 'Accademie, scienze e celebrazioni a Napoli nel primo Seicento'; Gianfrancesco, 'From propaganda to science'; Cocco, 'Locating the natural sciences in early modern Naples'. Dedicatory compositions in Donato D'Eremita's *Dell'Elixir Vitae* were written by members of three different academies (Gianfrancesco, 'From propaganda to science', 16). Tiberio Carafa, whose collection Major attempts to describe, was a member of both the Oziosi and the Infuriati academies.

## Bibliography


Aldrovandi, Ulisse. *Ornithologiae tomvs alter*. Bologna, 1600.

Aldrovandi, Ulisse. *Mvsaevm metallicvm in libros IIII distribvtvm*. Bologna, 1648.

Bartholin, Thomas. *De Vnicornv observationes Nouae*. Padua, 1645.

Bartholin, Thomas. *Epistolarum medicinalium a doctis vel ad doctos scriptarum, centuria 1–4*. Copenhagen, 1663.

Ben-Zaken, Avner. *Cross-Cultural Scientific Exchanges in the Eastern Mediterranean, 1560–1660*. Baltimore: The Johns Hopkins University Press, 2010.

Cardano, Girolamo. *De Svbtilitate libri XXI*. Basel, 1582.

Cardono, Girolamo. *The* De Subtilitate *of Girolamo Cardano*. Edited and translated by John M. Forrester. Tempe: Arizona Center for Medieval and Renaissance Studies (ACMRS), 2013.

Celano, Carlo. *Notizie del bello dell'antico e del curioso della Città di Napoli*. Atanasio Mozzillo, et al. (eds). Naples: Per le Edizioni scientifiche italiane, 1970.

Clericuzio, Antonio. 'Chemical medicine and Paracelsianism in Italy, 1550–1650'. In *The Practice of Reform in Health, Medicine and Science, 1500–2000*, edited by Margaret Pelling and Scott Mandelbrote, 59–79. Aldershot: Ashgate, 2005.

Clifton, James. 'Mattia Preti's *Madonna of Constantinople* and a Marian cult in seventeenth-century Naples'. In *Parthenope's Splendor: Art of the Golden Age in Naples*, edited by Jeanne Chenault Porter and Susan Scott Munshower, 337–63. Papers in Art History from The Pennsylvania State University 7. University Park, PA: Department of Art History, The Pennsylvania State University, 1993.

Cocco, Sean. 'Locating the natural sciences in early modern Naples'. In *A Companion to Early Modern Naples*, edited by Tommaso Astarita, 453–75. Leiden: Brill, 2013.

Colonna, Fabio. *Φυτοβασανος [Phytobasanos] sive plantarvm aliqvot historia in qva describvntur diversi generis plantae veriores, ac magis facie, viribusque respondentes antiquorum Theophrasti, Dioscoridis, Plinij, Galeni, aliorumque delineationibus, ab alijs hucusque non animaduersae.* Naples, 1592.
Colonna, Fabio. *Minus cognitarum Stirpium ac etiam rariorum nostro coelo orientium Stirpium Εκφρασιζ [Ekphrasis] … Item de Aquatilibus aliisque animalibus quibusdam libellus.* Rome, 1606.
Colonna, Fabio. *Purpura. Hoc est de Purpura ab Animale testaceo fusa, de hoc ipso Animali alijsq[ue] rarioribus testaceis quibusdam.* Rome, 1616.
Colonna, Fabio. *Opusculum de purpura*, edited by Johann Daniel Major. Kiel: Joachim Reumann, 1675.
De Cavi, Sabina. 'Il palazzo reale di Napoli (1600–1607): un edificio "spagnolo"?'. In *'Napoli è tutto il mondo': Neapolitan art and culture from humanism to the Englightenment, International Conference, Rome, June 19–21 2003*, edited by Livio Pestilli, Ingrid D. Rowland and Sebastian Schütze, 147–70. Pisa and Rome: Fabrizio Serra editore, 2008.
Della Valle, Pietro. *Viaggi.* Rome, 1650.
D'Eremita, Donato. *Dell'Elixir vitae libri qvattro.* Naples, 1624.
D'Eremita, Donato. *Antidotario, di Fra Donato d'Eremita dell Ordine de Predicatori, nel quali si discorre intorno all'osseruanza, che deue tenere lo speciale nell'elegere, preparare, componere, & co[n]seruare i medicamenti semplici, e composti.* Naples, 1639.
De Seta, Cesare. 'I viaggiatori stranieri e l'immagine di Napoli nel Seicento'. In *'Napoli è tutto il mondo': Neapolitan art and culture from humanism to the Englightenment, International Conference, Rome, June 19–21 2003*, edited by Livio Pestilli, Ingrid D. Rowland and Sebastian Schütze, 203–26. Pisa and Rome: Fabrizio Serra editore, 2008.
Dodero, Eloisa. *Ancient Marbles in Naples in the Eighteenth Century: Findings, collections, dispersals.* Studies in the History of Collecting & Art Markets 7. Leiden; Boston: Brill, 2019.
Farina, Antonio. *Compendio delle Cose piu Curiose di Napoli, e di Pozzuoli, con alcune notitie del Regno.* Naples, 1679.
Ferrari, Giovanni Battista. *De florvm cvltvra libri IV.* Rome, 1633.
Findlen, Paula. 'The Museum: Its classical etymology and Renaissance genealogy', *Journal of the History of Collections* 1 (1989): 59–78.
Findlen, Paula. *Possessing Nature: Museums, collecting, and scientific culture in early modern Italy.* Berkeley, Los Angeles; London: University of California Press, 1994.
Fischer, Alois. *Die literarische Tätigkeit des Johann Heinrich von Pflaumern 1584–1671, Doktors beider Rechte, Anwalts, Kaiserlichen Rats und Bürgermeisters der freien Reichsstadt Überlingen am Bodensee.* Bonn: P. Hauptmann'sche Buchdruckerei, 1909.
Fulco, Giorgio. 'Per il "museo" dei fratelli Della Porta'. In *Rinascimento Meridionale e altri studi in onore di Mario Santoro*, edited by Maria Cristina Cafisse et al., 105–75. Naples: Società Editrice Napoletana, 1987.
Gabrieli, Giuseppe. 'Il "Liceo" di Napoli', *Rendiconti della Classe di Scienze morali, storiche e filologiche. Accademia Nazionale dei Lincei*, 6th ser., 14 (1938): 535–41.
Gianfrancesco, Lorenza. 'Accademie, scienze e celebrazioni a Napoli nel primo Seicento', *Quaderni di Simbolon* 5 (2010): 175–209.
Gianfrancesco, Lorenza. 'From propaganda to science: Looking at the world of academies in early seventeenth-century Naples', *California Italian Studies* 3 (2012): 1–31.
Gianfrancesco, Lorenza. 'Books, gold, and elixir: Alchemy and religious orders in early modern Naples', *Ambix* 65 (2018): 250–74.
Imperato, Ferrante. *Dell'Historia Natvrale … Libri XXVIII nella qvale ordinatamente si tratta della diuersa condition di miniere, e pietre. Con alcune historie di Piante, & Animale, sin'hora non date in luce.* Naples, 1599.
Imperato, Francesco. *De Fossilibvs opvscvlvm … In quo miro ordine continentur Naturalis disciplinae scitu dignissima, eiusque Professoribus omnino necessaria; ab alijs minmie excogitata. Accesserunt. Multa, quae Hieroglyphicè per Fossilia nouiter interpetramur [sic]; non nullaeque Icones fideliter ad viuum delineate.* Naples, 1610.
Imperato, Francesco. *Discorsi intorno a diverse cose natvrali … opera non meno cvriosa, che vtile, e necessaria à Professori della natural Fiosofia.* Naples, 1628.
Imperato, Ferrante. *Historia Natvrale … nella qvale ordinatamente si tratta della diuersa condition di Minere, Pietre pretiose, & altre curiosità. Con varie Historie di Piante, & Animali, sin'hora non date in luce.* 2nd ed. Venice, 1672.

Jencquel, Kaspar Friedrich [alias Caspar Friedrich Neickelius]. *Museographia oder Anleitung zum rechten Begriff und nützlicher Anlegung der Museorvm, oder Raritäten-Kammern*. Leipzig; Breslau, 1727.

Jonstonus, Johannes. *Historiae Naturalis de Auibus Libri VI*. Frankfurt am Main,1650.

Keller, Vera. 'Professionalising doubt: Johann Daniel Major's observation "On the horn of the bezoardic goat", curiosity collecting, and periodical publication'. In *The Institutionalization of Science in Early Modern Europe*, edited by Giulia Giannini and Mordechai Feingold, 199–235. Scientific and Learned Cultures and Their Institutions 27. Leiden; Boston: Brill, 2020.

Keller, Vera. 'Johann Daniel Major (1634–1693) and the experimental museum', *Journal of the History of Collections* 33 (2021): 459–68.

Keller, Vera. 'Vernacular knowledge, learned medicine and social technologies in the Leopoldina, 1670–1700, or, How to publish on sirens, dragons and basilisks'. In *Collective Wisdom: Collecting in the early modern academy*, edited by Anna Marie Roos and Vera Keller, 127–54. Turnhout: Brepols, 2022.

Kircher, Athanasius. *Mundus subterraneus, in XII Libros digestus …* 2 vols. Amsterdam, 1665.

Kramer, Fabian. 'The *Curiosi* as *Collectores*: The publications of the Academia naturae Curiosorum, c.1652–1706'. In *Collective Wisdom: Collecting in the early modern academy*, edited by Anna Marie Roos and Vera Keller, 107–26. Turnhout: Brepols, 2022.

Lachmund, Friedrich. *De Ave Diomedea dissertatio, cum vera eijs effigie ari incisa, ex Museo D. Friderici Lachmund Hildesheim*. Amsterdam, 1674.

Licetus, Fortunius. *De Lvcernis antiqvorvm reconditis libb. sex …* Udine, 1652.

Major, Johann Daniel. *Index Alphabeticus, aliter in posterum (iuxta Situs loterum) disponendus, deutende unterdeß auf uhr-alt und neue, groß- und kleine, vollkommen- oder erst angefangene Kunst-Antiquitäten-Schatz- und fürnehmlich Naturalien-Kammern, Conclavia, Musea, Repositoria, oder auch nur kleinere Scrinia, Rerum Naturalium Selectiorum, bey Stands- und Privat-Personen, inn- und außerhalb Europae*. n.p. 1674.

Major, Johann Daniel. *Unvorgreiffliches Bedencken von Kunst- und Naturalien-Kammern ins gemein*. Kiel, 1674.

Major, Johann Daniel. *Vorstellung etlicher Kunst- und Naturalien-Kammern in Italien zu Neapolis und Alt-Rom*. Kiel, 1675.

Major, Johann Daniel. *Musei Cimbrici cum contentis in eo Rebus selectioribus, privatim declarandi Aditus II*. Kiel, 1689.

Major, Johann Daniel. 'Vorstellung etlicher Kunst- und Naturalien-Kammern in Africa und an Gräntzen Europa.' In *Museum Museorum, oder vollstëndige Schu-Bühne aller Materialien und Specereijen nebst deren natürlichen Beschreibun, Election, Nutzen und Gebrauch, aus andern Material- Kunst-und Naturalien-Kammern, Oost- und West-Indischen Reisz-Beschreibungen, Curiosen Zeit- und Tag-registern, Natur- und Artzney-Künigern, wie auch selbst-eigenen Erfahrung, zum Vorschub der Studirenden Jugend, Materialisten, Apothecker und deren Visitatoren, Wie auch anderer Künstler, als Jubelirer, Mahler, Färber, u.s.w.*, edited by Michael Bernhard Valentini, vol. 1, 40–60. Frankfurt am Main, 1704.

Marggraf, Georg. *Historiae Rervm Natvralivm Brasiliae, libri octo*. In *Historia Naturalis Brasiliae … In qua non tantum Plantae et Animalia, sed et Indigenarum morbi, ingenia et mores describuntur et Iconibus supra quingentas illustrantur*. Leiden; Amsterdam, 1648.

Meganck, Tine Luk. *Erudite Eyes: Friendship, art and erudition in the network of Abraham Ortelius (1527–1598)*. Studies in Netherlandish Art and Cultural History 14. Leiden; Boston: Brill, 2017.

Moller, Johann. *Cimbria Literata, sive scriptorum ducatus utriusqve Slesvicensis et Holsatici, qvibus et alii vicini qvidam accensentur*. 3 vols. Copenhagen, 1744.

Olearius, Adam. *Gottorffische Kunst-Kammer, worinnen allerhand ungemeine Sachen, sotheils die Natur, theils künstliche Hände hervor gebracht und bereitet. Vor diesem aus allen vier Theilen der Welt zusammen getragen, und vor einigen Jahren beschrieben, auch mit behörigen Kupffern gezieret*. Schleswig, 1674.

Olmi, Giuseppe. 'La colonia lincea di Napoli'. In *Galileo e Napoli*, edited by Fabrizio Lomonaco and Maurizio Torrini, 23–58. *Acta Neapolitana* 7. Naples: Guida editori, 1987.

Olmi, Giuseppe. '"Molti amici in varij luoghi": Studio della natura e rapporti epistolari nel secolo XVI', *Nuncius* 6 (1991): 3–31.

Pflaumer, Johann Heinrich von. *Mercurius Italicus, hospiti fidus per Italiae praecipuas regiones & vrbes dux, indicans explicans quaecumque in iis sunt visu ac scitu digna*. 2nd ed. Augsburg, 1650.

Radcliffe, Anthony. 'Ricciana', *Burlington Magazine* 124 (1982): 412–21; 423–4.

Rak, Michele. 'L'immagine di Napoli nel Seicento europeo'. In *'Napoli è tutto il mondo': Neapolitan art and culture from humanism to the Englightenment, International Conference, Rome, 19–21 June 2003*, edited by Livio Pestilli, Ingrid D. Rowland and Sebastian Schütze, 271–94. Pisa; Rome: Fabrizio Serra editore, 2008.

Recchi, Nardo Antonio, et al. *Rerum medicarum novae Hispaniae thesaurus, seu plantarum animalium, mineralium mexicanorum historia, ex Fr. hernandez … relationibus*. Rome, 1651.

Reinbacher, W. Rudolph. *Leben, Arbeit und Umwelt des Arztes: Johann Daniel Major (1634–1693), Eine Biographie aus dem 17. Jahrhundert, mit neuen Erkenntnissen*. Linsengericht: M. Kroeber, 1998.

Reinke, J. *Der älteste Botanische Garten Kiels. Urkundliche Darstellung der Begründung eines Universitäts-Instituts im siebzehnten Jahrhundert*. Kiel: Kommissions-Verlag der Universität Kiel, Lipsius & Tischer, 1912.

Sachs von Lewenheimb, Philipp Jacob. *Gammarologia sive gammarorum vulgo cancrorum consideratio physico-philologico-historico-medico-chymica: in qua, praeter gammarorum singularem naturam, indolem & multivarium usum non minus reliquorum crustatorum institutur tractatio; ad normam Collegii Naturae …* Frankfurt, 1665.

Schultz, Simon. 'De Ovo gallinaceo serpentifero', *Miscellanea curiosa medico-physica Academiae Naturae Curiosorum sive Ephemeridum medico-physicarum Gemanicarum* 3 (1672): 303–10.

Speciale, Gabriella Belloni. 'La ricerca botanica dei Lincei a Napoli: Corrispondenti e luoghi'. In *Galileo e Napoli*, edited by Fabrizio Lomonaco and Maurizio Torrini, 58–79. *Acta Neapolitana* 7. Naples: Guida editori, 1987.

Tognoni, Federico. 'Nature described: Fabio Colonna and natural history illustration', *Nuncius* 20 (2005): 347–70.

Vaiani, Elena. 'Alle origini della ricerca sulle lucerne antiche: Il Seicento (1621–1691)'. In *Lumina: Convegno Internazionale di Studi, Urbino 5–7 giugno 2013*, edited by Maria Elisa Micheli and Anna Santucci, 11–32. Pisa: Edizioni ETS, 2015.

Valentini, Michael Bernhard. *Museum Museorum, oder vollstëndige Schu-Bühne aller Materialien und Specereijen nebst deren natürlichen Beschreibun, Election, Nutzen und Gebrauch, aus andern Material- Kunst-und Naturalien-Kammern, Oost- und West-Indischen Reisz-Beschreibungen, Curiosen Zeit- und Tag-registern, Natur- und Artzney-Künigern, wie auch selbsteigenen Erfahrung, zum Vorschub der Studirenden Jugend, Materialisten, Apothecker und deren Visitatoren, Wie auch anderer Künstler, als Jubelirer, Mahler, Färber, u.s.w.* 3 vols. Frankfurt am Main, 1704.

Vesling, Johann. *De plantis Aegiptiis observationes et notae*. Padua, 1638.

Volgnad, Heinrich. 'De globis vitulinis', *Miscellanea curiosa medico-physica Academiae Naturae Curiosorum sive Ephemeridum medico-physicarum Gemanicarum Curiosarum* 2 (1671): 398–401.

Worm, Ole. *Museum Wormianum, seu Historia rerum rariorum, tam Naturalium, quam Artificialium, tam Domesticarum, quam Exoticarum, quae Hafniae Danorum in aedibus Authoris servantur*. Amsterdam, 1655.

Zeiller, Martin. *Itinerarium Italiae Nov-Antiquae: Oder, Raiß-Beschreibung durch Italien: Darinn Nicht allein viel vnderschiedliche Weg durch das Welschland selbsten, vnd in dasselbe … verzeichnet, vnd das alte Italien mit dem jetzigen Welschland verglichen … Deßgleichen allerhand Erinnerungen, von den jetzigen Potentaten in Italia … biß auff vnsere jetzige Zeit deducirt werden*. Frankfurt, 1640.

7

# Chinese medicine as a mirror of Neapolitan experimentalism: a case study of *Parere* (1681) by Leonardo Di Capua (1617–1695)

Daniel Canaris

## Introduction

Leonardo Di Capua (1617–1695)[1] was one of the most prominent figures in the Accademia degli Investiganti, a seventeenth-century association of Naples-based intellectuals who sought to promote the reform and advancement of science.[2] His main claim to fame was the *Parere* (opinion or judgement), first published in 1681 and republished three times in 1689, 1695 and 1714.[3] Di Capua wrote the *Parere* in response to a request of a committee set up in 1678 by the viceroy of Naples and the Collateral Counsel (Consiglio Collaterale) to solicit the views of Naples' leading physicians on abuses and errors in medical practice. The viceroy felt compelled to act after a close friend of his died from a new chemical remedy administered by a Galenist who had been improperly trained in the therapeutic application of chemistry (iatrochemistry).

Surprisingly, Di Capua resisted the viceroy's call for further regulation. Instead he argued that precisely the opposite was needed: the foundations of medical science were so uncertain that it would be impossible to stipulate the essence of best medical practice with laws. In essence, Di Capua articulated a libertine manifesto, opining that all medical practitioners should be free to reject authority and to use their own sensorial experience to improve therapeutic methods. Di Capua's thesis was formulated in eight reasonings ('ragionamenti') that surveyed the history of medicine from earliest antiquity to the present, then analysed the deleterious effects of blind obedience to authority (especially the indemonstrable presuppositions of Galenic medicine) for medical practice.

Di Capua linked his position on the uncertainty of medicine to the scepticism of the Greek Pyrrhonist philosopher Sextus Empiricus (fl. mid to late second century AD) and made radical claims about the impossibility of ever resolving the disputes between medical sects. However, Di Capua's contemporary critics found his scepticism unconvincing. The Jesuit Giovanni Battista De Benedictis (pseudonym Benedetto Aletino, 1622–1706), one of the most outspoken opponents of the Neapolitan freethinkers in the late 1680s and 1690s, thought Di Capua's scepticism a fig leaf for Cartesianism:

> René must be therefore your Philosopher, and although you have the appearance of a sceptic, you are in truth Cartesian.[4]

Even Di Capua's defender Francesco D'Andrea (1625–1698) acknowledged that Di Capua was 'not purely sceptical' ('non puramente scettico').[5] Most modern scholars such as Michele Rak, Nancy Struever and Salvatore Serrapica reject the label of scepticism, whereas Andrea Carlino has argued that Di Capua's scepticism should not be summarily disregarded but understood in his seventeenth-Italian century context.[6] As Maurizio Torrini points out, scepticism in early modern Italy did not exhibit the radicalism that it assumed north of the Alps. More specifically, scepticism in Naples was manifest in opposition to unified systems of knowledge and an emphasis upon sense perception as the primary yet problematic means through which limited knowledge can be acquired:

> Science and truth changed: truth became a construct of time, identified with the search for truth, not with the result. It became the product of successive and progressive accretions.[7]

This chapter argues on the basis of Di Capua's treatment of Chinese medicine in the third 'ragionamento' that his views on the uncertainty of knowledge should be understood in terms of limited scepticism – namely a methodological tendency to challenge received authority without concluding that medical knowledge is impossible to acquire.

Di Capua's interest in Chinese medicine has been neglected by scholars.[8] That Di Capua should consider the testimony of Chinese medicine is not surprising, given that increasing information on Chinese medicine was entering Europe. This occurred primarily through the works of Jesuit missionaries, but also through the writings of Dutch and other European traders. When Di Capua published the *Parere* in 1681, the interested European reader could access information about Chinese medicinal herbs, pulse theory, acupuncture and moxibustion. Notably, scientists in the Royal Society such as Robert Boyle, whom the Investiganti considered their intellectual peers,[9] also turned their attention to the properties and efficacies of Chinese remedies such as ginseng.[10]

Westerners who encountered Chinese medicine in the early modern period were prone to 'galenise' it by drawing parallels between the humoural model of balance and imbalance and vitalist concepts in Chinese medicine such as *qi* 氣 (stuff/air) and *yinyang* 陰陽.[11] Yet Di Capua resists this interpretative trend: his interest in Chinese medicine stems precisely from his perception that Chinese medicine served as an antithesis and antidote to the 'Galienisti filosofanti' who had muddied medicine with their incessant and abstruse metaphysics. For Di Capua, China offered the example of a medical tradition that never practised bloodletting, prioritised the discovery of efficacious 'simples' (i.e. medicinal herbs) and understood the importance of diet in both preventative and curative medicine. Di Capua stresses the resemblances between the experimental methodologies of the first Chinese physicians and his own iatrochemistry to configure the ancient Chinese physicians – and, by extension, the first physicians of all ancient peoples – as proto-chemists.

Yet Di Capua neglects to identify his sources. Only once does Di Capua mention the name of a Jesuit missionary, Matteo Ricci (1552–1610), one of the co-founders of the Jesuit China mission.[12] In all other instances he refers only ambiguously to the testimony of unnamed writers. However, since Di Capua paraphrases significant extracts of his sources, it is possible to identify most of them. They were are primarily Jesuit, and the manner in which Di Capua employs and critiques his Jesuit sources reveals the limits of his professed scepticism.

While Di Capua is critical of the information he received from the Jesuits, he does not dismiss this information *tout court*, but rather filters it through his own historiographical and iatrochemical presuppositions. As much as Di Capua considers the origins of medicine obscure because of the mythologising tendencies of ancient sources, he broadly accepts the claims made in Jesuit sources about the experimental and rational processes through which the earliest ancient Chinese physicians first came to understand the therapeutic properties of various plants. Ancient Chinese medicine thus confirmed Di Capua's own hypotheses about the experimental origins of medicine and served to reflect his own proposals for medical reform.

## Chinese 'materia medica' and medical practices

Throughout the seventeenth century, the most authoritative sources on China were produced by the Jesuits. They sought to publish news about their mission in China primarily to promote their mission, but also in part to bring knowledge about China into dialogue with the European Republic of Letters. In the writings produced in the first half of the century, there was scattered information about Chinese 'materia medica' and medical practices, though this would be later supplemented with more systematic treatments. The first influential Jesuit publication on China was the 1615 *De Christiana expeditione apud Sinas* (*On the Christian Mission to China*), a Latin translation of Matteo Ricci's (1552–1610) Italian manuscript writings by the Flemish Jesuit Nicolas Trigault (1577–1627).[13] Trigault's edition enjoyed significant readership; it was reprinted in four Latin editions (1616, 1617, 1623, 1648), translated into French (1616, 1617, 1618), German (1617), Spanish (1621), Italian (1622) and partially into English (1625).[14] Notably, the Italian edition of this work was published in Naples by the Neapolitan Lazzaro Scoriggio (*c.*1580s–1638), who published a number of works for the Jesuits in Naples. The translator of the work, Antonio Sozzini (dates unknown), specifically addressed the translation to the regent of the Consiglio Collaterale, Juan Enriquez, with the express intention that the Neapolitan government may draw inspiration from Chinese governance and moral philosophy.[15] This work was most likely read by Di Capua either in Trigault's Latin or Sozzini's Italian. Indeed, on one occasion, Di Capua explicitly referred to Ricci's authority in describing Chinese medicine as ancient, efficacious and based on the administration of herbal remedies alone.[16] These positive reports of Chinese herbal medicine would have been of great interest to Di Capua, who believed that physicians needed a solid foundation in herbal remedies ('simples'). In particular, he praised the Neapolitan apothecarist Giuseppe Donzelli (1596–1670) who cultivated a vast array of medicinal herbs in the garden surrounding his villa (still standing in Arenella).[17]

However, Ricci was rather negative about the state of medicine in contemporary China, concluding that the Chinese held medicine in low esteem and that Chinese medical graduates were of poor quality.[18] Personal experience seems to have played a role in forming Ricci's opinion. When his confrère António De Almeida (1557–1591) fell ill in Shaozhou at the end of 1590, Ricci found all the local doctors 'ignorant' ('non sanno niente') and insisted that he be sent back to Macau to receive Western medicine.[19] While Di Capua remains on the whole positive about the state of Chinese medicine, he echoes Ricci's remark that the most intelligent and ambitious Chinese are discouraged from embarking on medical careers because of the low prestige of medicine in China. Instead they dedicate themselves to the study of moral philosophy, the focus of the imperial examination system.[20]

Di Capua's positive assessment of Chinese medicine thus appears to have been informed by later, primarily Jesuit texts, which were much more favourable in their treatment of Chinese medicine. One text that he may have consulted was the *Imperio de la China* (1642) by the Portuguese Jesuit Álvaro De Semedo (1585/6–1658). This work was immediately translated into Italian (1643), and the Italian translation was even reprinted in 1678 – around the time Di Capua was preparing the *Parere*.[21] Whereas for Ricci the medical sciences are 'hardly flourishing' ('fioriscono assai puoco'), for Semedo Chinese medicine is in 'an excellent state' ('in ottimo posto'), boasting numerous superb works on the subject that have survived from antiquity.[22]

Significantly, Semedo stresses the dissimilarities between Chinese and European medicine. He claims that Chinese herbal medicine developed independently of European medicine because Chinese medical works were all by Chinese authors, who never read European writings. For Semedo, the differences between European and Chinese medicine can be seen in the fact that Chinese doctors do not draw blood, practise cupping or use syrups, laxatives, pills or even cauterisation.[23] For Di Capua, some of these differences are important for corroborating his critique of the absurdity of Galenic medicine. In the *Parere*, Di Capua repeatedly denounces the Galenic practice of bloodletting ('salasso'), a treatment that he considered inconsistent in its application and unsupported by empirical evidence.[24] The absurdity of this practice is proven by the fact that from earliest antiquity Chinese physicians had never used bloodletting, but their patients nonetheless recovered much better and more quickly from illnesses compared to their European counterparts. Di Capua thus implores his European colleagues to consider the Chinese experience in order that this injurious practice can be definitively abolished.[25]

Semedo's account is also notable for providing the first European description of ginseng.[26] Semedo praises the root, which he reports is taken by both healthy and sick alike to increase or restore vigour respectively. Although Di Capua mentions ginseng only fleetingly, he seems to share Semedo's admiration for the root, describing it as 'precious' ('preziosa') and so efficacious that even the moribund are seen to recover after taking it.[27] However, Semedo's indication of price suggests that Di

Capua supplemented this information about ginseng by using another Jesuit source. Whereas Semedo claimed that ginseng was worth 'twice its weight in silver' ('che al tempo della mia partita era il suo Prezzo, pesarsi due volte a peso d'argento'),[28] Di Capua valued a pound of ginseng at 'no less than three pounds of silver' ('non val meno di tre libre d'argento').[29] This was precisely the value given by the Jesuit cartographer and historiographer Martino Martini in his *Atlas Sinensis*; it was then cited in the *China illustrata* (1667) by the Jesuit encyclopedist Athanasius Kircher (1602–1680).[30]

Another important – but unacknowledged – source of Di Capua's understanding of Chinese medicine was *La Cina* (1663), by the Jesuit historiographer Daniello Bartoli (1608–1685). Bartoli was tasked with compiling a comprehensive history of the Jesuit missions.[31] He never went to China and could not read Chinese, but he meticulously collated and translated into Italian texts sent to him by missionaries in China and elsewhere.[32]

Although Bartoli's writings were closely tied to the promotion of the Jesuit order, they were immensely influential in seventeenth-century Italy, even among secular circles. In particular, Bartoli was widely admired for his mastery of prose style. His work on Italian diction, *Il torto e il diritto del 'Non si può'* (1655), which sought to refute the pedantry of the Accademia della Crusca with a renewed attention to the literary authorities of the Trecento (Dante, Petrarch and Boccaccio), was printed numerous times in the seventeenth-century in Naples. As Fausto Nicolini attests, Bartoli was the only 'modern' author whom Di Capua and his followers (the 'capuisti') emulated in formulating their own principles for creating a pure prose style to defeat the excesses of the baroque.[33]

Di Capua's basic definition of Chinese 'materia medica' was effectively lifted word for word from *La Cina*, with only slight variations in phraseology. Di Capua describes Chinese medicine as being based on the use of seeds ('semi'), foliage ('frondi') and bark ('corteccie'), as well as stones ('pietre'). He mentions that images of this 'materia medica', along with explanations of their therapeutic properties, can be found in books.[34] All of these details can be found in *La Cina* in almost exactly the same order.[35] However, whereas Bartoli attributes the formulation of these maxims and aphorisms to the 'Hippocrates of China', the Yellow Emperor ('Ienti = Huangdi'), Di Capua omits this attribution. Instead he emphasises the commonalities between Chinese and European medical precepts ('e le loro virtù narrate ne' precetti, e nelle regole medicinali, non guari da noi eglino ne van lontani'). In this case, it is probable that Di Capua drew this observation about commonalities between European and Chinese medical precepts from Trigault's *De christiana expeditione apud Sinas*.[36]

Like Bartoli, Di Capua then proceeds to discuss the importance of dietary practices in Chinese medicine. While Semedo also mentions Chinese diets,[37] the details and phraseology of Di Capua's description are almost identical to those in Bartoli's account, according to which the Chinese subjected their patients to strict diets lasting up to 20 days. During these diets patients were not allowed to eat anything other than pear juice three or four times per day (water could be taken without limitation). For Bartoli, the effectiveness of this treatment was demonstrated by the experience of a moribund Chinese Jesuit brother who made a spectacular recovery after being subject to this regime; however, Bartoli cautioned against adopting such therapies in Europe out of fear that 'European stomachs would not be able to cope with it'.[38] Here Di Capua shows his willingness to use his clinical experience and historical knowledge to critique his unnamed Jesuit source for 'imagining that such a diet could not be tolerated by our people' ('immagina alcuno, che tal dieta non potrebbe sofferirsi da' nostri huomini'). He points out that a similar dietary regime was practised by the ancient Greeks and adds – with a tinge of Orientalism – that the Chinese were 'more tender and much more delicate than us', the implication being that Bartoli's concern was baseless in light of stronger European constitutions.[39]

However, Di Capua's admiration for Chinese medicine was not unguarded; in a number of instances his distrust of textual authority leads him to dispute or reinterpret the claims of his sources. This is especially seen in his treatment of pulse diagnosis, which European accounts of Chinese medicine generally regarded in a favourable light. Jesuit missionaries such as Semedo were immensely impressed by the ability of Chinese physicians to arrive at precise and accurate diagnoses without inquiring about or even examining any symptoms.[40] In the decade prior to the publication of the *Parere*, these skeletal descriptions of pulse diagnosis were enriched with much more detailed information from Jesuit sources. A treatise, probably based on the writings of the Polish Jesuit Michał Boym (1612–1659), was published in French translation under the title *Les Secrets de la médecine des Chinois* (1671); this in turn was translated into Italian by Pietro Francesco D'Amphous as *Secreti svelati della medicina de' Chinesi* (1676).[41] Yet Di Capua displays no knowledge of these more recent sources, instead remaining wedded to Bartoli for his basic understanding of this Chinese medical practice. For Bartoli, Chinese pulse diagnosis demonstrated the excellence of Chinese physicians in their understanding of symptomatology:

> There is no doubt that through study and experience they have acquired incomparable skill in recognising intimately the nature and the concomittent properties of every type of illness, their peculiar signs and their true indications. In this respect they far surpass European doctors. They never ask a patient about his current state or medical history – to ask such questions would be an open confession of the physician's ignorance. Instead, the physician will seat himself beside the patient and observe most attentively for around half an hour the pulse, and from the diversity of the irregular palpitations which they discern most perceptively (it must be added that they know the characteristics of every various internal disposition, which is interpreted by the heart, both for movement and for experiencing pain), and they understand and explain to the patient what happened up to that day while making predictions about the future.[42]

Di Capua's description of pulse diagnosis is essentially a paraphrasis of this passage. After relating the consensus ('dicono tutti') that 'Chinese doctors were incomparably more skilful than our doctors in treating illnesses' ('i Cinesi medici sono senza alcun paragone assai più de' nostri, valenti in guarire i mali'), Di Capua then includes almost exactly the same details about pulse diagnosis as Bartoli. Curiously, whereas for Bartoli pulse diagnosis demonstrates the excellence of Chinese medicine, for Di Capua it reveals evidence of medical fraud:

> But nevertheless deceit and trickery still accompany medicine; although they understand a lot about pulses, to appear even more skilled they stay for up to half an hour, pretending to observe in minute detail the changes while feeling the wrist, and they assert that with such diligence they may arrive at the knowledge of every various internal disposition, including the most hidden, as well as the nature and true cause of even the most unusual illnesses.[43]

Bartoli also discusses fraud in Chinese medicine below this passage, but not in connection to pulse diagnosis. Rather, for Bartoli, the chief fraud in Chinese medicine is the claim made by an ever-increasing number of itinerant charlatans ('ciurmadori') that certain secret concoctions can rejuvenate the elderly or bestow immortality.

Di Capua's tendency to challenge untested information is also evidenced in his discussion on the purported therapeutic properties of tea. Unlike ginseng, which would have been difficult to obtain in

Naples, tea was readily available, thanks to a thriving trade conducted principally by the Dutch.[44] This availability allowed Di Capua to test the claims in his written sources using an experimental method, which was the epistemic requirement of the scientific methodology promoted by the Investiganti. Jesuit sources invariably raved at the marvellous therapeutic properties of tea. For instance, in the *China illustrata* Kircher, again citing Martini's *Atlas Sinensis*, listed manifold health benefits of tea: it is diuretic, pleasant tasting, helps scholars to concentrate for long hours and prevents kidney stones and gout.[45]

At the time, it was also common to contrast the effects of tea with those of coffee and chocolate, the first stimulants to enter the European market. A large literature sought to examine the medicinal properties of these beverages using Galenic humoural theory, as well as the chemical and mechanical theories of physiology which were becoming increasingly prominent.[46] For example, Kircher used Galenic humoural theory to argue for the superiority of tea over coffee and chocolate because tea has no negative side effects, whereas coffee causes bile to ascend and chocolate causes the body to overheat in summer.[47] In contrast, the Danish physician Simon Paulli (1603–1680) contested the medicinal properties of these exotic beverages, arguing that more readily available – and cheaper – European ingredients were more effective.[48]

Di Capua makes a tentative foray into these debates, refuting 'some of our writers' ('alcuni de' nostri scrittori') who claim that tea can cause apoplexy ('apoplessia') – a side effect which he argues experience has demonstrated to be false ('noi ben sappiamo per pruova, esser ciò falso'). However, he seems doubtful about the utility of tea as a treatment in the European context. While Di Capua does not directly challenge reports about its marvellous therapeutic effects in China, he has not observed such properties in Europe and provides a chemical explanation: the potency of tea is weakened through the loss of ammonia ('volatile Alcali') during transport from China to Europe. Interestingly, unlike Martini and Kircher, Di Capua does not attribute the absence of gout ('podagra') and kidney stones ('il male della pietra') in China to tea, but rather to the Chinese practice of drinking hot water.[49] Overall, Di Capua cautions against the fetishisation of exotic medicines and seems to agree with Paulli in prioritising European herbal remedies ('erba nostrale') over tea in his own medical practice. The value of European remedies was reflected in the fact that the Dutch traded Chinese tea for European sage, which the Chinese valued at three times the price of tea.[50]

## Rewriting the history of Chinese medicine

Di Capua's mixture of admiration and scepticism for Chinese medicine is further reflected in his discussion on its origins. Although he does not give his source, Di Capua undoubtedly derived his information from the *Sinicae historiae decas prima* (Munich, 1658) by the Jesuit missionary Martino Martini (1614–1661).[51] He makes an elliptical reference to Martini as 'he who took care and employed his intellect to translate the chronicles of the Chinese into Latin' ('colui che si diè cura, e impiegò il suo ingegno a traslatare in latino idioma le croniche de' Cinesi') and cites various information about the ancient Chinese emperor Shennong, the legendary founder of Chinese medicine, that could only have been taken from Martini's work. Unlike the aforementioned works, this was not translated into Italian at the time, but it was nonetheless read in Naples, as evidenced by the citation of it by Giambattista Vico (1668–1744) in the 1730 and 1744 editions of the *Scienza nuova*.[52] Martini's observations about Shennong were adapted from a variety of Chinese historiographical sources, possibly including the thirteenth-century work *Zizhi tongjian xubian* 資治通鑑續編 (*Continuation to the comprehensive mirror to aid in government*) by Chen Jing 陳桱 (Yuan dynasty).[53]

Despite Shennong's mythical status in Chinese folklore, Martini presented the emperor as an historical figure and did not question the veracity of the fabulous narrative. According to Martini, Shennong tested on his own body the therapeutic properties of herbs and explored the effect that they had on the digestion of food, as well as their potential to cause and heal illnesses. Martini related (erroneously) that the prodigious Shennong managed to catalogue 60 types of poisonous herbs and their antidotes.[54] Martini used this historical discussion to make a general statement about the excellence of contemporary Chinese medical practice, arguing that the Chinese possessed medical books of such quality that they had no need for either ancient or modern Western medicine.[55] He cites one such Chinese book – possibly Li Shizhen's (1518–1593) *Ben cao gangmu* 本草綱目 – which catalogued numerous plants and roots with lifelike depictions.[56]

According to Martini, Shennong's experiments were so successful that he was 'able to penetrate all the secrets of the human body with a vision sharper than that of lynxes' ('ut eum vulgo oculis plusquam lynceis crederent omnia hominum corporumque arcana perspicere').[57] Martini's reference to the lynx's vision is interesting because the eyes of the lynx had been used by Giambattista Della Porta (1535–1615) as a metaphor for the experimental method:

with lynx-like eyes, examining those things which manifest themselves, so that having observed them, he may zealously use them (lynceis oculis perpendens, quae se sibi demonstrant, ut re inspecta sedulo operetur)[58]

In turn, the lynx became the eponymous emblem of the Accademia dei Lincei, the celebrated group of experimental scientists among whom was Galileo Galilei. Martini was possibly drawing a comparison between Shennong's methodology and modern experimental European science.

Di Capua's interest in Shennong stems from his discussion about the historiographical and epistemological problem of how the first human beings, who had limited rational faculties, came to acquire medical knowledge.[59] One prominent theory, promoted by Francis Bacon (1561–1626), was that medicine was discovered out of necessity or by chance.[60] For Di Capua, animal behaviours can provide clues as to how necessity and chance interacted with knowledge obtained from the senses to construct primitive medical knowledge. The premise of Di Capua's discussion is the conventional Aristotelian distinction between rational human souls and animal souls: unlike people, animals are devoid of a rational soul, yet they able to fine-tune their instinct using their senses, which help them understand how to respond to and manipulate even the slightest sensory impressions to the benefit of their self-preservation.[61] For instance, animals learn to abstain from certain foods after bad experiences, whereas animals such as wolves, cats and dogs treat indigestion by consuming grass which stimulates vomit.[62] Yet Aristotle's categoric distinction between the rational and animal souls falters as Di Capua perceives in animal responses to sensory phenomenon a primitive form of 'philosophising' ('filosofare') that is restricted to the perception and analysis of external impressions – namely 'the bark of things' ('la prima sola corteccia delle cose').[63]

Di Capua therefore draws an implied parallel between the sensory knowledge of animals and the experimental methodology of the Investiganti, attributing to animals the ability 'to investigate in a rough and imperfect way' ('con rozzo, ed imperfetto modo investigare').[64] Hence the contrast between rational and sensory faculties is not ontological; it is rather an inverse relationship in which higher level rational faculties blunt sensory acuteness: 'where reason is lacking the senses are in abundance' ('dove manca ragione il senso abbonda').[65]

The insight that even in brute beasts 'some shadow of Rational medicine' ('qualche ombra di Razional medicina') can be found leads Di

Capua to the hypothesis that even earliest primitive people ('que' primi rozzi huomini') possessed rational faculties. They thus primarily acquired their first medical knowledge not through chance, but through rational investigations.[66] Such a hypothesis, of course, required historiographical corroboration through a survey of early medical history. Di Capua's chief difficulty was that he considered the extant sources about early medical history unreliable. Elsewhere, he expresses extreme scepticism about the cures supposedly discovered by figures in Greco-Roman mythology,[67] and he does not exempt the Chinese from this charge of mythologising early history: he believes that the stories about Shennong are 'the stuff of fable' ('favolose') and that it would have been impossible for Shennong to have performed all the experiments by himself in the timeframe indicated (that of one day). In the Chinese histories Di Capua sees further evidence that the custom of promoting 'lies' ('menzogne') and 'boasts' ('millanterie') about medicine was practised by all peoples in every period of history.[68]

For Di Capua, the unreliability of the Chinese records about early medicine had been further exacerbated after the burning of 'no less than all' ('poco men che tutti') Chinese books 2,000 years ago by the 'Chinese emperor' (i.e. Qin Shihuang, 259–210 BC). This event was frequently related in Jesuit sources, but was invoked primarily to explain corruptions and lacunas in the Confucian textual tradition.[69] Di Capua's claim that this affected the integrity of medical texts seems unprecedented in both Jesuit accounts and even Chinese sources, which specifically exempted medical and other technological texts from the flames.[70] Di Capua's next claim that whatever fragments survived the conflagration were essentially unreadable was also a gross overstatement; it could not be supported by the information provided by Jesuit sources, which had already summarised significant content from these pre-Qin texts. Di Capua's hasty conclusions seem to be prompted by his superficial reading of Chinese sources and lack of historical perspective in his understanding of them.

Be that as it may, Di Capua evidently believes that the Chinese records must have some historical value because he still relies upon Martini's narrative to make inferences about Shennong's investigative methodology. For Di Capua, Shennong's classification of so many poisonous plants and antidotes would have been impossible without a deliberate methodology, providing further evidence that the origins of human medicine were not rooted in necessity or chance discovery. The Chinese, like the ancient Egyptians, Scythians, Arabians, Phoenicians, the Druids of Gaul and ancient Greeks, must have been proto-chemists who used their rational faculty to investigate the therapeutic properties of medicines.[71] This rational process was not necessarily sophisticated, but was rather 'uncouth' ('rozzo') and 'imperfect' ('imperfetto'); it had a certain epistemic continuity with the rational processes of modern science. As a result, the ancient Chinese arrived at a significant understanding of geometry, philosophy and other sciences, especially chemistry, as reflected by their explanation of natural things by reference to 'one principle of natural things' ('un solo essere il principio delle cose naturali') – a possible reference to *qi* 氣 (air) or *yinyang* – as well as to the 'secondary principles' ('secondi principj') of 'five substances' ('cinque sostanze'). These last were the five elements or 'wuxing' 五行 – that constitute bodies (fire, water, earth, metal and wood).[72]

In contrast, Di Capua's Jesuit sources approached both the Chinese first principle and the five elements from an Aristotelian perspective and were unambiguously critical. Trigault, for instance, perceived monist implications in the postulation of a single substance underlying the universe and called the Chinese 'foolish' ('inepte') for adding an element to Aristotle's Four Elements.[73] Similarly, Martini also begs his reader not to 'ridicule' ('ne … rideamus') the Chinese for having 'deviated from truth' ('a vero nimium aberrantes') in adding wood and metal to the 'elements'.[74] Yet Di Capua, like Boyle and many other 'novatores', rejected Aristotle's rigid enumeration of the elements; he thus did not see these deviations as inherently problematic.

The concept that the universe was made of a single stuff was also shared by Boyle, who called this universal stuff 'Catholick matter'.[75] In a similar vein, the chemist Thomas Willis (1625–1671) sought to update Aristotle in light of Paracelsus's concept of active particles (spirits or mercury, sulphur and salt) and Gassendi's atomism by proposing a theory of five elements: spirit, sulphur, salt, earth and water.[76] Although Di Capua also translates Willis's theory of the five elements as 'cinque sostanze' in the *Parere*,[77] his presentation of the *wuxing* (or five elements) is so elliptical that it is unclear to what 'principles of our chemists' ('principj de' nostri Chimici') Di Capua likens Chinese medicine. Nevertheless, the anti-Aristotelian/anti-Galenic and iatrochemical orientation of Di Capua's discussion is unmistakable, representing a significant departure from his Jesuit source texts.

At the same time, the Chinese histories also provide Di Capua with exemplars of how chemistry can be corrupted. Later in Martini's chronology is related the story of how the emperor Wu of Han (156–87 BC) was persuaded by Daoist chemists ('chimici') to pursue chemical elixirs for immortality. Paraphrasing Martini, Di Capua relates how this emperor concocted these elixirs from pearls soaked in dew, which was collected from a bronze bowl placed at the top of a tower.[78] Di Capua draws an interesting comparison between modern Chinese physicians, who inherited this tradition, and the Rosicrucians, who similarly sought to peddle elixirs of immortality with spurious claims of antiquity. Just as the Rosicrucians claim lineage from the medieval alchemist Arnaldus de Villa Nova (1240–1311), so the modern Chinese alchemists cite examples of the ancient Chinese who

> became immortal with powerful medicine, and dwell in the summits of the highest mountains, and thus they go or rather fly wherever they want, even in Heaven, *unbound from all human qualities*.[79]

Di Capua's citation of this passage reveals his anxiety to differentiate his own iatrochemistry from the pseudo-scientific principles of alchemy.

## Conclusion

By presenting Chinese medicine antithetically to the Galenic medical tradition, Di Capua made a significant contribution to the reception of Chinese medicine in early modern Europe. His comparison of Chinese medicine to his own experimental and iatrochemical orientation provided a form of legitimisation of ancient Chinese medicine, even as he and his fellow 'novatores' were undermining the authority of the Galenic tradition. Yet Di Capua was not abreast of the latest, much more detailed literature on Chinese medicine that was being published from the 1670s onwards. He reveals no knowledge of Chinese vitalism or of the metaphysical features of Chinese medicine which are not perhaps so amenable to an experimentalist interpretation. The significance of Di Capua's treatment of Chinese medicine lies less in the accuracy of his interpretative insights than in the role of Chinese medicine in clarifying his conception of scientific development and the nature of his professed scepticism.

First, Di Capua's willingness to learn from the methodologies and achievements of peoples who experimented with 'uncouth' ('rozzo') and 'imperfect' ('imperfetto') rational processes reflected his non-linear conception of scientific development. Di Capua saw no conflict between ancient and modern scientific theory and practice. Rather, as Struever elegantly puts it:

> not only may modern investigations, primarily in anatomy and chemistry, correct ancient theory and practice, but ancient theory may modify modern empirical initiatives.[80]

Di Capua's presentation of China was far from Sinophilic: while he extols the merits of Chinese medicine, he was also ready to critique claims which he believed were unsubstantiated, and to reprove those deviations from correct iatrochemical practice which he identified in both ancient and modern Chinese medicine.

Second, Di Capua's treatment of his (primarily Jesuit) sources on China sheds light on the import and limitations of his scepticism. While Di Capua cites Sextus Empiricus to assert the unreliability of mythological accounts of early medicine,[81] he nonetheless accepts a significant amount of the historical and medical information provided by his Jesuit sources, merely contesting those claims that he finds implausible in light of his own presuppositions and experience. Far from Pyrrhonist suspension of judgement, Di Capua uses his source material to construct positive claims about the origin and nature of primitive medical knowledge.

Scepticism provided Di Capua the methodological dictate to challenge and deconstruct systems of medical knowledge that cannot be substantiated with experimental methods, but it did not lead Di Capua to conclude that medical knowledge is impossible to acquire. However much Di Capua may lament the lack of an innate criterion with which humans can adjudicate between conflicting truths, he remains adamant about the infallibility of the senses, the soul's windows to its first rudiments of knowledge. For Di Capua, both the impressions which are represented to the soul and the soul that receives them 'can never lie' ('che i sensi né se medesimi, né l'anima mentir non possono giammai').[82] Appealing to Tertullian, Di Capua argues that to question the infallibility of the senses would undermine the foundations of 'all knowledge, all judgement, and all faith' ('ogni contezza, ogni giudicio, ogni fede').[83] Such a presupposition, which owes more to Epicureanism than to scepticism, would seem to provide the epistemological grounds for Di Capua's reflections on the role of the senses in both the construction of animal medicine and the experimentalism of primitive man and the ancient Chinese.

## Notes

1 His name is also written as 'Lionardo Di Capoa'. I follow the modernised spelling used by Silvano Scalabrella in his biography: Scalabrella, 'DI CAPUA, Leonardo', 712–15.

2 See Fisch, 'The Academy of the Investigators', 521–63.
3 Di Capua, *Parere*.
4 'Bisogna dunque, che Renato in sostanza sia il vostro Filosofo, e che comunque abbiate fatto sembiante di Scettico, siate per verità Cartesiano', Aletino, *Lettere apologetiche*, 118.
5 Carlino, 'La via dell'incertezza', 343.
6 Rak, 'Una teoria', 233–97; Struever, 'Lionardo di Capoa's *Parere*', 322–36; Serrapica, *Per una teoria dell'incertezza*; Carlino, 'La via dell'incertezza'.
7 Torrini, 'From Galileo to Vico', 337 (327–341).
8 Di Parere's interest in Chinese medicine is mentioned briefly by Carlino, 'La via dell'incertezza', 331; Conforti, 'La mutazione linguistic', 74.
9 Notably, in 1684, the first 'ragionamento' was published in English translation with a dedication to Robert Boyle. Di Capua, *The Uncertainty of the Art of Physick*.
10 Appleby, 'Ginseng and the Royal Society', 121–45.
11 Barnes, *Needles, Herbs, Gods, and Ghosts*, 19. A similar tendency can be found in Western approaches to other medical traditions even in the early eighteenth century. Giovannetti-Singh, 'Galenizing the New World', 59–72.
12 'Ed in quel vasto, e quasi immenso tratto di paese della China, come testimonia il Padre Matteo Ricci, si è medicato per molti, e molti secoli, e si medica tuttavia, ed assai felicemente coll'uso delle sole erbe', Di Capua, *Parere*, 519.
13 Trigault and Ricci, *De Christiana expeditione*. Ricci's original Italian text was only published in the twentieth century, first by Pietro Tacchi Venturi and then in a more Sinologically accurate edition by Pasquale M. D'Elia. See Venturi, ed., *Opere storiche*. The most recent edition of Ricci's original Italian-language manuscripts is Ricci, *Entrata della Compagnia di Giesù*.
14 Trigault and Ricci, *China in the Sixteenth Century*, xvii.
15 Trigault and Ricci, *Entrata nella China*, dedication (n.p.).
16 See above, n. 12.
17 Di Capua, *Parere*, 579. For this point, see Hanafi, *The Monster in the Machine*, 145–6.
18 Trigault and Ricci, *China in the Sixteenth Century*, 32.
19 D'Elia, *Fonti Ricciane*, 1: 304.
20 'Ma avvegnachè nella Cina i medici, quanto al fatto del medicare sien così fortunati, come divisato abbiamo: nondimeno avuti vi sono in pochissimo pregio, e stima. E quinci avvien poi, che tutti coloro, i quali sien d'alto ingegno, e di saggio avvedimento dalla natura forniti, nulla badandovi, alla moral filosofia studiosamente si volgano, onde a' primi onori del regno agevolmente poi pervengono', Di Capua, *Parere*, 181. Di Capua's phraseology and lexical choices might confirm that he was reading Trigault's text in Sozzini's Italian translation, which uses similar expressions such as 'niuna stima', 'ingegno' and 'i primi honori del Regno'. Trigault and Ricci, *Entrata nella China de' Padri della Compagnia del Gesù*, 25.
21 Semedo, *Relatione della grande monarchia della Cina*.
22 Semedo, *Relatione della grande monarchia della Cina*, 73.
23 Turk and Allen, 'Bleeding and cupping', 128–31. Semedo's description was not entirely accurate because cupping ('baguan') was very common in traditional Chinese medicine, and European cauterisation could find correspondence in Chinese moxibustion (the burning of mugwort leaves near or on the skin). Later missionary accounts of Chinese medicine reported cupping as being practised in China. Barnes, *Needles, Herbs, Gods, and Ghosts*, 58–9; 117.
24 'Così mordendosi l'un l'altro, e piatendo, niun l'imbrocca, e tutti a mal partito menano gli ammalati; volendo altri i salassi, ed altri vietandogli, ed altri una sol volta permettendogli, chi scarsamente, e chi sino a trar loro tutto il sangue, chi dalle vene delle braccia, e chi da quelle de' piedi, e chi anche da quelle parti, delle quali è bello il tacere, con appiccarvi le mignatte; altri a tutti costoro contrastando vogliono che dalla buccia solamente per coppette si tragga', Di Capua, *Parere*, 145–6.
25 'E quanto al non trar sangue, oltre al novero de' greci, e de' nostri medicanti, che seguono il medesimo istituto: la ben lunga prescrizione di quaranta, e più secoli, ne' quali han potuto guarir felicissimamente, ed in ispazio assai brieve le malattie, non gli rende degni, non dico di scusa, ma d'altissima loda? Ed a ciò vorrei, che ponesser mente tutti coloro che così di leggieri si lasciano a' medici trar sangue', Di Capua, *Parere*, 179–80. The aversion of Chinese physicians to bloodletting is also mentioned in Bartoli *La Cina* (1663), which, as discussed below, was also consulted by Di Capua. However, Bartoli merely claims that bloodletting was 'hardly' ('appena') administered in China, not, as Di Capua claims, 'never' ('mai').

26 Appleby, 'Ginseng and the Royal Society'. Semedo's original text was written in Portuguese, but was first published in Spanish. Semedo, *Imperio de la China*. For the Italian translation see Semedo, *Relatione della grande monarchia*.

27 'Usano frequentemente la preziosa radice, detta da loro Ginsen, dalla qual sovente si veggon guarir gl'infermi, eziandio morienti, e però una libra di essa, non val meno di tre libre d'argento', Di Capua, *Parere*, 181.

28 Semedo, *Relatione della grande monarchia*, 31.

29 Di Capua, *Parere*, 181.

30 'triplo argenti pondere appenso emitur libra', Kircher, *China monumentis* [...] *illustrata*, 179.

31 Bartoli, *Dell'historia della Compagnia di Giesù*.

32 The manuscript drafts of Bartoli's writings are held in Archivum Romanum Societatis Iesu [ARSI], *Hist. Soc.* 124. Unfortunately the sections relevant to China have yet to be analysed, and further work is required to establish the missionary sources that Bartoli used to construct his discussion of Chinese medicine.

33 Nicolini, *La giovinezza di Giambattista Vico (1668–1700)*, 86–7; Nicolini, *Commento storico*, 1: 232; Renaldo, 'Antecedents of Vico', 353; Canaris, *Vico and China*, 25.

34 'I moderni Cinesi medici non altrimenti, che gli antichi già si facessero, de' semi, delle frondi, delle corteccie d'alcune piante si vagliono, e d'alcune pietre altresì, e serban libri, ove son figurate l'immagini di tali piante, e pietre, e le loro virtù narrate l'immagini di tali piante, e pietre, e le loro virtù narrate ne' precetti, e nelle regole medicinali, non guari da noi eglino ne van lontani', Di Capua, *Parere*, 180.

35 'Tutti i loro rimedi son pietre, semi, radici, herbe, frondi, cortecce, e cotali altri semplici: e han libri che ne figuran le immagini, e ne divisano le virtù: seguendo in ciò massimemente i dettati, e gli aforismi d'un loro antichissimo Imperadore, tutto insieme herbolaio, e medico eccellente, per nome Ienti, il quale è l'Ippocrate de' Cinesi', Bartoli, *Dell'historia della Compagnia di Giesù. La Cina. Terza parte dell'Asia*, 62.

36 Trigault seems to contradict himself on whether Chinese medicine resembles or is dissimilar from European medicine. At the beginning of his description of Chinese medicine, he emphasises that the precepts of Chinese medicine 'differ significantly from European precepts' ('Artis medicae praecepta non parum a nostratibus discrepant'), but adds below that insofar as the administration of herbal remedies ('simples') is concerned, 'almost the entire art of Chinese medicine is contained in the precepts of our own herbal medicine' ('itaque ars tota Sinesnsi medica herbariae fere nostrae praeceptis continetur'), Trigault and Ricci, *De Christiana expeditione apud Sinas*, 31. This contradiction is not present in Ricci's original Italian text, in which Ricci unambiguously presents Chinese medicine as different from European medicine; he merely mentions a correspondence between the Chinese use of herbal medicine and European 'simples' ('risponde più tosto alla nostra herbolaria'), D'Elia, *Fonti Ricciane*, 1: 42.

37 'Non prohibiscono l'acqua, però ha da esser cotta, è vero *cha*. Prohibiscono bene il mangiare; di modo che se l'infermo ha fame, ha da mangiar leggiermente, e con molta dieta: se non ha fame, non si ammazzano che mangi, dicono che stando il corpo infermo, lo stomaco non fa bene l'officio suo, e così la concottione, che all'hora fa, tutt'è maligna, e contro la sanità', Semedo, *Relatione della grande monarchia*, 75.

38 'Ma ben sì il prescrivere un iquisito rigore in dieta, e tanto, che per avventura, non ad ognuno parrà possibile, non che vero: nè io mi ci arrischierei altrimenti, che havendone in casa testimoni, e di veduta in altrui, e prova in se stesso. Cioè, tener l'inferno sette, quattordici, e per sin anco venti dì, senza dargli una bricia di che che sia, per cibarsi. Bere acqua sì, quanta ne vuole, e due, tre, quattro volte al dì, sugo di pere. Così strettamente digiuno quattordici dì continuati, un nostro Fratel Cinese si compò d'una mortalissima infermità: ma gli stomachi Europei non reggerebbono a tanto', Bartoli, *Dell'historia della Compagnia di Giesù. La Cina. Terza parte dell'Asia*, 62.

39 'Prescrivono a' loro infermi sì rigorose diete, che alle volte lascian passar fino a venti dì senza dar loro altro cibo, che certo sugo di pere, tre, o quattro fiate il giorno, e ber quanto acqua richieggiono; e sì molte gravissime malattie a buono, e perfetto stato riducono. Immagina alcuno, che tal dieta non potrebbe sofferirsi da' nostri huomini; ma quanto egli vada errato, il può far vedere l'essere stata in uso appo gli antichissimi greci, e l'essere i Cinesi di noi più teneri, e dilicati assai', Di Capua, *Parere*, 180.

40 Semedo, *Relatione della grande monarchia*, 73.

41 *Secreti svelati della medicina de' Chinesi*; *Specimen Medicinae Sinicae*. See Hanson and Pomata, 'Travels of a Chinese pulse treatise', 23–57. Boym's translations and descriptions of pulse medicine would also appear in the *Specimen medicinae sinicae* edited by Andreas Cleyer (Frankfurt, 1682) and the *Clavis medica ad Chinarum doctrinam de pulsibus* (Nuremberg, 1686).

42 'Al che non ha dubbio che in gran maniera non conferisca una incomparabil peritia acquistata da essi, e per istudio, e per uso, di conoscere intimamente la natura, e le proprietà conseguenti qualunque sia specie di male; e i lor propri segni, e le lor vere indicationi: nel che avanzano a dismisura i medici Europei. Non chieggono essi mai all'infermo dell'essere suo presente, nè del succedutogli da che il prese il male: che cio sarebbe un confessarsi alla scoperta ignorante: ma sedutiglisi a canto, ne spiano per intorno a mezz'hora attentissimamente il polso, e dalla diversità de gl'irregolari suoi battimenti, che sottilissimamente discernono (e convien dire che sappiano d'ogni varia dispositione interna il suo proprio, di cui è interprete il cuore, tale al muoversi quale al patire) comprendono, e narrano all'infermo, quanto dì per dì gli è fino allora avvenuto, e soggiungono i presagi dell'avvenire', Bartoli, *Dell'historia della Compagnia di Giesù. La Cina. Terza parte dell'Asia*, 63.

43 'Ma nondimeno ancora ivi colla medicina s'accompagna l'inganno, e l'artificio; ed essendo eglino intendenti molto de' polsi, tuttavia per parere in ciò da più assai, s'intertengono sin' a mez'ora, fingendo d'osservar minutamente le lor mutazioni in toccandogli, e danno a diveder dapoi, che con una tal diligenza essi aggiungano a sapere d'ogni varia, e più occulta interna disposizione, e di qualunque più strana malattia la natura, e la vera cagione', Di Capua, *Parere*, 180.

44 Van Driem, *The Tale of Tea*, 290–367.

45 Kircher, *China monumentis [...] illustrata*, 180.

46 Albala, 'Stimulants and intoxicants in Europe', 53.

47 Kircher, *China monumentis [...] illustrata*, 180.

48 See Paulli, *Commentarius de abusu tabaci*.

49 'Ma lasciando sì fatte memorie da parte stare, si scorge quanto ben orniti fossero de' rimedj efficaci gli antichi Cinesi, dalle maravigliose cure, che con essi tuttavia fanno i moderni medici. Solamente potrebbesi levare incontro taluno, dicendo, che non siano giunti a saper quanto dilettevol sia il ber freddo, ne mai habbian messo in uso i salassi; ma tali apposizioni recar potrebbonsi eglino a somma loda; imperocchè col ber caldo si sono i Cinesi sottratti al male della pietra, alle podagre, e ad altre atrocissime malattie, che così frequenti, ed abbondevoli sono fra noi. E quanto al non trar sangue, oltre al novero de' greci, e de' nostri medicanti, che seguono il medesimo istituto: la ben lunga prescrizione di quaranta, e più secoli, ne' quali han potuto guarir felicissimamente, ed in ispazio assai brieve le malattie, non gli rende degni, non dico di scusa, ma d'altissima loda? Ed a ciò vorrei, che ponesser mente tutti coloro, che così di leggieri si lasciano a' medici trar sangue', Di Capua, *Parere*, 179–80. Cfr. Semedo, *Relatione della grande monarchia*, 75.

50 Other sources give a similar valuation. 'Les Chinois aiment tant la Sauge, qu'ils s'étonnent comment les Européens viennent chercher le Thé dans leurs pays, pendant qu'ils ont chez eux une plante aussi excellente, et qui réellement lui est préférable; aussi les Hollandois ont-ils grand soin d'enlever, à grand marché, presque toute la récolte de la Sauge qui croît sur nos côtes de Provence, et de la porter in Chine, on ils la vendent très cher, tant aux Chinois qu'aux Japonnois. On prétend que dans l'échange d'une caisse de Sauge, ils en obtiennent deux caisses et souvent trois de Thé vert', Jacques Christophe Valmont de Bomare, *Dictionnaire raisonné universel d'histoire naturelle*, vol. 5, 127.

51 This work was translated into French in 1692 and more recently into Italian. See Martini, *Histoire de la Chine*; Martini, *Opera omnia*, vol. 4.

52 Canaris, 'China in Giambattista Vico', 145–63.

53 For Martini's sources see Wu Liwei 吴莉苇, *Dang Nuoya fangzhou zaoyu Fuxi Shennong*, 95–116; Martini, *Opera omnia*, 4:xix–xx; Standaert, *Intercultural Weaving of Historical Texts*, 96–102.

54 In fact, the Han dynasty *Huainanzi* 淮南子 (19.1) stated that Shennong discovered 70 types of medicines: '一日遇七十毒'.

55 Martini, *Sinicae historiae decas prima*, 13–14.

56 Martini, *Opera omnia*, 4: 69, n. 53.

57 Martini, *Sinicae historiae decas prima*, 13.

58 Della Porta, *Magiae naturalis*, 14.
59 Di Capua, *Parere*, 172–5.
60 Conforti, 'Medicine, history and religion', 78.
61 For Di Capua's debt to Aristotelianism, see Struever, 'Lionardo Di Capoa's *Parere*', 322–36.
62 Di Capua, *Parere*, 173–4.
63 Di Capua, *Parere*, 174.
64 Di Capua, *Parere*, 174.
65 Di Capua, *Parere*, 175. There are strong parallels with Vico's epistemology, which similarly rests upon the inverse relationship between senses and reason. Cf 'La fantasia tanto è più robusta quanto è più debole il raziocinio' and 'Finalmante il niuno o poco uso del raziocinio porta robustezza de' sensi', Vico, *Opere filosofiche*, 258.
66 Di Capua, *Parere*, 172. Conforti seems to misinterpret Di Capua, claiming that Di Capua shared Bacon's view that medical knowledge was discovered 'primarily by chance'. Conforti, 'Medicine, history and religion', 78, n. 70. However, Di Capua states precisely the opposite view: 'E conciossiecosachè rari sien quei rimedj, che a caso ritrovar si possano; ne sembri verisimil punto, che le tante erbe, e radici, onde negli antichissimi tempi, non pur le ferite, ma gl'interni malori altresì medicavansi, venissero a sorte lor conosciute; *rimane adunque, che per la più parte dalla ragione i medicamenti stati sieno scoverti* [author's emphasis]', Di Capua, *Parere*, 172.
67 'But as for *Aesculapius*, he may well rest contented with that Glory, which, for his having been the first that taught the World the Art of Tooth-drawing, is rationally given him by the *Roman Orator*, when he saith; Aesculapius was the first that invented the drawing of Teeth: Altho' the Cures wrought by him are said to have been so rare and wonderful, and those in so many different ways related, that as *Sextus* the Empirick wisely reflects, they may hereby have been concluded to have been wholly Fabulous', Di Capua, *The Uncertainty of the Art of Physick*, 38.
68 'Ma chi non sarà per iscorgere anco a prima vista poi quanto sien favolose, ed inverisimili quelle pruove, che di Cinnungo si narrano, che egli facesse in se stesso lo esperimento delle piante nocevoli, e ristorative, e che nello spazio sì breve d'una sola giornata, tante ne provasse, e ne riprovasse; il che fa chiaramente conoscere, quanto la medicina, se acquistar vuole estimazione, in tutti i tempi, ed in tutti i luoghi abbia in costume di porre in opera le menzogne, e le millanterie', Di Capua, *Parere*, 178. There are strong parallels to these claims by Di Capua and Vico's 'borie delle nazioni', in which Vico, like Di Capua, pointedly includes the Chinese. Vico, *Opere filosofiche*, 432.
69 Meynard, *Confucius Sinarum Philosophus*, 108.
70 *Shiji*, 6.38: '所不去者，醫藥卜筮種樹之書'.
71 Di Capua, *Parere*, 176.
72 'Ma da qualche vestigio, che tuttavia ne rimane, si scorge apertamente, che i Cinesi nella geometria, nella filosofia, e nell'altre scienze molto furono addotrinati, e si valsero della Chimica, e conobbero un solo essere il principio delle cose naturali; e ser secondi principj le cinque sostanze dette da loro metallo, legno, acqua, fuoco, e terra; ma diversi da que' corpi, che comunemente con tal nome si chiamano, e non dissimili per avventura da' principj de' nostri Chimici', Di Capua, *Parere*, 178.
73 Trigault, *De Christiana expeditione apud sinas*, 106, 110.
74 Martini, *Sinicae historiae decas prima*, 9.
75 Lawrence M. Principe, *The Aspiring Adept*, 77.
76 O'Connor, 'Thomas Willis and the background to *Cerebri Antatome*', 139–43.
77 Di Capua, *Parere*, 438.
78 Di Capua, *Parere*, 179. Compare Martini, *Sinicae historiae decas prima*, 310.
79 'Che fattasi colla gran medicina immortali, dimorino nelle cime degli altissimi monti, e quindi vadano, anzi volino dove lor più sia a grado, ed anche in Cielo, *Sciolti da tutte qualitati umane*', Di Capua, *Parere*, 179. The words in italics are cited from Petrarch, *Rime sparse*, 15.14. These Petrarchan flourishes are a feature of Di Capua's literary style.
80 Struever, 'Lionardo Di Capoa's *Parere*', 325.
81 See above, n. 65.
82 Di Capua, *Parere*, 150.
83 Di Capua, *Parere*, 151.

# Bibliography

Albala, Ken. 'Stimulants and intoxicants in Europe, 1500–1700'. In *The Routledge History of Food*, edited by Carol Helstosky, 42–60. London: Routledge, 2015.

Appleby, John. 'Ginseng and the Royal Society'. *Notes and Records of the Royal Society of London* 37, no. 2 (1983): 121–45.

Barnes, Linda L. *Needles, Herbs, Gods, and Ghosts: China, healing, and the West to 1848*. Cambridge, Massachusetts: Harvard University Press, 2005.

Bartoli, Daniello. *Dell'historia della Compagnia di Giesù. La Cina. Terza parte dell'Asia*. Rome: Nella Stamperia del Varese, 1663.

Canaris, Daniel. 'China in Giambattista Vico and Jesuit accommodationism', *Intellectual History Review* 29, no. 1 (2019): 145–63. https://doi.org/10.1080/17496977.2019.1546443.

Canaris, Daniel. *Vico and China*. Oxford University Studies in the Enlightenment. Oxford: Voltaire Foundation, 2020.

Carlino, Andrea. 'La via dell'incertezza: una lettura scettica del Parere sull'incertezza della medicina di Lionardo di Capua'. In *Storia sociale e politica. Omaggio a Rosario Villari*, edited by A. Merola, G. Muto, E. Valeri and M. A. Visceglia, 323–51. Milan: Franco Angeli, 2007.

Conforti, Maria. 'Medicine, history and religion in Naples in the seventeenth and eighteenth centuries'. In *Medicine and Religion in Enlightenment Europe*, edited by Ole Peter Grell and Andrew Cunningham, 63–78. Hampshire, England: Ashgate, 2007.

Conforti, Maria. 'La mutazione linguistica nelle storie della medicina di età moderna'. In *Linguaggio, filosofia, fisiologia nell'età moderna: atti del Convegno Roma 23–25 gennaio 2014*, edited by Cristina Marras and Anna Lisa Schino, 65–83. Rome: Istituto per il Lessico Intellettuale Europeo e Storia delle Idee, 2015.

D'Elia, Pasquale M. *Fonti Ricciane: documenti originali concernenti Matteo Ricci e la storia delle prime relazioni tra l'Europa e la Cina (1579–1615)*. 3 vols. Rome: Libreria dello Stato, 1942–9.

Di Capua, Leonardo. *Parere [...] divisato in otto ragionamenti, ne' quali partitamente narrandosi l'origine, e'l progresso della medicina, chiaramente l'incertezza della medesima si fa manifesta*. Naples: Per Antonio Bulifon, 1681.

Di Capua, Leonardo. *The Uncertainty of the Art of Physick, Together with an Account of the Innumerable Abuses Practised by the Professors of That Art*. Translated by J. L. Gent. London: Printed by Fr. Clark, for Thomas Malthus at the Sun in the Poultrey, 1684.

Driem, George van. *The Tale of Tea: A comprehensive history of tea from prehistoric times to the present day*. Leiden: Brill, 2019.

Fisch, Max Harold. 'The Academy of the Investigators'. In *Science, Medicine and History: Essays on the evolution of scientific thought and medical practice*, edited by E. Ashworth Underwood, 1, 521–63. Oxford: Oxford University Press, 1953.

Giovannetti-Singh, Gianamar. 'Galenizing the New World: Joseph François Lafitau's "Galenization" of Canadian ginseng, *ca* 1716–1724', *Notes and Records* 75 (2021): 59–72.

Hanafi, Zakiya. *The Monster in the Machine: Magic, medicine, and the marvelous in the Time of the Scientific Revolution*. Durham, NC: Duke University Press, 2000.

Hanson, Marta and Gianna Pomata. 'Travels of a Chinese pulse treatise: The Latin and French translations of the *Tuzhu Maijue Bianzhen* 圖註脈訣辨真 (1650s–1730s)'. In *Translation at Work: Chinese medicine in the first global age*, edited by Harold J. Cook, 23–57. Leiden: Brill, 2020.

Kircher, Athanasius. *China monumentis, qua sacris qua profanis, nec non variis naturae et artis spectaculis, aliarumque rerum memorabilium argumentis illustrata*. Amsterdam: apud Jacobum a Meurs, 1667.

Martini, Martino. *Sinicae historiae decas prima: res a gentis origine ad Christum natum in extrema Asia, sive magno Sinarum imperio gestas complexa*. Munich: Typis Lucae Straubii, 1658.

Martini, Martino. *Histoire de la Chine*. Translated by Louis-Antoine Le Peletier. Paris: Barbin & Seneuze, 1692.

Martini, Martino. *Opera omnia: Sinicae historiae decas prima*. Edited by Federico Masini and Luisa M. Paternicò. Vol. 4 (of 5 vols). Trento: Università degli studi di Trento, 2010.

Meynard, Thierry. *Confucius Sinarum Philosophus (1687): The first translation of the Confucian classics*. Rome: Institutum historicum Societatis Iesu, 2011.

Nicolini, Fausto. *La giovinezza di Giambattista Vico (1668–1700): saggio biografico*. 2nd ed. Bari: G. Laterza & figli, 1932.

Nicolini, Fausto. *Commento storico alla seconda Scienza nuova*. 2 vols. Rome: Storia e letteratura, 1949.

O'Connor, James P. B. 'Thomas Willis and the background to *Cerebri Antatome*', *Journal of the Royal Society of Medicine* 96, no. 3 (2003): 139–43.

Principe, Lawrence M. *The Aspiring Adept: Robert Boyle and his alchemical question: Including Boyle's 'lost'* Dialogue on the Transmutation of Metals. Princeton, NJ: Princeton University Press, 1998.

Renaldo, John J. 'Antecedents of Vico: The Jesuit historians', *Archivum Historicum Societatis Iesu* 39, no. 2 (1970): 349–55.

Ricci, Matteo. *Della entrata della Compagnia di Giesù e Christianità nella Cina*. Macerata: Quodlibet, 2000.

Scalabrella, Silvano. 'DI CAPUA, Leonardo'. In *Dizionario Biografico degli Italiani*, 39: 712–15. Rome: Istituto della Enciclopedia Italiana, 1991.

Semedo, Álvaro. *Imperio de la China, i cultura evangelica en èl, por los Religios de la Compañía de Jesús*. Madrid: Juan Sánchez, 1642.

Semedo, Álvaro. *Relatione della grande monarchia della Cina del P. Alvaro Semedo*. Translated by Giovanni Battista Giattini. Rome: Sumptibus Hermanni Scheus, 1643.

Standaert, Nicolas. *The Intercultural Weaving of Historical Texts: Chinese and European stories about Emperor Ku and his concubines*. Leiden; Boston: Brill, 2016.

Struever, Nancy. 'Lionardo Di Capoa's *Parere*: A legal opinion on the use of Aristotle in medicine'. In *Philosophy in the Sixteenth and Seventeenth Centuries: Conversations with Aristotle*, edited by Constance Blackwell and Sachiko Kusukawa, 322–36. Ashgate: Routledge, 1999.

Tacchi Venturi, Pietro, ed. *Opere storiche del P. Matteo Ricci, S.I.* 2 vols. Macerata: F. Giorgetti, 1911–1913.

Trigault, Nicolas and Matteo Ricci. *De Christiana expeditione apud Sinas suscepta ab Societate Iesu*. Augsburg: Apud Christophorum Mangium, 1615.

Trigault, Nicolas and Matteo Ricci. *Entrata nella China de' Padri della Compagnia del Gesù*. Translated by Antonio Sozzini. Naples: Lazzaro Scoriggio, 1622.

Trigault, Nicolas and Matteo Ricci. *China in the Sixteenth Century: The journals of Matteo Ricci*. Translated by Louis J. Gallagher. New York: Random House, 1953.

Turk, J. L. and Elizabeth Allen. 'Bleeding and cupping', *Annals of the Royal College of Surgeons of England* 65 (1983): 128–31.

Vico, Giambattista. *Opere filosofiche*. Edited by Paulo Cristofolini and Nicola Badaloni. Florence: Sansoni, 1971.

Wu Liwei 吴莉苇. *Dang Nuoya fangzhou zaoyu Fuxi Shennong* 当诺亚方舟遭遇伏羲神农. Beijing: Zhongguo Renmin daxue chubanshe, 2005.

8

# Courtly exchange between Bourbon Naples and Hanoverian London after the French Revolution: culture, chemistry and the Herculaneum papyri

Frank A. J. L. James

## Introduction

O ye who patiently explore
The wreck of Herculaneum lore,
What rapture could ye seize
Some Theban fragment, or unroll
One precious, tender-hearted scroll
Of pure Simonides!

That were, indeed, a genuine birth
Of poesy; a bursting forth
Of genius from the dust:
What Horace gloried to behold,
What Maro loved, shall we enfold?
Can haughty Time be just![1]

Thus wrote the poet William Wordsworth (1770–1850), expressing in the final stanzas of his poem 'Upon the Same Occasion', composed in September 1819, the hopes of many at the time that much lost classical literature might be discovered by unrolling the papyri excavated from Herculaneum, south-east of Naples. Written mostly in Greek, these 1,800 or more papyri had lain buried in what is now known as the Villa dei Papiri since Vesuvius erupted in August AD 79.[2] Herculaneum had been rediscovered early in the eighteenth century and the Villa in the 1750s.[3] The papyri aroused European-wide excitement about what long-lost classical texts might be found if only they could be safely unrolled.

Such interest was particularly intense during the opening decades of the nineteenth century, when Europe was embroiled in a seemingly endless war. No cultural work of any kind is conducted in a social or political vacuum, especially if significant resources are deployed. This chapter is concerned with understanding how the papyri, as material objects of cultural, social, political and scientific value, came to such prominence. To do this we need to interpret their historical contexts in at least three interlocking levels.

The first is operational. The papyri went originally to the royal palace at Portici near Herculaneum; it quickly became the Museum, housing objects excavated from the buried town. Since the 1750s enormous effort, resource and expertise has been – and continues to be – devoted to unrolling the papyri by various methods and transcribing, reading and translating their content.[4] These processes may be seen in terms of solving purely technical problems, whether, of understanding their chemistry, for instance, or of reading the Greek (and sometimes Latin). The methods involved can be, and indeed have been, discussed in the vast literature on the Herculaneum papyri in such operational terms. These accounts are invariably (and legitimately) written from the perspective of seeking to understand the history of particular papyri, when were they unrolled, who transcribed them and how they came to be in their present location or destroyed etc.[5]

The second interpretive level is cultural. It suggests that the importance attached to the papyri throughout Europe, by figures such as Wordsworth, was some sort of recompense for the destruction of the continent's Republic of Letters that came about directly due to the communication breakdown brought about by the war. The statistical study by Elise Lipkowitz showed that the correspondence networks that had sustained the Republic during the eighteenth century gradually ceased during nearly a quarter century of war.[6] The third interpretative strand thus relates, as we shall see, to the constantly changing political, diplomatic and military situation in Europe in the decades following 1789.

Within this interpretive framework are some specific themes or commonalities. Soon after their discovery Camillo Paderni (*c.*1715–1781), the Director of the Portici Museum, wrote to the London physician and book collector Richard Mead (1673–1754) to inform him that papyri with legible writing had been discovered in Herculaneum.

Mead regarded this news as so important that he had the letter translated into English and published in the Royal Society of London's *Philosophical Transactions*,[7] a journal that had previously published several articles on the spectacular finds from Herculaneum.[8] In his letter Paderni also reported that he had been commanded by the Bourbon King of Naples, Carlo VII (1716–1788),[9] to attempt opening the papyri, despite their extreme fragility.

This very early report on the papyri referenced many issues that recurred subsequently in their story. Their extreme brittleness (a quality that words really cannot convey), and the consequent difficulty in opening them, became a constant refrain. The initial methods used were purely physical – even, one has to say, brutal (cutting them in half or mechanically unrolling them). Over the decades, however, as we will see, ideas developed of chemically understanding the papyri and treating them accordingly before unrolling. In addition, and beginning early in the eighteenth century, issues relating to cultural ownership and its protection were always present[10] and continued with the excavated papyri. In his letter to Mead, Paderni supplied with a transcription of a Latin text, but 'my fidelity to the king not permitting me to send you any more' meant that he provided only two lines.[11] As Paderni made clear, the Neapolitan state, in the person of the king, claimed cultural ownership of all archaeological discoveries. State involvement, especially by royalty and senior diplomats (thus entailing significant expenditure), continued well into the nineteenth century.

Another key issue concerned who actually should do the unrolling, transcribing, translating etc. Throughout, with some shortish gaps, there existed in Naples a team of unrollers and interpreters. However, so painfully slow was their work, not helped by the Neapolitan culture of secrecy, that complaints and calls for more rapid progress became common, particularly in England. After briefly examining their early unrolling up to 1800, I will then discuss three schemes during the following two decades, all funded and managed by the British government, which avowedly intended to increase drastically the number of papyri unrolled. These projects will be discussed in some detail as they reveal the multiplicity of interactions between state actors and bureaucracies and the limits of their authority, the internal politics of cultural institutions, public interest (reflected in extensive newspaper coverage) and national aspirations and prestige.

All of this led to the papyri becoming prominent as objects of exchange in the cultural politics of the courts of Naples and London. In turn, the process suggests that in addition to endeavouring to recover texts that might be central to European culture, it also attempted to continue courtly practices fractured by the impact of the French Revolution.

## Early unrolling

Unrolling these papyri to allow their text to be read was a formidable conservation task. Very quickly chemical methods, with no empirical or other justification, such as dipping the papyri in wine or in mercury and then cutting them longitudinally in half, were tried but without success.[12] In mid-1753 Antonio Piaggio (1713–1796) arrived from the Scuole Pie in Rome to tackle the problems. To him the most obvious approach was mechanical, but it took him 10 years before he refined a machine to unroll the papyri successfully. Its use also required skill, so the first papyri to be unrolled, the treatise *De Musica* by the Epicurean philosopher Phildoemus (*c*.110–*c*.35 BC), took Piaggio four years; the next, *De Rhetorica* by the same author, took a year. By 1800 only 18 papyri had been unrolled;[13] Elizabeth, Lady Webster, later Lady Holland (1771–1845), visiting Naples in 1792 and not the most patient of people, blamed 'the indolence of the Neapolitans' for the lack of progress.[14] Such painfully slow progress in the physical unrolling was matched by slowness in publication. The decision to publish was not taken until 1787; from then it took six years before the first text, *De Musica*, appeared,[15] some 40 years after being excavated. The first copy to reach England arrived in the autumn of 1795 and was sold to Christ Church, Oxford.[16]

By this time, on his father becoming King Carlo III of Spain in 1759, Ferdinando IV (1751–1825)[17] had ascended the Neapolitan throne aged eight, with a regency until his majority. The effective head of the government was an Englishman, John Acton (1736–1811), who commanded the Neapolitan armed forces. Through him the British Ambassador to Naples from 1764, the antiquary and volcanologist William Hamilton (1731ns–1803), engaged in dealings with the government. Hamilton's early interest in the papyri is evident in a long letter written to him by Frederick Hervey (1730–1803)[18] in 1788; it expressed how unfortunate it was that, despite his influence, Hamilton had been unable to have the papyri unrolled. They might well contain, Hervey opined, unknown or lost classical texts, some possibly relating to the rise of Christianity. In view of their potential importance he suggested, in an early indication of impatience with the speed of work, that the unrolling should be undertaken by foreigners and the king paid for allowing access.[19]

Coinciding with a major eruption of Vesuvius in August 1779, from September that year Hamilton paid Piaggio to make daily observations of the volcano until 1795.[20] Not only must this have been welcomed additional income,[21] but it also put Hamilton into close contact with Piaggio and his other occupation of papyri unrolling. From an early date until just before his death, Hamilton considered publishing on Herculaneum. He indeed brought to London what appears to have been a large quantity of Piaggio's papers on papyri, which subsequently disappeared.[22]

Hamilton's continuing interest in Herculaneum is most noticeable in his correspondence with his close friend Joseph Banks (1743ns–1820), President of the Royal Society of London from 1778 until just before his death. In the 1780s he turned to Banks to help find a superintendent for the gardens at the royal palace at Caserta (north of Naples). Banks provided so much help that Hamilton hinted to Acton that Ferdinando IV should give Banks 'a set of the Herculaneum'.[23] This probably referred to the magnificent eight-volume *Le Antichità di Ercolano Esposte* illustrating the spectacular finds from Herculaneum. The volumes appeared very slowly between 1757 and 1792, and in the ensuing years Hamilton reassured Banks that he would receive a copy.[24]

Charles Blagden (1748–1820), a Secretary of the Royal Society of London from 1784 to 1797 and close to Banks, visited the Portici Museum early in 1793.[25] Reporting to Banks, he gave a long description of a text called the

> trattato di Fisica ... the work of that eternal book-maker Philodemos & intended to prove that it was unsafe to form general conclusions from particular facts.[26]

The tone of this passage, followed by a long description of the procedures adopted to transcribe, translate and interpret the papyri, once unrolled, indicates the contemporary disappointment felt at the (in)significance of what was being revealed after so much time and effort.

By this time, however, the wars that followed the French Revolution had commenced. Although there are further references in the 1790s to the papyri, as in Blagden's diary, they became perfunctory as the conflict continued and spread. A very confused political situation arose in Naples, which resulted in Ferdinando IV fleeing the city to Palermo in Sicily for the first half of 1799, during the establishment of the very short-lived Parthenopean Republic. This state was under French protection, and arrangements began for the transfer to Paris of objects from the Portici Museum. These items did not include the papyri as they had been taken to Palermo, an indication of their cultural value even at the worst of times.

With the aid of the Royal Navy under Horatio Nelson (1758–1805), Ferdinando IV and his government were restored in mid-1799. They instituted savage reprisals on the Republic's supporters, not least because his queen, Maria Carolina (1752–1814), never forgave the Jacobinism that had sent her sister, the French Queen Marie Antoinette (1755–1793), to the guillotine. Though the Bourbon restoration prevented the French from looting Neapolitan cultural property, the peace agreement negotiated between Naples and France provided for transferring many antiquities from Pompeii and Herculaneum to Paris. These objects, including six papyri, arrived in Paris in the autumn of 1802. The French military dictator, Napoleon Buonaparte (1769–1821, styled from 1799 First Consul and from 1804 Emperor), ordered the papyri to be placed in the Institut de France. Everything else went to Malmaison, the private palace near Paris of Buonaparte and his wife Joséphine Buonaparte (1763–1814).[27] While Buonaparte would have doubtless viewed this as courtly exchange, it did have the appearance of *force majeure*, rather than voluntarily gifting objects, albeit usually for reasons of diplomatic influence.

## John Hayter

From early 1800 the Prince of Wales (1762–1830, from 1811 Prince Regent and from 1820 King George IV), began taking a strong interest in the papyri that lasted for 20 years. A major cultural patron,[28] the Prince doubtless chose to involve himself specifically with the Herculaneum papyri to associate himself potentially with the (re)discovery of culturally valuable texts. It has been suggested that Hamilton facilitated his involvement.[29] However, the timing does not quite work as the Prince started to take an interest in them shortly after Hamilton's recall from Naples, as shown in a letter dated 20 December 1799 (though Hamilton did not return to London until November the following year). Naples had assumed a far more prominent role in British strategic thinking and required professional diplomacy rather than the attentions of a dilettante connoisseur. Furthermore, Hamilton had been disgraced because of his mounting debts and the affair between his wife, Emma Hamilton (bp.1765–1815), and Nelson, as well as the involvement of all three in the reprisals against the Parthenopean Republic's leaders.[30]

What is possible, however, is that with Hamilton's imminent departure from Naples, the Prince realised that there would be no English interest in the papyri. This may have been drawn to his attention by the Prince's Private Secretary between 1795 and 1803, Thomas Tyrwhitt (1762–1833), a Devonshire MP.[31] Unusually for a servant of the Prince's, Tyrwhitt never lost favour and acted for him on papyri matters throughout. Nephew of a famous classicist of the same name, Tyrwhitt was also interested in the classics, although not particularly successfully. It is possible that his Devonshire connections led to his acquaintance with the classically trained clergyman John Hayter (1755–1818), with whom he had much in common.[32]

However the Prince of Wales became aware of Hayter, it was he whom he appointed to speed up the unrolling of the Herculaneum papyri. Hayter had discussed the problems with Banks, who suggested that unspecified chemical processes might be of use.[33] On 18 March 1800 Tyrwhitt wrote to Hayter giving him the Prince's instructions to proceed to Naples, find out what needed to be done and commence operations; he also authorised Hayter to draw £150.[34] Two days later, in a pamphlet addressed to the Prince, Hayter reviewed the situation. His idea for increasing speed was

> that if the Frames should be multiplied to the proposed extent, several Pages of Thirty different Manuscripts might be disclosed, and transcribed within the Space of one Week.[35]

Never having seen a Herculaneum papyrus before, this passage illustrates Hayter's profound ignorance about the problems involved. A few days later the project's 'proposed extent' became clear in a newspaper account. This reported that the Prince had commissioned the Neapolitan Consul-General in London, Francesco Sastres (d.1822), to ask Ferdinando's permission to undertake the project 'to set 30 [frames] at work at once, by which means the whole 600 volumes may be in due time transcribed'.[36] It would thus appear that the project had been planned in London during the previous weeks or months without any consultation with the owners. Possibly the assumption in London was that because the Royal Navy had restored the Bourbons, the Neapolitan government would be happy to co-operate.

In April Hayter left England, leaving his wife and five children with a curate in charge of his parish. Sailing, courtesy of the Royal and Neapolitan navies, via Minorca, Genoa and Palermo, he reached Naples in late July.[37] There he found, much to his surprise, that because of war and illness the papyri had not yet been returned from Palermo to Naples.

Hayter thus returned to Palermo and eventually located the papyri boxes in the dockyard magazine where, in the confusion of evacuation, they had been deposited. With aid from Arthur Paget (1771–1840), Britain's relatively new Minister Plenipotentiary in the city, he 'procured a Royal Order for the placing [of] them under my superintendency, and for developing them'.[38] Furthermore, in June 1801 Paget secured a further £1,200 from the Foreign Secretary Robert Banks Jenkinson, Lord Hawkesbury (1770–1828, later (1808) 2nd Earl of Liverpool).[39]

In the meantime Paget had left for Vienna, which meant that Robert Waldron temporarily became the senior British diplomat in Palermo as *Chargé d'Affaires*. He negotiated some rooms in Palermo's former Jesuit College of San Francesco Saverio where Hayter could begin work unrolling the papyri, though he encountered significant obstruction in his work.[40] In August William Drummond (1770?–1828), a classicist, arrived as envoy extraordinary. He soon drew the money approved the previous year by the Foreign Secretary[41] and arranged for the papyri to be returned to the Portici Museum, directed from 1802 by Carlo Maria Rosini, Bishop of Pozzuoli (1748–1836).[42]

The papyri arrived in Portici on 23 January 1802 and Hayter, presumably at Drummond's instigation, was given a nearby house.[43] Hayter began working on the papyri and, although friction arose between Rosini and Hayter, the substantial sums given to the latter enabled him during the next 18 months to increase the number of unrolling machines to seven, operated by 13 staff.[44] The texts revealed would then be copied simply as images, his workers not knowing Greek. Due to the brittleness of the papyri these *disegni*, as they were named, are often the most complete, or sometimes the only, surviving record of those papyri.[45]

The Preliminaries (30 September 1801) of what became the Peace of Amiens (25 March 1802) made communication between London and Naples easier for nearly 20 months. The results of Hayter's work thus became known in England where they were favourably reported, for example to the Society of Antiquaries and (especially) in the provincial press.[46] This doubtless helped to support the Prime Minister and acting Chancellor of the Exchequer, Henry Addington (1757–1844, later (1804) Viscount Sidmouth), in his moving a motion in the House of Commons that a further £1,700 be authorised for Drummond to spend on the papyri, that is for Hayter's use.[47] During October the provincial press published very similar reports on the project, praising Hayter and the Prince of Wales for the project's successes.[48]

Reports of Hayter's work then ceased for nearly two years. This might be partially attributed to the resumption of war following the breakdown of the Peace of Amiens (18 May 1803), but it was also due to the arrival soon after of Hugh Elliot (1752–1830) as the new British envoy to Naples replacing Drummond. Very quickly Hayter met Elliot, who informed him that he had no instructions about continuing payments to Hayter. The following day Hayter provided Elliot with an account of the project detailing payments to staff and stating that since 1802 the project had unrolled 41 papyri; a further nine were still being processed. Publication, he wrote, was being delayed by the Neapolitan authorities who refused to let Hayter pay for the *disegni* to be engraved[49] – an indication of the unresolved issues over who actually owned the contents of the papyri.

A month later Elliot referred the matter to Hawkesbury, enclosing Hayter's letter and asking for instructions. He noted that the last bill authorised by Drummond had not been honoured, which 'will I fear materially contribute to stop the further progress of the unrolling of the Manuscripts'.[50] Elliot's delay in referring the matter to London may have been due to Hayter working to obtain the King's permission (granted 18 July)[51] to give the Prince of Wales some papyri, possibly intended as a justification for continuing the funding. Elliot ended his letter with news of the gift of six papyri which he had entrusted to the Legation's Secretary, William A'Court (1779–1860), to take to England.[52] Their imminent arrival prompted Tyrwhitt to muse about the possibility of offering a prize for inventing a better unrolling machine, but nothing came of that.[53]

Although Elliot had still not received further instructions by April, he continued to authorise Hayter's expenditure – though he then said that would be his final payment until he heard from London.[54] It would appear that Elliot found himself in something of a dilemma. On the one hand, he had no authorisation from the government (who, after all, were then more concerned with preparing to meet a possible French invasion of Britain) for further expenditure on unrolling the papyri. On the other, Elliot was also well aware of the Prince of Wales's personal interest and was conscious that he might not be pleased if the project came to an abrupt end. A wrong decision would further impact Elliot's already somewhat lacklustre diplomatic career, during which he had already received two formal reprimands. At the start of July 1805, having still received no instructions, Elliot thus advanced Hayter 500 ducats to cover both back wages and payments until October. He also expressed his severe displeasure that Hayter had not, as he had strongly advised, written to any government minister asking for the instructions to be sent.[55]

It may have been in this context that a list of 177 papyri opened between 1802 and November 1805, the vast majority of which were fragments, was handed to Samuel Taylor Coleridge (1772–1834) – then in Naples on his way north, having until recently been Secretary of the Government of Malta. That list went no further, however, and remained in Coleridge's papers.[56] Hayter also wrote to Tyrwhitt detailing what he had so far achieved, doubtless with the hope that this would be passed onto the Prince of Wales. Unfortunately, that letter has not been found, but early in 1806 Hayter wrote again, informing Tyrwhitt about his work in the intervening months and including another list of the papyri opened.[57]

Despite all the financial problems, together with those of communicating with Britain, Hayter did his best to ensure that the project remained before the Prince and the public. For instance, in May 1804, writing to an unknown correspondent though published widely, Hayter, while emphasising the difficulties of the work, described his successes so far in unrolling by then nearly 90 papyri in two years. He admitted, however, that seven Latin papyri had suffered a 'fatal accident which reduced them to powder'.[58]

The other thing that kept the project in the news was the arrival in London during November 1804, after more than two years, of Ferdinando's gift of six papyri.[59] Although the newspapers suggested that scholars would vie with one another to work on the papyri, the Prince of Wales immediately sent for Banks to work on them – a decision that pleased his friend, the recently retired President of the Board of Trade, the 1st Earl of Liverpool (1729–1808).[60] Banks made extracts from papers on Herculaneum published in the *Philosophical Transactions* and also transcribed the account by the elder Pliny (*c.*AD 23–79) on growing the papyrus plant.[61] It took a little time to prepare the apparatus that he needed, including a mercury bath, steam jackets and an unrolling machine of some sort. Around 10 December Banks immersed a papyri in mercury; a week later, he began working on that and another with Taylor Combe (1774–1826), the newly appointed superintendent of the coins and medals collection at the British Museum. The papyri were heated in a steam jacket but by 20 December, after three days work, all they had achieved was detaching a fragment five by two inches (13 × 5 cm).[62] Neither Banks nor Combe had been to Naples to witness the papyri being unrolled, so during those days a couple of men who had observed the process, Mr Brown and Edward Ash (*c.*1765–1829),[63] visited to give their opinions. Both said that the rolls that Banks and Combe worked on were darker than those they had seen in Naples.[64]

At some point Thomas Young (1773–1829), formerly Professor of Natural Philosophy at the Royal Institution and from 1804 Foreign Secretary of the Royal Society of London, became involved. He worked on the same two papyri and initially used an anatomical blow-pipe, finding that the dampness of the breath helped to soften the papyri. Like Banks, he tried various chemicals, including immersing one papyri in water for six months.[65] Unsurprisingly these two papyri did not survive these processes and only a single letter was recovered.[66]

The letter that Banks wrote to Tyrwhitt reporting the 'little success' achieved, stated he had decided that 'after this disappointment we determined not to proceed farther till new means of unrolling more likely to succeed should be devised'. He added, however, that he had obtained (possibly from Hamilton's papers?) Piaggio's description of his unrolling method, which he would have translated, and so hoped to learn more about how best to proceed.[67] Tyrwhitt responded smoothly, saying that were it not for the war he would go to Naples to bring back two or three of the Neapolitans who had unrolling experience.[68] He emphasised the Prince of Wales's full confidence in Banks and asked him to take full charge of all the papyri. However, Tyrwhitt stressed that, though he remained interested in unrolling the other four papyri, there was no intention of sacrificing them.[69] Pleading gout and the cold weather Banks declined the invitation 'even if H.R.H. was in person to delegate it to me'.[70]

On 8 February 1806 the French army invaded the Kingdom of Naples. While fighting continued for a few months in a number of places, a week later Joseph Buonaparte (1768–1844) – appointed King of Naples by his younger brother, the now Emperor Napoleon – triumphantly entered the city. By then Ferdinando IV and Maria Carolina had once again fled to Palermo with the rest of the court, including Elliot and Hayter, who took the *disegni* with him.[71] In the chaos the papyri were left in Portici, and furthermore Hayter appears to have given his facsimiles to the Neapolitan government.[72] When he learned about all this, Tyrwhitt admonished Hayter severely and refused him permission to return overland to London, although he wrote that he might return on a Royal Navy ship. He hoped that Elliot would help Hayter to recover the facsimiles, adding that the new Foreign Secretary, Charles James Fox (1749–1806), had written with that instruction.[73]

Regaining the facsimiles probably prompted Hayter initially to remain in Palermo, doubtless helped by Drummond's return as envoy in October 1806. Hayter, with Drummond's help, succeeded in early September 1807.[74] As a consequence, he received a conciliatory letter from Tyrwhitt, written in early 1808, asking Hayter to arrange for them to be sent to London as soon as possible.[75] But in July Drummond took charge of the facsimiles since, shortly after their return by the government, Hayter seems to have suffered a serious breakdown. In mid-December Drummond, about to be replaced as envoy by William, 2nd Baron Amherst (1773–1857), sent Hayter a long and severe reprimand. He accused him of an '*entire and utter* neglect of' his employment 'ever since the *papyri* were restored by the Sicilian government'.[76] Indeed, he and the classicist Robert Walpole (1781–1856), then in Naples, had offered to help Hayter; the implication is that this had been rejected. Such a move might not have been entirely disinterested, as Drummond and Walpole were writing a collection of essays, dedicated to the Prince of Wales, on Herculaneum – including a couple stemming from the papyri.[77] But it was Hayter's personal behaviour that most drew Drummond's ire. The Neapolitan foreign minister Tommaso Somma, Marquis de Circello (1737–1826), had sent him a letter of

> remonstration … on the broils, quarrels, and riots, occasioned by persons living in my house. The stories of your battles in brothels, &c, so made much noise, and were multiplied so fast, that I found it impossible to apologise for your conduct.[78]

Matters got worse in April when Hayter was accused of attempting to abduct a young girl from a convent.[79]

Consequently, nine years after first arriving in Naples Hayter was recalled to England.[80] There he seems to have benefited from what appears to have been a political cover-up designed to minimise any embarrassment that knowledge of his behaviour might provoke. He was appointed a Chaplain in Ordinary to the Prince of Wales (possibly to provide him with a salary)[81] and as such was presented to the Prince at the second levée of the Regency with the title of HRH's 'Superintendent of the Herculaneum Manuscripts'.[82] In 1811 Hayter claimed £1,156 in backpay for his Herculaneum work.[83] In either May or June he published a self-serving account of his time in Naples and Palermo in the form of a (long) letter to the Prince Regent.[84]

In March 1810 the Prince of Wales gave Hayter's facsimiles and the remaining four papyri to Oxford University, where they remain in the Bodleian.[85] The presentation had been brokered by William, Lord Grenville (1759–1834), who served briefly as Prime Minister in the 'Ministry of all the talents' between 1806 and 1807 and who, for the following 10 years, led the Whig opposition.[86] It would appear that the gift was part of Grenville's successful campaign to be elected University Chancellor; certainly the papyri were referenced in the verse spoken at his installation in the Sheldonian.[87] Doubtless these gifts helped motivate the University to confer an honorary doctorate on the Prince in May 1810.[88]

In 1811 Oxford University formed a committee to deal with the Prince's gifts. Because of Hayter's extensive knowledge they offered him £200 to edit, engrave and publish the facsimiles.[89] The attitude of the Prince and Tyrwhitt to this was not entirely straightforward and it is very hard to understand their thinking. According to Hayter, they first refused him permission to undertake the work, then insisted that he did. Finally Tyrwhitt compelled Hayter to resign from the project on the grounds of ill health.[90] In the end he worked in Oxford from February to August 1812, but in October the committee noted that Hayter had 'quitted [*sic*] the University under circumstances which seem to preclude all idea of his returning to fulfil his engagements';[91] nothing was published until the mid-1820s.[92]

Despite all the manoeuvrings in 1815 Hayter was working, in London, on one of 1804 papyri yet to be transferred to Oxford.[93] He then went to Paris, allegedly to work on the papyri there (with no effect). He got into financial difficulty and was obliged to ask the British ambassador for money.[94] Hayter died of an apoplexy in Paris towards the end of 1818, aged 63.

## Friedrich Sickler

Meanwhile in Naples, in 1808 the Emperor replaced Joseph as king with their brother-in-law Joachim Murat (1767–1815), who reigned until May 1815. The allies then once again restored Ferdinando IV as King of Naples; he soon had Murat executed by firing squad. From the end of 1816 the crowns of Naples and Sicily were united under the restored King, reigning as Ferdinando I of the Two Sicilies. In spring 1816 the British government saw an opportunity for a major cultural coup which, had it come off, would have been in the same rank (and controversy) as the transfer of the Elgin Marbles from Athens to London. Tyrwhitt asked A'Court, now back in Naples as envoy, if the government would transfer all the Herculaneum papyri to England; the answer, maintaining Neapolitan cultural integrity, was a resounding no, as A'Court told the Foreign Secretary Robert Stewart, Viscount Castlereagh (1769–1822). He added, however, that Ferdinando, presumably as a sign of gratitude for his second Restoration, would gift some papyri;[95] of these, 12 were sent to the Prince Regent and arrived in England in July.[96]

Towards the end of 1816 Tyrwhitt travelled to Paris, doubtless to see how Hayter's work there was progressing – slowly – and possibly with the idea of re-employing him to deal with the newly arrived London papyri. While there, however, Tyrwhitt heard about the work of Friedrich Sickler (1773–836), a philologist from Hildburghausen in Saxony.[97] Sickler had spent time in Naples during a six-year visit to Italy, where he seems to have become acquainted with Hayter, or at the very least with his work. Sickler had access to at least one papyrus and developed, in his view, a better unrolling method – involving using a liquid to soften the papyri – before subjecting them to a machine similar to Piaggio's.[98]

Sickler had presented his results to the *Königliche Gesellschaft der Wissenschaften* at Göttingen. Their *Göttingische gelehrte Anzeigen* 1814 volume published a positive report on his work, though not Sickler's memoir itself.[99] He seems to have sent that document to Tyrwhitt, who received it in Paris. Evidently impressed by Sickler's memoir, although he kept the precise details secret, Tyrwhitt, seemingly on his own initiative, invited him to come to Paris (it seems unlikely that he did) and then London to test his unrolling method.[100] By return, Sickler replied to Tyrwhitt setting out his terms. These included a payment of £3,000, as well as travelling and living expenses for him and his wife, and thereafter a £100 annuity.[101] Possibly indicating some concerns in London, Banks, Grenville and the classicist Charles Burney (1757–1817) discussed Sickler's Greek.[102]

Nevertheless, it was agreed that a high-level committee to oversee the work should be formed. The possible members were Banks, Burney, Grenville, Tyrwhitt, George Hamilton-Gordon – fourth Earl of Aberdeen (1784–1860) and President of the Society of Antiquaries – and Humphry Davy (1778–1829), England's foremost chemist and one of Europe's. All these individuals were named in a letter of early April that Castlereagh, on the Prince Regent's behalf, wrote to Tyrwhitt. In it he authorised Tyrwhitt to visit Sickler in Hildburghausen to make the necessary arrangements.[103] Following Sickler's arrival in London on 12 June 1817,[104] Tyrwhitt invited the former House of Commons Speaker, Charles Abbot, Lord Colchester (1757–1829), to join the committee; he noted it in his diary.[105] The final member, William Richard Hamilton (1757–1817), Under Secretary of State at the Foreign Office, had been responsible for ensuring that the Rosetta Stone and the Elgin Marbles were transported safely to the British Museum. Judging by the tone of their correspondence, he and A'Court were good friends. A Parliamentary official, James Pulman (1783–1859), served as the committee's secretary.

Just over a fortnight after his arrival in London, Sickler first met the committee (present: Colchester, Tyrwhitt, Burney, Davy and Hamilton) in Tyrwhitt's apartment in the Houses of Parliament.[106] Throughout the summer and into October the committee and Sickler continued work, but without success. They received advice from Blagden, Combe and Drummond, as well as A'Court, who had seen the papyri unrolled in Naples.[107] At their third meeting Davy examined the liquid to be used by Sickler and on that basis the committee authorised Sickler to continue work;[108] in the committee's report, Davy commented that the liquid would not damage the papyri, though he added that it would not help in their unrolling.[109] This appears to have been Davy's sole contribution to the committee's work, since in mid-July he left London[110] to tour Scotland and north-east England, returning in mid-October. The committee report concluded 'that Dr. Sickler has totally failed in his endeavours to satisfy the Committee'.[111] Sickler left London on 26 October, the whole project having cost the large sum of £1,117.7s.[112]

## Humphry Davy

Before he left London for the north-east, Davy had suggested to Colchester 'a mode … for unrolling the papyrus, by immersing it in a gelatinous fluid of moderate heat, and then suddenly cooling it, which might detach the laminae'.[113] Following Sickler's failure, it would seem that the next step would be to see whether Davy's ideas were workable. It is not clear where – or precisely when in early 1818 – Davy undertook this research;[114] as with his development of the miners' safety lamp, no contemporary laboratory notes of his experiments have been found. Davy later wrote that Tyrwhitt secured for him papyri samples previously used by Sickler and by Hayter. Furthermore Young, Davy's former colleague at the Royal Institution, sent him some papyri pieces that he had been unable to unroll himself.[115]

Davy concluded that the nature of the papyri had been misunderstood. It had been supposed that they had been carbonised by the heat of volcanic material. Davy argued that they were like peat which had been formed by fermentation and chemical reactions due to the vegetable nature of the papyri. A chemical method to reverse the process should thus be possible. Davy proposed that by exposing the papyri to small quantities of chlorine or iodine and then heating in a controlled manner, the leaves became detachable and legible,[116] which Banks opined showed that the chemistry worked.[117] But as Sickler had doubtless taken his unrolling apparatus when he departed, there was no possibility of Davy undertaking the next stage of physically unrolling the London papyri. These survived and are now in the British Library.[118]

To show that Davy's process could be used to unroll papyri, there was no option but for him to go to Naples. On 24 May 1818 Davy attended an audience with the Prince Regent who commanded him to continue work in Naples, though it is not clear whether his expenses would be met. Davy, together with his wife, Jane, Lady Davy (*c*.1780–1855), left London the following day.[119] Both seasoned travellers, the couple did not hurry south, reaching Rome only in mid-October; later that month the Regent raised Davy from a knight to a baronet.[120]

The protocols followed by the Prince and his advisors in sending Davy to Naples had remarkable similarities to those which sent Hayter there in 1800. The project must have been planned in the weeks and months before and, as with Hayter, the letters to the Neapolitan authorities were sent well after Davy had left London. In August Castlereagh wrote to A'Court, emphasising the Prince's personal interest in the matter to 'open to the Public these valuable stores of ancient Literature'. He enclosed a copy of his letter to Davy and the Regent's letter to Ferdinando, which A'Court presented at the end of September.[121]

Naples responded positively, but Davy did not learn this until mid-November while in Rome.[122] He eventually arrived in Naples at the start of 1819, only to find that the King and A'Court would leave two days later for Persano (another Bourbon palace, south-east of Salerno). However, before A'Court departed Davy talked to him, reviving the suggestion (dismissed by A'Court) of transferring the papyri to London where they might be worked on by the 'great Grecian' Peter Elmsley (1774–1825), then at Rome.[123]

Davy, initially given every facility in Naples, began by inspecting the Villa dei Papiri, the condition of which confirmed his views of the chemical nature of the damage to the papyri.[124] He also spent 10 days working on the papyri in the museum, determined to keep his method secret.[125] That appears to have been a major source for the tension that emerged with the director of the Museum, Carlo Rosini, and may account for Davy spending less than a month there before returning to Rome in early February.[126] Davy's concern with secrecy throughout the first half of 1819 was probably connected with his bruising experiences in defending his priority over the invention of the miners' safety lamp; news of its design had circulated before its formal publication, and this he claimed (probably incorrectly) had allowed others to develop their own lamps.[127] The secrecy may also account for A'Court's perception of Davy being 'very, very lukewarm' about the papyri, especially when in April, owing to Lady Davy's poor health, they headed north from Rome rather than return to Naples.[128]

Once back in Rome, Davy worked quickly. He experimented on some papyri that he had brought with him and wrote a long report, dated 12 February, parts of which had been drafted before he left London.[129] Sending this to William Richard Hamilton at the Foreign Office, Davy accompanied it with a letter expressing his view that the reason for the slowness of unrolling was due to the Museum staff wanting 'a job for life'; he added that mentioning such views in Naples would earn him a stiletto.[130] Davy, now confident that his (still secret) method would enable the unrolling, recommended in both the report and letter employing a chemist and a Greek scholar to manage the work. Though Davy still hoped that some 'celebrated long-lost works of antiquity' might yet be found, he had resigned himself to the 'probability' that only texts on materialism would be found. Nevertheless, he believed that continuing the project would be worthwhile.[131]

A few days later Davy wrote to the former laboratory assistant at the Royal Institution, Michael Faraday (1791–1867), whom he clearly trusted as Davy disclosed his process:

> I find that by raising to heat very slowly ie taking 5 or 6 hours to raise it to 600°ft the separation of the leaves is effected without any fracture or injury to the MSS. In the brown MSS a low atmosphere of Chlorine seems to assist the effect of heat. Of course I mention this in confidence to you & I wish for the present to have everything relating to *the nature of the process* kept a secret. I have sent a report on the state on the MSS to our government with a plan for the undertaking of unrolling.[132]

Davy then hinted that he might suggest Faraday as the chemist to come to Naples to continue the work, depending on the Prince Regent's reaction to the report. Faraday's reply has not been found, but in view of his difficulties in accompanying the Davys on their 1813–15 continental tour, it may be easily imagined.[133]

Clearly Davy's report had to be widely distributed and it would appear that one of its recipients, Joseph Banks, sent it to the *Quarterly Journal of Science* for publication. Davy's successor at the Royal Institution, William Thomas Brande (1788–1866), edited this journal, but Faraday frequently deputised for him. Hence Davy, now in Florence, wrote urgently to Faraday requesting him not to publish it.[134] However, that had already occurred; it was reported extensively in the press and reprinted in the widely read weekly *The Literary Gazette*.[135] Probably the report, with its optimistic tone, prompted Wordsworth, who knew Davy fairly well, to reference the project a few months later in 'Upon the Same Occasion'.[136]

The report's publication vexed Davy 'more than I can well express', since he thought it would damage his chances of obtaining support for continuing the work.[137] He need not have worried. Later in May the Prime Minister, now the Earl of Liverpool, and Castlereagh authorised A'Court to allow Davy to draw up to £1,500 for the project.[138] Hamilton wrote directly to Davy, at Bagni di Lucca, with the good news, to which he responded enthusiastically two months later (having been in Carniola).[139] On staffing, Davy accepted that Faraday would not be the chemical operator and so was looking for one 'accustomed to do even the dirty work at Lectures'.[140] For expertise in 'Calligraphy' and 'Greek',[141] he would write respectively to William Gell (1777–1836), a prominent Naples-based antiquarian, whom he had suggested in the letter to Hamilton that accompanied his report, as well as to Elmsley, with whom he had discussed the project in Florence,[142] offering them engagement on the project.[143] Davy had Castlereagh's consent for both appointments.[144]

Unfortunately, Elmsley, who had been in Italy, was returning to England and by the end of July had reached Paris. There a couple of weeks later he received news of the offer in a letter, dated 21 June, from a friend in London which had followed him from Florence.[145] Around the same time Elmsley heard further details about the project from his friend William Sotheby (1757–1833).[146] However, it was not until towards the end of August, when he received the Foreign Office offer of the position, that he decided to visit England briefly before returning to Italy towards the end of the year.[147] Banks's view, based on a letter from Davy, was it would have been better to wait until the rolls had been transcribed before appointing a Grecian.[148]

Despite the government support he had received, Davy did not hurry back to Naples. From Bagni di Lucca he went first to Florence and then north again to Bologna, before returning to Rome in mid-October. Leaving there after six weeks, he finally arrived again in Naples on 4 December where he found Elmsley, who had reached Naples on 25 November.[149]

Davy's problem was that the Neapolitan government had not yet granted him permission to work on the papyri. On 12 December he explicitly stated to A'Court precisely what he wanted, enclosing an extract from his report to the British government.[150] Instead, Davy spent much of his time studying Vesuvius, again erupting, ascending it on 5 December.[151] Access to the papyri was granted around the end of the year and Davy began work. The secrecy that had pertained during his visit in early 1819 vanished and Davy explained the process to visitors such as Elizabeth Fremantle (1778–1857) and Sydney Morgan (bp.1783–1859).[152]

In a letter to Banks at the end of January 1820 Davy, though he claimed 'great success' with the project, mostly discussed Vesuvius, adding that he intended to return to England in the spring.[153] He also mentioned that almost everything unrolled was 'of the Epicurean shop [i.e. by Philodemus]'. A'Court, in a mid-February letter to Hamilton, made clear that the issue was the contents of the papyri. He reported that Davy and Elmsley were making 'great progress' with unrolling and transcribing the papyri, but unfortunately little of interest had been found, adding that both Davy and Elmsley doubted the value of continuing the project.[154] As the younger John Soane (1786–1823) put it in a letter to his father written from Naples in early 1820, the contents of the papyri were 'interesting to neither artists nor scholars'.[155]

During the two months Davy worked on the papyri, he employed six people at the Museum (in addition to Gell and Elmsley). They managed partially to unroll 23 papyri and to make limited transcriptions.[156] At first things seem to have gone well, but in late January Elmsley reported tensions with the Museum[157] and in early February Rosini claimed that Davy's processes were destroying letters on the papyri. Davy strongly rebutted that charge in two letters to A'Court, where he peremptorily ended the project and settled the financial matters.[158] Elmsley departed for Florence,[159] telling a friend that A'Court refused to intervene without instructions from London and that his own presence, as 'a ninth [interpreter] from Oxford', caused more hostility than Davy's.[160]

In his letter to Hamilton describing what had gone wrong, A'Court provided the same explanation. He added that the problems lay with the staff of Museum who were fearful for their jobs if Davy – and particularly Elmsley – were too successful. Casting his own role in a favourable light, A'Court declared that 'no obstacles [were] thrown in the way of the undertaking by this Govt. If there had been any, it would have been my business to remove them'. (Even on the day he ended the project Davy found the Neapolitan foreign minister, Circello, 'aimable & courteous'.)[161] A'Court thought that if Davy had spent more time in

Naples he would have unrolled more papyri successfully. However, he took the view that Davy had done enough to show that 'very little hope can be entertained that anything of value exists in the Collection'.[162]

By 7 March Davy had returned to Rome where he spent a month or so[163] before leaving for London, arriving just before Banks's death on 19 June 1820. Davy presented George IV with a magnificent bound volume of drawings that Gell had made of the Herculaneum papyri 'unrolled by Chemical means',[164] and in November was elected to Banks's Presidential chair. Davy then wrote his paper on the Herculaneum papyri, read to the Society on the Ides of March the following year.[165] In this he used Elmsley's presence as the explanation of the tensions with the Neapolitan authorities,[166] never being one to accept blame himself.

## Conclusion

Davy's imperious, not to say impetuous, behaviour in suddenly leaving Naples suggests two things. First, that owing to the breakdown of the Republic of Letters during the wars[167] Davy, unlike Hamilton, did not know the etiquette of dealing with other European cultures; he thus lacked the necessary cultural sensitivity.[168] Second, that lack led to tensions with the Neapolitan authorities which came to a head at the end of February 1820. While both of those factors were certainly present, and did not help the situation, simply to blame the failure on Davy's tactlessness in his dealings with others (of which there are many examples)[169] misses the longer historical and cultural perspective.

What happened was not just the outcome of the very few months Davy spent in Naples, but the culmination of problems, also present during Hayter's time in Naples, that had existed during 20 years of British involvement with the papyri. Such friction may have originated with the initial secrecy to which the Herculaneum finds were subjected from the early eighteenth century onwards. Hamilton's interest in the papyri, on the other hand, seems not to have caused any problems in Naples; perhaps his close connection with the Royal Family prevented that. Only after his recall and the beginning of direct British government involvement did problems arise.

They would have been exacerbated throughout the period by the protection of the Bourbon monarchy by British armed forces, who twice restored it to the Neapolitan throne. That demonstrated clearly that Naples could not defend itself properly, an asymmetry of power (reflected in this chapter) that made the Neapolitan government firmly indebted to Britain. But Naples played to its cultural strengths and gifted only a few papyri, perhaps with the idea that there were always more which could be used should the need arise. The fact that papyri were sent to the Prince of Wales in 1802 and again in 1816 (soon after both Restorations) indicates Ferdinando's recognition of Britain's strong cultural interest in them – a view confirmed when the British government attempted to acquire the entire collection. The Prince Regent, trying to continue the cultural reciprocity of the *ancien régime*, as opposed to the outright looting or *force majeure* employed by the French government to acquire cultural material, responded to the 1816 papyri gift by sending five kangaroos from the Royal Menagerie to Naples – playing to Britain's cultural (and imperial) strengths.[170]

For two decades after 1800 the British government at the highest levels initiated three well-funded projects to unroll the Herculaneum papyri owned by the government of Naples. Despite fighting a global war for much of that time, heads of state, leading ministers and diplomats spent appreciable amounts of time and effort on these projects. The reaction of George IV to one outcome, the Gell album, has not been found; however, so far as the evidence allows, he appears to have taken no further interest in the papyri. This perhaps reflected the broader failure of the project to recover from them anything then regarded as culturally significant; only a certain kind of culture, expressed by Wordsworth, was of interest.

Neither England's foremost chemist nor a leading Grecian had found any justification for continuing work in those circumstances. So, it could be argued, George IV had made the wrong decision in 1800 to concentrate so much effort on the papyri.[171] Nevertheless this episode serves to illustrate the central importance played by the possession of material cultural objects on both European and national scales during the conflict. That role continues to be a feature of warfare, as has been recently demonstrated so graphically in both Syria and Ukraine.

## Appendix: two *dramatis personae*

*John Hayter*. The son of an Anglican clergyman, he studied at Eton and King's College, Cambridge. There he won the gold medal for the best Greek ode in imitation of Sappho (c.630–c.570 BC) and for three years from 1776 was a Fellow of the College. From 1779 he served as Rector of Chagford, Devonshire, a village about 10 miles south-east of

Okehampton and 13 miles north-east of Tor Royal. In 1810, following his return to England, he was presented with the Rectory of Hepworth, Suffolk, though he left the parish in the care of a succession of curates. (Methold, 'The parish of Hepworth', 404–6).

*Thomas Tyrwhitt*. The son of an Anglican clergyman, educated at Eton and Christ Church, Oxford, he practised as a lawyer before election as MP for Okehampton in 1796. He later sat for Plymouth, ceasing to be an MP in 1812 when he became Black Rod, a position he held until shortly before his death. Tyrwhitt's long connection with Devonshire included enclosing land (which he named Tor Royal), where he produced very high quality flax, and helping to establish Dartmoor prison (originally called Princeton) for prisoners of war. He did not merit an entry in ODNB.

## Notes

1 Wordsworth, 'Upon the same occasion' in Wordsworth, *The River Duddon*, 189–2, on 192 and Ketcham, *Shorter Poems*, 284–6.
2 There is some uncertainty about the precise number of papyri excavated because pieces became detached in the process; a single papyri roll might therefore comprise two or more fragments. Gigante, *Catalogo dei Papiri Erconlanesi*, listed 1,826 items while Agnese Travaglione, *Catalogo descrittivo dei Papiri Ercolanesi*, recorded somewhat more. For a general overview see Mario Capasso, *Manuale de Papirologia Ercolanese*.
3 Sider, *The Library of the Villa dei Papiri*, 16–24.
4 For a brief overview see Maresca, 'Early attempts to open and read the papyri'. For recent work see Seales *et al.*, 'Technology and the quest to unlock the secrets of the Herculaneum scrolls'. Most recently non-destructive CT scans and AI have been used to transcribe two papyri held by the Institut de France. See Marchant 'First passages'. I am grateful to Christopher James for drawing my attention to this article.
5 For example, McIlwaine, 'Davy in Naples', though it has a promising title, focuses upon the identification of which papyri Davy worked on, as does Auricchio, 'L'esperienza napoletana del Davy'. Though it does not reference the papers in The National Archives and some other collections used here and contains some minor errors, more along the narrative line of this paper is McIlwaine, 'British interest in the Herculaneum papyri'.
6 Lipkowitz, 'The sciences are never at war?'.
7 Paderni, 'Extract of a letter'.
8 D'Amore, *The Royal Society [of London]*, esp. 142–3.
9 He reigned also as Carlo V of Sicily.
10 D'Amore, *The Royal Society [of London]*, 119–21 and the sources cited.
11 Paderni, 'Extract of a letter', 73. For the precautions taken by the Museum to prevent images being made of any of the objects there see Descamps-Lequime, 'The Ferdinand IV donation', 143–4.
12 McIlwaine, *Herculaneum*, 1: 65.
13 Hayter, *A Report upon the Herculaneum Manuscripts*, 45.
14 Elizabeth Webster, *Diary*, late 1792, British Library (hereafter cited as BL) add MS 51927, f.18r.
15 *Herculanensium voluminum quæ supersunt tomus 1*.
16 'Account of the MSS. found at Herculaneum', *The Scots Magazine*, September 1795, 556–7.
17 Also Ferdinando III of Sicily.
18 Bishop of Derry from 1768 until his death.

19 Frederick Hervey to William Hamilton, 2 July 1778, National Library of Ireland MS 2262, quoted in Thibaudeau, *Catalogue*, 55.
20 Jenkins and Sloan, *Vases & Volcanoes*, 44. The eight volumes of observations now form Royal Society of London (hereafter cited as RSL) MS 2/1–8. See also William Hamilton to Joseph Banks, 31 May 1797, BL add MS 34048, f.86–91.
21 Charles Townley in his diary said as much. Quoted in Jenkins and Sloan, *Vases & Volcanoes*, 45.
22 William Hamilton to Charles Greville, 23 October 1781, Huntington Library MS HM 34460, quoted in Thibaudeau, *Catalogue*, 77; Jenkins and Sloan, *Vases & Volcanoes*, 44–5; Constantine, *Fields of Fire*, 279–81.
23 William Hamilton to Joseph Banks, 8 January 1788, Banks, *Scientific Correspondence*, 3: 367–8.
24 William Hamilton to Joseph Banks, 6 April 1790 and 11 June 1793, BL add MS 34048, f.60–2 and f.72–3.
25 Blagden, *Diary*, 29 January 1793, RSL MS CB/3/2.
26 Charles Bladgen to Joseph Banks, 30 March 1793, BL add MS 33272, f.115.
27 Denoyelle and Descamps-Lequime, 'Josephine's Greek and Roman antiquities', 18–19.
28 Heard and Jones, *George IV*.
29 See the references given in Janko, *Philodemus*, 9–10.
30 Most of this paragraph is derived from Constantine, *Fields of Fire*, 244–6.
31 See appendix. Tyrwhitt's possible role, though not explaining the timing, was suggested by Knight, 'Canguri e papiri', 305–6. Janko, *Philodemus*, 9 came to the same conclusion.
32 See appendix.
33 John Hayter, *The Herculaneum and Pompeian Manuscripts*, 15.
34 Thomas Tyrwhitt to John Hayter, 18 March 1800, Bodleian MS Gr Class c.10, f.45–6; published in Lapatin, *Buried by Vesuvius*, 234.
35 Hayter, *The Herculaneum and Pompeian Manuscripts*, 15–16.
36 *The Star*, 24 March 1800, 3b.
37 Hayter, *A Report upon the Herculaneum Manuscripts*, 80–2.
38 Hayter, *A Report upon the Herculaneum Manuscripts*, 85; John Hayter to Thomas Tyrwhitt, 29 October 1800, Bodleian MS Gr Class c.10, f.50–5 (draft).
39 Hayter, *A Report upon the Herculaneum Manuscripts*, 85.
40 Hayter, *A Report upon the Herculaneum Manuscripts*, 87–8.
41 William Drummond to Lord Hawkesbury, 18 January 1802 (£1000) and 22 June 1802 (the remaining £200), The National Archives (hereafter cited as TNA) FO 70/17, f.62–3 and 262 respectively.
42 On Rosini see Cerasuolo, 'Carlo Maria Rosini'.
43 Hayter, *A Report upon the Herculaneum Manuscripts*, 89–90.
44 John Hayter to Hugh Elliot, 27 June 1803, TNA FO 70/21, f.123–4; the previous year newspapers such as *Ipswich Journal*, 9 October 1802, 4b reported 10. For further details Auricchio, 'John Hayter'; Indelli 'John Hayter'; Capasso, 'Carlo Maria Rosini', 166–7.
45 Sider, *The Library of the Villa dei Papiri*, 54–5.
46 Bennett, 'Account of the ancient rolls of papyrus' (read 2 December 1802). For the provincial press see, for example: *Hampshire Chronicle*, 24 May 1802, 3c; *The Scots Magazine*, July 1802, 589; *Ipswich Journal*, 31 July 1802, 2b–c.
47 *The Times*, 26 June 1802, 2d, reporting the previous day's proceedings.
48 *The Scots Magazine*, October 1802, 66; *Ipswich Journal*, 9 October 1802, 4b; *Hampshire Chronicle*, 11 October 1802, 2e.
49 John Hayter to Hugh Elliot, 27 June 1803, TNA FO 70/21, f.123–4.
50 Hugh Elliot to Lord Hawksbury, 26 July 1803, TNA FO 70/21, f.121–2.
51 Francesco La Vega to John Hayter, 18 July 1803, Bodleian MS Gr Class c.10, f.93.
52 Hugh Elliot to Lord Hawksbury, 26 July 1803, TNA FO 70/21, f.121–2. La Vega's letter of 18 July made it explicit – the number of papyri chosen provided parity with those sent to Paris around the same time.
53 Thomas Tyrwhitt to Earl Spencer, 27 September 1803, BL add MS 76091 (no foliation).
54 Hugh Elliot to John Hayter, 30 April 1804, Bodleian MS Gr Class c.10, f.95–6.
55 Hugh Elliot to John Hayter, 4 July 1805, Bodleian MS Gr Class c.10, f.97–9.
56 Victoria University of Toronto, S MS F14.15.
57 John Hayter to Thomas Tyrwhitt, 6 January 1806, BL RP 9680 (photocopy). There is a contemporary copy of the letter, though not the list which added four further papyri unrolled

since Coleridge's list, in Natural History Museum (hereafter cited as NHM) MS DTC 16, f.179–82. The letter was published, in part, three years later as 'Fac-simile of a line of a Latin poem found at Herculaneum'.

58 Quoted in *The Bath Chronicle*, 12 July 1804, 2b.

59 *The Morning Chronicle*, 23 November 1804, 2d–3a; *The Times*, 24 November 1804, 3d.

60 Earl of Liverpool to Joseph Banks, 27 November 1804, Lincolnshire Archives HILL 22/2/3/15.

61 BL add MS 56299, f.18–19 and f.20–22 respectively.

62 Banks's report on the experiments, 17, 19 and 20 December 1804, BL add MS 56299, f.1–2. This report was published in McIlwaine, 'Sir Joseph Banks', 204–5.

63 Ash, a physician, visited Italy between 1792 and 1794. Ingamells, *A Dictionary of British and Irish Travellers in Italy*, 30.

64 Banks's report, BL add MS 56299, f.1–2; McIlwaine, 'Sir Joseph Banks', 204–5.

65 [Young], '[Review of] William Drummond and Hugh Walpole', 18–19, written in the third person.

66 Thomas Tyrwhitt to John Hayter, 5 June 1806, Bodleian MS Gr Class c.10, f.106–7.

67 Joseph Banks to Thomas Tyrwhitt, 21 December 1804, NHM DTC XV, f.205–6 (copy).

68 He later requested Hayter, unsuccessfully, to arrange this. Thomas Tyrwhitt to John Hayter, 5 June 1806, 30 January 1808, Bodleian MS Gr Class c.10, f.106–7, 112.

69 Thomas Tyrwhitt to Joseph Banks, 23 December 1804, BL add MS 33981, f.178.

70 Joseph Banks to Thomas Tyrwhitt, 29 December 1804, BL add MS 33981, f.179.

71 Hayter, *A Report upon the Herculaneum Manuscripts*, 103.

72 William Drummond to John Hayter, 13 December 1808, Bodleian MS Gr Class c.10, f.113–16.

73 Thomas Tyrwhitt to John Hayter, 5 June 1806, Bodleian MS Gr Class c.10, f.106–7.

74 'Catalogue of transcripts of papyri delivered to Mr. Hayter, Sept. 2, 1807', Bodleian MS Gr Class c.10, f.35–43; see also William Drummond to John Hayter, 13 December 1808, Bodleian MS Gr Class c.10, f.113–16.

75 Thomas Tyrwhitt to John Hayter, 30 January 1808, Bodleian MS Gr Class c.10, f.112; published in Lapatin, *Buried by Vesuvius*, 234.

76 William Drummond to John Hayter, 13 December 1808, Bodleian MS Gr Class c.10, f.113–16.

77 Drummond and Walpole, *Herculanensia*.

78 William Drummond to John Hayter, 13 December 1808, Bodleian MS Gr Class c.10, f.113–16. Needless to say none of this appeared in Hayter, *A Report upon the Herculaneum Manuscripts*. He did note that the facsimiles had come into Neapolitan possession (105–6), however, but did not explain how that had happened.

79 Auricchio, 'John Hayter', 181.

80 Thomas Tyrwhitt to Lord Amherst, 19 or 20 June 1809, Yale University MS Osborn fc112 (draft).

81 Noted in his obituary in *Gentleman's Magazine* 89(1), (1819): 179.

82 *The Morning Herald*, 13 March 1811, 3a.

83 *Journals of the House of Commons*, January to July 1812, 822–3.

84 Hayter, *A Report upon the Herculaneum Manuscripts*, dated 20 April 1811; *Gentleman's Magazine* 81(1), (1811): 508–9 published in its June issue a glowing puff by 'B' on Hayter's work.

85 *Jackson's Oxford Journal,* 24 March 1810, 3a. Scott, *Fragmenta Herculanensia* (on 3–4 this is slightly confused about the chronology of events); Auricchio, 'Sui disegni oxoniensi'; Lindsay, 'The Bodleian facsimiles'.

86 Prince of Wales to Lord Grenville, 9 March 1810, Bodleian MS Gr Class c.10, f.11–15; published in Lapatin, *Buried by Vesuvius*, 235.

87 *The Morning Chronicle*, 21 July 1810, 3b.

88 *Gentleman's Magazine* 80(1), (1810): 439.

89 Oxford University Herculaneum Committee minutes, 13 May 1811, Bodleian MS Top Oxon c.349, f.4.

90 John Hayter to John Cole, 20 March 1813, Bodleian MS Top Oxon c.349, f.40–2. For his resignation see John Hayter to John Cole, 6 December 1812, Bodleian MS Top Oxon c.349, f.38.

91 Oxford University Herculaneum Committee minutes, 30 October 1812, Bodleian MS Top Oxon c.349, f.16.

92 *Herculanensium voluminum* (1824–5).

93 Thomas Tyrwhitt to William A'Court, 10 November 1815, BL add MS 41537, f.101–3.

94 John Hayter to Charles Stuart, 22, 27 February and 17 March 1817, National Library of Scotland (hereafter cited as NLS) MS 21270, f.125, 141 and 21271, f.25. See also Charles Stuart to Thomas Tyrwhitt, 30 November 1818, NLS MS 21277, f.90 which reported Hayter's debts after his death.

95 William A'Court to Lord Castlereagh, 4 April 1816, BL add MS 41517, f.178 (copy). The idea of acquiring all the papyri had been floated a few months before in Thomas Tyrwhitt to William A'Court, 10 November 1815, BL add MS 41537, f.101–3.

96 Thomas Tyrwhitt to William A'Court, 16 July 1816, BL add MS 41517, f.241.

97 On Sickler, see Steiner, *Die Sphinx zu Hildburghausen*.

98 Friedrich Sickler to Thomas Tyrwhitt, 3 January 1817 in *Herculaneum Rolls*, 7–9.

99 *Göttingische gelehrte Anzeigen*, 1814, 1993–8; translated into English in *Herculaneum Rolls*, 22–7.

100 Thomas Tyrwhitt to Friedrich Sickler, 24 December 1816, *Herculaneum Rolls*, 5–6.

101 Friedrich Sickler to Thomas Tyrwhitt, 3 January 1817, *Herculaneum Rolls*, 7–9.

102 Joseph Banks to Charles Blagden, 1 and 2 April 1817, Banks, *Scientific Correspondence*, 6: 228–9.

103 Lord Castlereagh to Thomas Tyrwhitt, 9 April 1817, in Herculaneum Rolls Committee minutes, 27 June 1817, Royal Collection, Windsor Castle, inventory number 1196789, pp.1–3 (copy); published in Sickler, *Die Herkulanensischen Handschriften in England*, 105–6. See also 'Report of the committee appointed to superintend the experiments of Dr. Sickler, for the purpose of proving the efficacy of a method, proposed by him, for unrolling and decyphering the Herculaneum manuscripts', *Parliamentary Papers*, 1818, 15.23, p.2; Joseph Banks to Charles Blagden, 1 and 2 April 1817, Banks, *Scientific Correspondence*, 6: 228–9.

104 'Report of the committee', 2.

105 Colchester's diary is in TNA PRO 30/9/36. Much of it was published in Abbot, *The Diary and Correspondence*, entry for 25 June 1817, 3: 9.

106 Abbot, *The Diary and Correspondence*, 27 June 1817, 3: 11; Herculaneum Rolls Committee minutes, 27 June 1817, Royal Collection, Windsor Castle, inventory number 1196789, 1–11.

107 'Report of the committee', 3.

108 Herculaneum Rolls Committee minutes, 4 July 1817, Royal Collection, Windsor Castle, inventory number 1196789, 16.

109 'Report of the committee', 4.

110 *The Morning Post*, 11 July 1817, 3b.

111 'Report of the committee', 5.

112 'Report of the committee', 7. The report was not laid before the Commons until March. It was then reported extensively in newspapers including *The Times*, 1 April 1818, 3a and *The Caledonian Mercury*, 6 April 1818, 4b–c.

113 Charles Abbot, *The Diary and Correspondence*, 3 July 1817, 3: 12–13.

114 Davy, 'Some observations and experiments', 192 stated that the experiments were made in 1818.

115 Davy, 'Report of the state of the manuscripts', 154; Davy, 'Some observations and experiments', 193.

116 Davy, 'Some observations and experiments', 193–4.

117 Joseph Banks to Edward Daniel Clarke, 16 May 1818, University College London MS Gilbert, box 1, file 2, enclosure B (typescript).

118 Presented by Queen Victoria in 1865.

119 Reported in *The Morning Chronicle*, 26 May 1818, 3a.

120 *The London Gazette*, 20 October 1818, issue 17410, 1875.

121 Viscount Castlereagh to William A'Court, 20 August 1818, BL add MS 41519, f.101; William A'Court to Viscount Castlereagh, 29 September 1818, TNA FO 70/84, f.304.

122 Humphry Davy to Teodoro Monticelli, 19 November 1818, Davy, *Letters*, 3: 665.

123 William A'Court to William Richard Hamilton, 28 March 1819, BL add MS 41520, f.31–2 (retained copy). While a very distinguished classicist, 'great' here might also refer to Elmsley's physical size, frequently referenced in contemporary correspondence.

124 Humphry Davy, 'Report of the state of the manuscripts', 155–7.

125 Humphry Davy, 'Report of the state of the manuscripts,' 158; Humphry Davy to William Richard Hamilton, 12 February 1819, Davy, *Letters*, 3: 673.

126 Humphry Davy to Michael Faraday, 18 February 1819, Faraday, *Correspondence*, 1: 95 and Davy, *Letters*, 3: 674 noted he had been back in Rome for about a fortnight.
127 James, 'How big is a hole?'.
128 William A'Court to William Richard Hamilton, 16 April 1819, TNA FO 70/86, f.109.
129 Davy, 'Report of the state of the manuscripts' (159 for his Roman experiments). The draft, including pages copied in London by Faraday and amended by Davy – something quite typical of his practice – is in Royal Institution MS HD/10, 235–67. A photocopy of a duplicate of the final manuscript report is in BL RP 9680, along with a note recording its arrival at the Foreign Office on 22 March 1819.
130 Humphry Davy to William Richard Hamilton, 12 February 1819, Davy, *Letters*, 3: 673.
131 Davy, 'Report of the state of the manuscripts,' 159–60.
132 Humphry Davy to Michael Faraday, 18 February 1819, Faraday, *Correspondence*, 1: 95; Davy, *Letters*, 3: 674.
133 James, *Michael Faraday*, 35–7.
134 Humphry Davy to Michael Faraday, 7 May 1819, Faraday, *Correspondence*, 1: 101; Davy, *Letters*, 3: 684.
135 *The Literary Gazette*, 1 May 1819, 280–2.
136 For a contextual discussion of these stanzas see Porter, 'Hearing voices'.
137 Humphry Davy to Michael Faraday, 15 May 1819, Faraday, *Correspondence*, 1: 102; Davy, *Letters*, 3: 685.
138 William Richard Hamilton to William A'Court, 20 May 1819, BL add MS 41520, f.63.
139 Humphry Davy to William A'Court, 5 September 1819, Davy, *Letters*, 3: 688.
140 Humphry Davy to William Richard Hamilton, 23 July 1819, Westminster School (hereafter cited as WS) MS 5/ELM/2/2/4 (copy), quoted in Nicholas Horsfall, 'Classical studies in England', 474–5.
141 Humphry Davy to William A'Court, 5 September 1819, Davy, *Letters*, 3: 688.
142 Peter Elmsley to Robert Finch, 20 May 1819, Bodleian MS Finch d.5, f.333–4. Partly quoted in Nitchie, *The Reverend Colonel Finch*, 69.
143 Humphry Davy to William Richard Hamilton, 23 July 1819, WS MS 5/ELM/2/2/4, quoted in Horsfall, 'Classical studies in England', 474–5. Well-known as a prominent supporter of Princess Caroline, Gell's appointment suggests that his excellence as an antiquarian may have overcome any opposition in London to his appointment. Jason Thompson, *Queen Caroline and Sir William Gell*.
144 Humphry Davy to William A'Court, 5 September 1819, Davy, *Letters*, 3: 688.
145 Peter Elmsley to Robert Finch, 20 Augst 1819, Bodleian MS Finch d.5, f.335–6.
146 Peter Elmsley to Mary Elmsley (his widowed sister-in-law), 12 August 1819, WS MS 5/ELM/2/5/15.
147 Peter Elmsley to Mary Elmsley, 28 August 1819, WS MS 5/ELM/2/5/16.
148 Joseph Banks to Charles Blagden, 11 October 1819, Banks, *Scientific Correspondence*, 6: 360–2.
149 Peter Elmsley to Robert Finch, 7 December 1819, Bodleian MS Finch d.5, f.337–8.
150 Humphry Davy to William A'Court, 14 December 1819, Davy, *Letters*, 3: 701.
151 Davy, 'On the phænomena of volcanoes'; Humphry Davy to Michael Faraday, 10 December 1819, Faraday, *Correspondence*, 1: 108; Davy, *Letters*, 3: 700.
152 Elizabeth Fremantle, *Diary*, 18 and 19 February 1820, MS in private possession. I am grateful to Professor Elaine Chalus for providing me with a copy of this entry. Morgan, *Italy*, 2: 350–1.
153 Humphry Davy to Joseph Banks, 28 January 1820, Davy, *Letters*, 3: 703.
154 William A'Court to William Richard Hamilton, 15 February 1820, TNA FO 70/90, f.24.
155 John Soane jr to John Soane sr, 5 January 1820, Sir John Soane Museum MS Private Correspondence XIV.D.2.13. I am grateful to Sue Palmer for drawing my attention to this letter.
156 Davy, 'Some observations and experiments', 203.
157 Peter Elmsley to Robert Finch, 25 January 1820, Bodleian MS Finch d.5, f.339–40. Partly quoted in Nitchie, *The Reverend Colonel Finch*, 69–70.
158 Humphry Davy to William A'Court, 29 February 1820, Davy, *Letters*, 3: 704 and 705.
159 Charles Wynn to Peter Elmsley, 2 April 1820, WS MS 5/ELM/2/15/19.
160 Peter Elmsley to Robert Finch, 25 January 1820, Bodleian MS Finch d.5, f.333–4.
161 Humphry Davy to William A'Court, 29 February 1820, Davy, *Letters*, 3: 705.
162 William A'Court to William Richard Hamilton, 16 March 1820, TNA FO 70/90, f.58–9.

163 Humphry Davy to Teodoro Monticelli, 7 March 1820, Davy, *Letters*, 3: 708.
164 'Copies of specimens of papyri, unrolled under the auspices of His Sacred Majesty King George the Fourth', Royal Collection, Windsor Castle, inventory number 1076170.
165 Davy, 'Some observations and experiments'. Although in that paper Davy failed to acknowledge that Gell executed the drawings for it, this was noted in the Royal Society of London's journal book, published in *Abstracts of the Papers Printed in the Philosophical Transactions*, 145–6.
166 Davy, 'Some observations and experiments', 204.
167 Lipkowitz, 'The sciences are never at war?'.
168 White, *Italian Cultural Lineages*, 23 made a similar point, but seems not to have been aware of Elmsley's involvement.
169 Young privately attributed the débâcle entirely to Davy. Thomas Young to Macvey Napier, 4 October 1820, BL add MS 34612, f.387.
170 Unfortunately, Castlereagh forgot about the matter and they were not sent, via Malta, until the summer of 1819. William Parker to Joseph Planta, 18 May 1819, TNA FO 70/87, f.63. For this episode see White, *Italian Cultural Lineages*, 9–13, 16–18. On the exchange see also Knight, 'Canguri e papyri'.
171 Work did not seriously resume on them until nearly 40 years after Davy and Gell left Naples – when, significantly, another of Europe's leading chemists, Justus Liebig, became involved. Maresca, 'Early attempts to open and read the papyri', 34.

## Bibliography


*The Scientific Correspondence of Joseph Banks*, 6 vols, edited by Neil Chambers. London: Pickering, 2007 (NB this does not publish all Banks's letters). The work is cited here as Banks, *Scientific Correspondence*, followed by volume and page number.

*The Collected Letters of Sir Humphry Davy*, 4 vols, edited by Tim Fulford and Sharon Ruston, advisory editors Jan Golinski, Frank A. J. L. James and David Knight, assistant editor Andrew Lacey. Oxford: Oxford University Press, 2020. The work is cited here as Davy, *Letters*, followed by volume and letter number.

*The Correspondence of Michael Faraday*, 6 vols, edited by Frank A. J. L. James. London: Institution of Electrical Engineers / Engineering and Technology, 1991–2012, The work is cited here as Faraday, *Correspondence*, followed by volume and letter number.

Abbot, Charles. *The Diary and Correspondence of Charles Abbot, Lord Colchester: Speaker of the House of Commons 1802–1817*, 3 vols. London: John Murray, 1861.

*Abstracts of the Papers Printed in the Philosophical Transactions of the Royal Society of London, From 1800 to 1830 inclusive*, vol. 2. London: Richard Taylor, 1833.

Auricchio, Francesca Longo. 'John Hayter nella Officina dei Papiri Ercolanesi'. In *Contributi alla Storia della Officina dei Papiri Ercolanesi*, edited by Marcello Gigante, 159–215. Naples: Industria Tipografica Artistica, 1980.

Auricchio, Francesca Longo. 'L'esperienza napoletana del Davy', *Proceedings of the XIXth International Congress of Papyrology*, Cairo (1989): 189–202.

Auricchio, Francesca Longo. 'Sui disegni oxoniensi dei papiri Ercolanesi', *Cronache Ercolanesi* 22 (1992): 181–4.

Bennett, Henry Grey. 'Account of the ancient rolls of papyrus, discovered at Herculaneum, and the method employed to unroll them', *Archaeologia* 15 (1806): 114–17.

Capasso, Mario. 'Carlo Maria Rosini e i papiri Ercolanesi'. In *Carlo Maria Rosini (1748–1836): un unmanista flegreo fra due secoli*, 129–92. Pozzuoli: Azienda autonoma di cura, soggiorno e turismo, 1986.

Capasso, Mario. *Manuale de Papirologia Ercolanese*. Galatina: Congedo, 1991.

Cerasuolo, Salvatore. 'Carlo Maria Rosini Studioso e Unmanista'. In *Carlo Maria Rosini (1748–1836): un unmanista flegreo fra due secoli*, 13–60. Pozzuoli: Azienda autonoma di cura, soggiorno e turismo, 1986.

Constantine, David. *Fields of Fire: A life of Sir William Hamilton*. London: Weidenfeld and Nicholson, 2001.

D'Amore, Manuela. *The Royal Society [of London] and the Discovery of the Two Sicilies: Southern routes in the Grand Tour*. Cham: Palgrave Macmillan, 2017.

Davy, Humphry. 'On the phænomena of volcanoes', *Philosophical Transactions* 118 (1828): 241–50.

Davy, Humphry. 'Report of the state of the manuscripts of papyrus, found at Herculaneum', *Quarterly Journal of Science* 7 (1819): 154–61.

Davy, Humphry. 'Some observations and experiments on the papyri found in the ruins of Herculaneum', *Philosophical Transactions* 111 (1821): 191–207.

Denoyelle, Martine and Sophie Descamps-Lequime. 'Josephine's Greek and Roman antiquities: A rediscovered collection'. In *The Eye of Josephine: The antiquities collection of the Empress in the Musée du Louvre*, edited by Martine Denoyelle and Sophie Descamps-Lequime, 11–25. Atlanta, GA: High Museum of Art, 2007.

Descamps-Lequime, Sophie. 'The Ferdinand IV donation to the First Consul and his wife: Antiquities from the Bay of Naples at Malmaison'. In *Rediscovering the Ancient World on the Bay of Naples, 1710–1890*, edited by Carol C. Mattusch, 141–60. Washington DC: National Gallery of Art, 2013.

Drummond, William and Robert Walpole. *Herculanensia; or Archaeological and Philological Dissertations, containing a Manuscript found among the Ruins of Herculaneum*. London: Cadell and Davies, 1810.

'Fac-simile of a line of a Latin poem found at Herculaneum', *Repository of Arts* 1 (1809): 180–1.

Gigante, Marcello. *Catalogo dei Papiri Erconlanesi*. Naples: Bibliopolis, 1979.

Hayter, John. *A Report upon the Herculaneum Manuscripts in a Second Letter, addressed, by permission, to His Royal Highness the Prince Regent*. London: Richard Phillips, 1811.

Hayter, John. *The Herculaneum and Pompeian Manuscripts*. London: [no publisher], 1800.

Heard, Kate and Kathryn Jones, eds. *George IV: Art & spectacle*. London: Royal Collection Trust, 2019.

*Herculanensium voluminum*, 2 vols. Oxford: Clarendon, 1824–5.

*Herculanensium voluminum quæ supersunt tomus 1*. Naples: Royal Typographer, 1793.

*Herculaneum Rolls. Correspondence relative to a proposition made by Dr. Sickler of Hildburghausen upon the subject of their developement*. London: Barfield, 1817.

Horsfall, Nicholas. 'Classical studies in England, 1810–1825', *Greek, Roman and Byzantine Studies* 15 (1974): 449–77.

Indelli, Giovanni. 'John Hayter e i Papiri Ercolanesi'. In *Contributi alla Storia della Officina dei Papiri Ercolanesi*, edited by Marcello Gigante, 217–25. Naples: Industria Tipografica Artistica, 1980.

Ingamells, John. *A Dictionary of British and Irish Travellers in Italy 1701–1800*. New Haven: Yale University Press, 1997.

James, Frank A. J. L. 'How big is a hole?: The problems of the practical application of science in the invention of the miners' safety lamp by Humphry Davy and George Stephenson in late Regency England', *Transactions of the Newcomen Society* 75 (2005): 175–227.

James, Frank A. J. L. *Michael Faraday: A very short introduction*, Oxford: Oxford University Press, 2010.

Janko, Richard, ed. *Philodemus On Poems, Books 3–4 with the fragments of Aristotle On Poets*. Oxford: Oxford University Press, 2011.

Jenkins, Ian and Kim Sloan. *Vases & Volcanoes: Sir William Hamilton and his collection*. London: British Museum Press, 1996.

Ketcham, Carl H. ed. *Shorter Poems 1807–1820 by William Wordsworth*. Ithaca, NY: Cornell University Press, 1989.

Knight, Carlo. 'Canguri e papiri', *Cronache Ercolanesi* 32 (2002): 305–20.

Lindsay, W. M. 'The Bodleian facsimiles of Latin papyri from Herculaneum', *The Classical Review* 4 (1890): 441–5.

Lipkowitz, Elise S. 'The sciences are never at war?': The scientific republic of letters in the era of the French Revolution, 1789–1815'. Northwestern University, PhD thesis, 2009.

McIlwaine, I. C. 'British interest in the Herculaneum papyri', *Proceedings of the XVIIIth International Congress of Papyrology*, Athens, 1 (1986): 321–9.

McIlwaine, I. C. 'Davy in Naples: The British viewpoint', *Proceedings of the XIXth International Congress of Papyrology*, Cairo (1989): 107–13.

McIlwaine, I. C. *Herculaneum: A guide to printed sources*, 2 vols. Naples: Bibliopolis, 1988.

McIlwaine, I. C. 'Sir Joseph Banks and the Herculaneum papyri', *Atti del XVII Congresso Internazionale di Papirologia*, Naples, 1 (1984): 197–205.

Marchant, Jo. 'First passages of rolled-up Herculaneum scroll revealed', *Nature*, 5 February 2024, https://www.nature.com/articles/d41586-024-00346-8 (accessed 9 February 2024).

Maresca, Sofia. 'Early attempts to open and read the papyri: 1750s–1990s'. In *Buried by Vesuvius: The Villa dei Papiri at Herculaneum*, edited by Kenneth Lapatin, 28–36. Los Angeles: The J. Paul Getty Museum, 2019.

Methold, Thomas Tindal. 'The parish of Hepworth and its rectors', *Proceedings of the Suffolk Institute of Archæology and Natural History* 8 (1894): 380–407.

Morgan, Sydney. *Italy*, 2 vols. London: Henry Colburn, 1821.

Nitchie, Elizabeth. *The Reverend Colonel Finch*. New York: Columbia University Press, 1940.

Paderni, Camillo. 'Extract of a letter ... to Dr. Mead, concerning the antiquities dug up from the ancient Herculaneum, dated from Naples, Nov. 18, 1752. Translated from the Italian', *Philosophical Transactions* 48: (1753): 71–3.

Porter, James I. 'Hearing voices: The Herculaneum papyri and classical scholarship'. In *Antiquity Recovered: The legacy of Pompeii and Herculaneum*, edited by Victoria C. Gardner Coates and Jon L. Seydl, 95–113. Los Angeles: The J. Paul Getty Museum, 2007.

Scott, Walter. *Fragmenta Herculanensia: A descriptive catalogue of the Oxford copies of the Herculanean rolls together with the texts of several papyri accompanied by facsimiles*. Oxford: Clarendon Press, 1885.

Seales, W. Brent and Christy Chapman with Vito Mocella. 'Technology and the quest to unlock the secrets of the Herculaneum scrolls'. In *Buried by Vesuvius: The Villa dei Papiri at Herculaneum*, edited by Kenneth Lapatin, 124–32. Los Angeles: The J. Paul Getty Museum, 2019.

Sickler, Friedrich. *Die Herkulanensischen Handschriften in England und meine, nach erhaltenem Ruf und nach Auftrag der englischen Regierung, im Jahre 1817 zu ihrer Entwickelung gemachten Versuche*. Leipzig: Brockhaus, 1819.

Sider, David. *The Library of the Villa dei Papiri at Herculaneum*. Los Angeles: The J. Paul Getty Museum, 2005.

Steiner, Gerhard. *Die Sphinx zu Hildburghausen: Friedrich Sickler, ein schöpferischer Geist der Goethezeit*. Weimar: Böhlau, 1985.

Thibaudeau, Alphonse Wyatt. *Catalogue of the Collection of Autograph Letters and Historical Documents formed by Alfred Morrison (Second Series, 1882–1893). The Hamilton & Nelson papers. Volume I. 1756–1797*. [London]: For Private Circulation, 1893.

Thompson, Jason. *Queen Caroline and Sir William Gell: A study in royal patronage and classical scholarship*. Cham: Palgrave Macmillan, 2019.

Travaglione, Agnese. *Catalogo descrittivo dei Papiri Ercolanesi*. Naples: Biblioteca Nazionale 'Vittorio Emanuele III', 2008.

White, Jonathan. *Italian Cultural Lineages*. Toronto: University of Toronto Press, 2007.

Wordsworth, William. *The River Duddon, A series of Sonnets: Vaudracour and Julia: and other poems*. London: Longman, 1820.

[Young, Thomas]. '[Review of] William Drummond and Hugh Walpole, Herculanensia; or archaeological and philological dissertations: containing a manuscript found among the ruins of Herculaneum', *Quarterly Review*, 3 (1810): 1–20.

# Part IV

# Neapolitan history in the present age

9

# 'The art of healing, healing through art': art, medicine and science at the Hospital of the Incurables of Naples

Gennaro Rispoli

On 23 March 1522 the venerable Maria Lorenza Longo (1463–1539) founded the hospital of the Real Santa Casa di Santa Maria del Popolo degli Incurabili, which became known across Europe as the Hospital of the Incurables (Fig. 9.1). It was the first hospital to be founded in the Kingdom of Naples, and its broad wards welcomed up to 2,000 patients at a time drawn from all regions of the kingdom. Situated on the ancient hill of Caponapoli, it became a key centre of medical assistance for Neapolitans. Important figures such as Giovanni Filippo Ingrassia (1509–80), Marco Aurelio Severino (1580–1656), Giovanni Battista Quadri (1780–1851), Leonardo Bianchi (1848–1927), Antonio Cardarelli (1831–1927) and Giuseppe Moscati (1880–1927) practised the healing arts within these ancient walls. The Hospital of the Incurables has also functioned as a centre of scientific scholarship not only on a local level, but also for other regions in the south of the Italian peninsula. Since its foundation, this institution has hosted the Academy of the Oziosi. Over the years it has also welcomed a number of the leading exponents of seventeenth-century medical and philosophical culture, Enlightenment rationalists of the eighteenth century and members of the nineteenth-century Neapolitan Medical School.

Medicine and charity, myth and pragmatism are intertwined in the history of the hospital complex, giving birth to a microcosm bustling with life and energy that, as we have seen, hosted a great variety of figures. These elements have been united by a desire to advance scientific progress, an ambition that has marked the history of the hospital ever since it was founded. The hospital was also devoted to academic education, especially in the fields of surgery, obstetrics

**Figure 9.1** Courtyard of the Hospital of the Incurables, from the magazine *Poliorama pittoresco*, 1839. © Collection of the Museum of Sanitary Arts of Naples.

and anatomy. Indeed, the complex actually possessed four anatomy theatres.

The hospital complex has recently rediscovered its former vitality thanks to the intervention of the cultural association *Il Faro di Ippocrate*. Its volunteers have reopened these historical buildings to the public and founded the Museum of Medical Arts and History of Medicine. They constantly devote their time to research activities.[1]

## The Hospital of the Incurables: its wards and its cloisters

The Hospital of the Incurables is an institution built to express human solidarity. Its name reflects the fact that it was instituted with the aim of welcoming those suffering from syphilis, a disease for which at the time of its foundation there was no known cure. The hospital also welcomed those who could not afford to pay for medical care, reflecting its strong sense of charity and desire to help others. From the beginning of its activity, the hospital practised a form of social medicine. This activity was strongly supported by both lay and religious organisations, such as the confraternity of the Bianchi della Giustizia, the brotherhood of the Theatines and the order of San Giovanni di Dio. A further important later example was the Sisters of Charity, an order founded by Saint Jeanne-Antide Thouret (1765–1826). In 1813 Thouret was sent to Naples by Letizia Bonaparte, the mother of Napoleon. After settling with some other nuns in the convent of Regina Coeli, she started to undertake activities that supported the work of the hospital doctors. The order remains active today.[2]

In the following sections I will escort you on a virtual tour of the hospital complex. Visitors can access the courtyard around which the hospital is structured through one of two doors, the first situated on the northern side and the second on the southern side of the courtyard. Upon entering the hospital, it is still possible to see traces of the original sixteenth-century wards. In the principal atrium stands a statue of the Virgin Mary, who welcomes under her cloak the sick and their prayers for healing. Her symbolic presence underlines the strong relationship between medical activity and religion in this institution.

The current structure of the complex is the result of renovations dating from the first half of the eighteenth century. They were designed to improve its exposure to the flow of the winds, believed to make the location healthier. During the course of these renovations a medical garden was established, in order to cultivate plants that could be used as ingredients for medicines. The garden was initially proposed by Domenico Cirillo (1739–1799) and his student Michele Tenore (1780–1861); it was implemented by Vincenzo Stellati, a member of the Medical College of Barber Surgeons (Collegio Medico Cerusico), in 1811. The garden was also used by medical students.[3]

In recent years the garden has been restored, thanks to the work of the volunteers of *Il Faro di Ippocrate*. After studying historical botanical texts, they have managed to identify the plants that were initially present in the garden and to replant them. One that deserves particular attention is the *Camellia japonica*, an extremely important plant for the kings of the Spanish Bourbon dynasty.[4] The garden also features a majestic, ancient camphor tree (*Cinnamomum camphora*) (Fig. 9.2). Planted in the eighteenth century, it stands approximately 35 metres tall and has a canopy that extends 20 metres across. Camphor was rare in Italy at the time this tree was planted. It was considered to have numerous medical properties and was used to treat a variety of illnesses, especially respiratory diseases. Moreover, it was believed that the smell of camphor would purify the air and facilitate the healing of the patients.

Just outside the garden is the hospital's astonishing cloister, known as the cloister of Santa Maria delle Grazie. Its archways are decorated with frescoes filled with alchemical symbols. There is an image of a bird-woman, for example, which in Greek iconography is a symbol for a mermaid. It is a clear reference to the myth of the siren Parthenope, the mythical founder of the city of Naples; her remains, according to legend, are buried on the same hill where the hospital is located. Another interesting image is the figure of a mask with three faces, which represent the tripartite division of time between past, present and future (Fig. 9.3).

Members of the Oziosi academy used to meet in this cloister. Their name was derived from the Latin word 'otium', meaning productive leisure, which reflected the academicians' rejection of an active life in order to pursue scientific and cultural research. The academy's

**Figure 9.2** Camphor tree in the Medical Garden of the Hospital of the Incurables. © Collection of the Museum of Sanitary Arts of Naples.

**Figure 9.3** 'Affresco a grottesca' of the cloister of Santa Maria delle Grazie in the Hospital of the Incurables. © Collection of the Museum of Sanitary Arts of Naples.

members included Giambattista Della Porta (1535–1615). A genius with multiple talents, his interests ranged from the natural sciences to astrology.[5]

Upon arriving at the end of the cloister, the visitor encounters a large passage through which it is possible to reach the hospital's obstetrics ward. The strong female presence that dominates the whole complex finds its clearest expression in this ward. Maria Lorenza Longo decided to devote this part of the complex to pregnant women, to support them during the last trimester of their pregnancies until the moment of delivery (Fig. 9.4).

Most importantly, the women who gave birth here could choose to do so while wearing a veil. This expedient enabled them to hide their identity, and then if they wished to give up the baby, either because the child was born from an illicit affair or because they lacked the financial means to keep it. A commemorative plaque on the side of the door states:

> Every woman, rich or poor, noble or peasant, native or foreigner, as long as she is pregnant, can knock, and the door will be opened (Qualunque donna, ricca o povera, patrizia o plebea, indigena o straniera, purché incinta, bussi e le sarà aperto)

**Figure 9.4** Affresco dalla Sala del Governatore in the Hospital of the Incurables with San Gaetano da Thiene and Maria Lorenza Longo, unknown artist, sixteenth century. © Collection of the Museum of Sanitary Arts of Naples.

According to historical sources, this phrase was once engraved on the door of the ward, but this has now unfortunately been lost. Attributed to Maria Lorenza Longo herself, it conveys a strikingly modern message concerning the universality of medical treatment.

## The Pharmacy of the Incurables: when art heals

Leaving the hospital's corridors, the visitor will encounter a splendid eighteenth-century structure, enclosed by two curving staircases (Fig. 9.5). This structure is the hospital's pharmacy.[6] It represents an extraordinary combination of art and science, capable of lifting the spirits of doctors who struggled with the terrible suffering and death that they saw every day in the hospital's anatomy theatres and wards. The pharmacy was built according to a 1729 design produced by the painter, sculptor and architect Antonio Vaccaro (1678–1745). A unique building, it is characterised by a carefully organised sequence of spaces, both functional and decorative. The realisation of the project was made possible thanks to a donation from Antonio Magiocca, one of the eighteenth-century governors of the Hospital of the Incurables.

After ascending the stairs, the visitor arrives in front of a majestic door. It is overseen by a monstrous figure with a devilish grin, the

**Figure 9.5** Exterior of the Historical Pharmacy of the Hospital of the Incurables. © Collection of the Museum of Sanitary Arts of Naples.

symbol of pharmaceutical products. This image represents the early modern medical–alchemical concept that opposites were united, so that everything good could turn into something bad and everything bad could be turned into something good. This made poisons and remedies complementary and intertwined: many contemporary medicines did indeed contain traces of poison, which is why an erroneous prescription could lead to death.

The first room that the visitor sees upon entering the pharmacy is the 'controspezieria'. This was the space in which medical remedies were produced and sold. It is dominated by a large wooden counter, in front of which are situated golden niches containing glass flasks; these still contain some of the ingredients once used to produce medicines. However, the most impressive piece of art in this room is undoubtedly a gilded wooden sculpture of a uterus located behind the counter (Fig. 9.6). Above this sculpture, one can once again find the figure of a beautiful, winged woman: it is Parthenope, guarding a mystical knowledge that could only be accessed by a few.[7]

The next room, known as the Sala Grande, is the most majestic in the pharmacy (Fig. 9.7). It was a reception hall that once welcomed the nobility of Europe and hosted conferences and medical meetings (Fig. 9.8). The shelves in this room display several vases, decorated

**Figure 9.6** Allegory of the virginal uterus, gilded wooden sculpture in the 'controspezieria' of the Historical Pharmacy, G. Di Fiore, eighteenth century. © Collection of the Museum of Sanitary Arts of Naples.

**Figure 9.7** Sala Grande of the Historical Pharmacy. © Collection of the Museum of Sanitary Arts of Naples.

**Figure 9.8** Majolica floor of the Sala Grande of the Historical Pharmacy, Brothers Massa, eighteenth century. © Collection of the Museum of Sanitary Arts of Naples.

in shades of blue, yellow and green – colours that represent the city of Naples. The same colours are also found on the decoration of the floor. These decorations were produced by the atelier of the brothers Massa. These famous artisans had also worked on the decorations of the cloister of Santa Chiara, another of the city of Naples's most famous monuments.

Once again, the figure of a uterus dominates this room (Fig. 9.9). This image of a uterus, however, is different from the one found in the *controspezieria*: it is more anatomically precise and distinguished by

**Figure 9.9** 'Allegory of a caesarean section', gilded wooden sculpture in the Sala Grande of the Historical Pharmacy, G. Di Fiore, eighteenth century. © Collection of the Museum of Sanitary Arts of Naples.

the presence of a vertical cut. This detail is a reference to the Caesarean section, a procedure practised at this hospital. Under the direction of the governor, Antonio Magiocca, the hospital upheld the idea that the foetus was an independent juridical entity. Consequently it had to be protected and saved, even in the case of complications during labour that could cost the mother her life.[8]

Another part of the pharmacy, where the pharmacists' laboratories were once located, now hosts the permanent collection of the artistic masterpieces owned by the historic hospitals of Naples. The Hospital of the Incurables used to own numerous paintings; the collection was intended to help alleviate the pressure that doctors experienced due to their exhausting job. It is perhaps surprising to the modern eye that the patrons of the hospital not only recognised the danger of burnout among medical professionals, but also appreciated that it could be treated through the beauty of art. In these halls, magnificent paintings produced by artists such as Giovanni da Nola (1478–1559), Marco Pino (1521–1583), Battistello Caracciolo (1578–1635), Belisario Corenzio (1558–1646), Francesco De Mura (1696–1782) and Pietro Bardellino (1728–1806) are displayed, along with more technical objects, scientific and medical instruments that remind the visitor of these spaces' original scientific purpose (Fig. 9.10).

**Figure 9.10** Pietro Bardellino, 'Machaon heals wounded Menelaus', ceiling cloth of the Sala Grande of the Historical Pharmacy, eighteenth century. © Collection of the Museum of Sanitary Arts of Naples.

The Hospital of the Incurables now also houses an extraordinary historical and cultural heritage. It is composed of pieces of art collected from all of the historic hospitals of Naples that fall under the aegis of Local Sanitary Agency (ASL) Napoli 1 Centro. These include the hospital Elena d'Aosta, the hospital Gesù e Maria and the hospital Santa Maria della Pace. The artworks have been assembled over the last 20 years by the volunteers of *Il Faro di Ippocrate*, thus fulfilling one of their main goals – namely, to create a unified collection designed to give voice to the history of medicine in Naples. The collection consists of more than 2,000 pieces, ranging from frescoes, canvases and paintings to statues, religious clothes and liturgical items, all of which date from between the sixteenth and the twentieth centuries (Figs 9.11 and 9.12).[9]

The core of the collection is formed by paintings that once decorated the lateral chapels of the church of Santa Maria del Popolo. They were created by some of the greatest artists active in Naples between the sixteenth and the eighteenth centuries, including Marco

**Figure 9.11** Angelo Viva, marble portrait of Domenico Cotugno, eighteenth century, located in the Museum of Sanitary Arts and History of Medicine in the Hospital of the Incurables. © Collection of the Museum of Sanitary Arts of Naples.

**Figure 9.12** Tommaso Solari, bronze portrait by Domenico Cirillo, 1884, located in the Museum of Health Arts and History of Medicine in the Hospital of the Incurables. © Collection of the Museum of Sanitary Arts of Naples.

Pino, Carlo Sellitto (1581–1614), Francesco De Mura and Francesco Solimena (1657–1747). Some of the most valuable paintings are undoubtedly *La Pietà* by Giuliano Bugiardini (1475–1555), dating back to the 1530s, and *The Crucifix*, dating back to 1577 and signed by Marco Pino. The latter painting is surmounted by a cymatium representing *The Resurrection of Christ*. The visitor's eye is also drawn to *The Mystic Press*, which represents Christ's suffering on a cross that is also a wine press.

At the end of this route is a marble vase made by Crescenzio Trinchese. It was designed to contain one of the most distinctive products of the time: theriac. The recipe of this potion was said to date back to Mithridates VI, King of Pontus. It was brought to Italy by the Romans, which started a history of prolific production that would continue until the twentieth century. This product, made out of vipers' skin, was considered to be both a powerful aphrodisiac and a panacea.

## The Museum of Medical Arts and History of Medicine: the art of healing, healing through art

Having left the pharmacy, the visitor can proceed to another building located to the south of the square. There a piperno stairway leads to a building that was part of the original sixteenth-century complex, once home to the Hospital of the Incurables' psychiatric ward. On the stairway is located the 'Pozzo dei pazzi' ('the well of the crazy ones') (Fig. 9.13). This well was never intended to be used simply to provide water; it was instead devoted to curing the mentally ill and, in some instances, even punishing them. The mentally ill were treated with the strangest remedies, such as being locked in the well during violent episodes. Other methods included feeding 100 egg yolks to those deemed to be inactive. Such cures were attributed to the legendary figure of Giorgio Cattaneo, also known as *Mastro Giorgio*, who in local popular culture is considered a symbol of authority.[10]

Climbing the stairs, one arrives at the convent of the Repented, known as 'Pentite'. This institution was a home for former prostitutes who had been converted by Maria Lorenza Longo. After taking religious vows, these women participated in the life of the hospital, assisting those who were ill with syphilis. They were given this role because they

**Figure 9.13** 'Il Pozzo dei pazzi', sixteenth century, located on the access staircase to the former convent of the 'Pentite', now the seat of the Museum of Sanitary Arts and History of Medicine of Naples. © Collection of the Museum of Sanitary Arts of Naples.

were considered to be immune to the illness, having survived previous exposure. Since at the time there was no precise information about the nature of the disease, it was believed to be spread by direct contact. All other sanitary operators were thus frightened of getting too close to sufferers, as they feared becoming infected themselves.[11]

Currently part of the building hosts the Museum of Sanitary Arts and History of Medicine, overseen by the current director, Professor Surgeon Gennaro Rispoli. The museum forms the engine room of the projects and cultural activities that take place in the complex. Founded on 23 March 2010, the anniversary of the institution of the hospital, it contains a rich collection of ancient surgical materials, prints and books gathered from private collections, which narrate the medical history of the south of Italy.

The opening of the museum marked the realisation of a project first proposed in the nineteenth century. In these years Gennaro de Rosa, the librarian of the Royal Medical–Surgical Academy of Naples, decided to move his library into a building close to the pharmacy, uniting it with a pre-existing library that had belonged to Luigi Tortora, an eighteenth-century surgeon who worked in the hospital (Fig. 9.14).

The idea of creating a museum was outlined in a speech delivered at the academy at some point between June and July 1885, during which a collection of prints, paintings and busts of important doctors was shown.[12] The academy had been founded in 1818 and it hosted the initial nucleus of the medical collection.

An article written by the surgeon Amedeo Lepre, 'Lacing of the external iliac artery following a crural injury', published in 1881, reveals that the project of creating a museum was not always welcomed by the Hospital's surgeons. It was published in the *Annali Clinici dell'Ospedale degl'Incurabili*, an important journal produced by the Hospital for almost 150 years.[13] In this article Lepre complained that he could not use a Cooper needle for an operation because that 'beautiful and brilliant instrument has been locked into the collection of art instruments, which, for simple spectacle are exposed in the Saint Maison'.[14] Lepre's aim was to obtain the academy's permission to use the large collection of surgical instruments for medical purposes, rather than simply keeping them for show.

These testimonies allow us to trace the existence of an earlier museum of the Hospital of the Incurables. It was probably an institution designed to support scientific and medical research, which displayed the collections of anatomical preparations and surgical materials. The museum of those years found its vitality through the presence of several academies and scientific publications of historical and medical content;

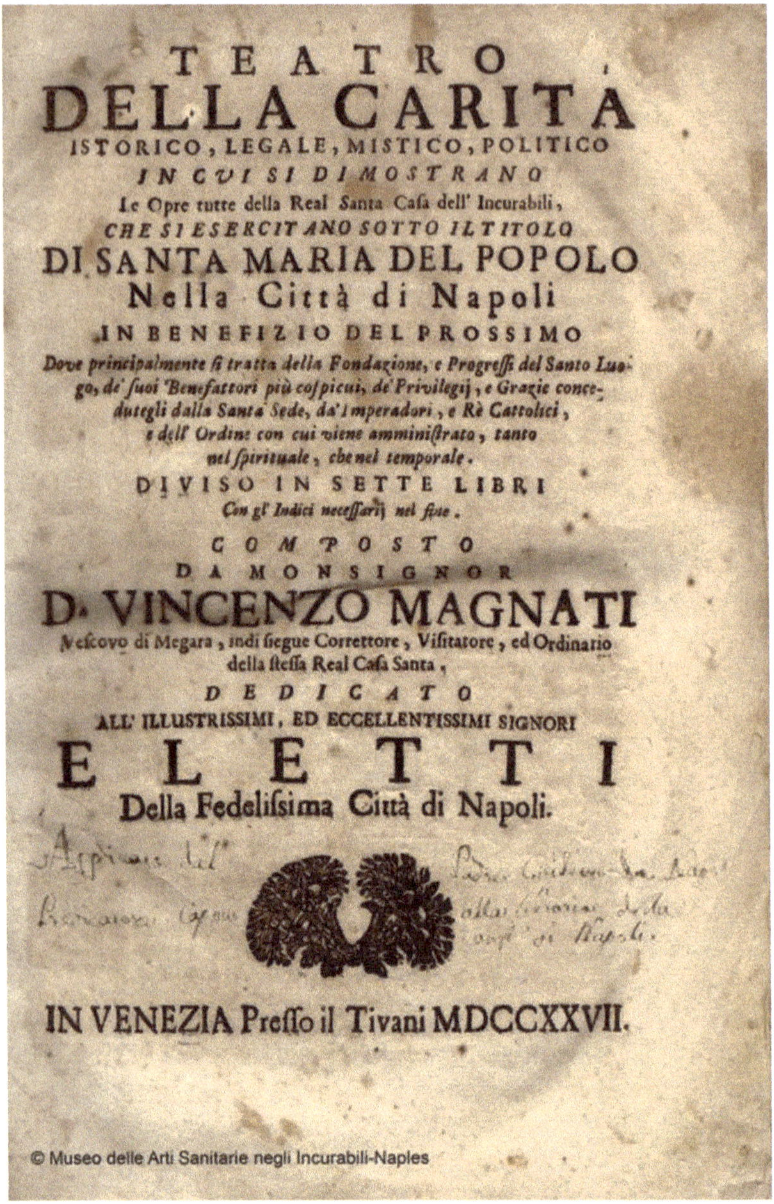

TEATRO
DELLA CARITÀ
ISTORICO, LEGALE, MISTICO, POLITICO
*IN CUI SI DIMOSTRANO*
Le Opre tutte della Real Santa Casa dell' Incurabili,
*CHE SI ESERCITANO SOTTO IL TITOLO*
DI SANTA MARIA DEL POPOLO
Nella Città di Napoli
IN BENEFIZIO DEL PROSSIMO
*Dove principalmente si tratta della Fondazione, e Progressi del Santo Luogo, de' suoi Benefattori più cospicui, de' Privilegj, e Grazie concedutegli dalla Santa Sede, da' Imperadori, e Rè Cattolici, e dell' Ordine con cui viene amministrato, tanto nel spirituale, che nel temporale.*
DIVISO IN SETTE LIBRI
*Con gl' Indici necessarij nel fine.*
*COMPOSTO*
DA MONSIGNOR
D. VINCENZO MAGNATI
Vescovo di Megara, indi siegue Correttore, Visitatore, ed Ordinario della stessa Real Casa Santa,
*DEDICATO*
ALL' ILLUSTRISSIMI, ED ECCELLENTISSIMI SIGNORI
ELETTI
Della Fedelissima Città di Napoli.

IN VENEZIA Presso il Tivani MDCCXXVII.



**Figure 9.14** Title page of the volume *Teatro della carità istorico, legale, mistico, politico: in cui si dimostrano le opre tutte della Real Santa Casa ... composto da monsignor Vincenzo Magnati* (Venice: presso il Tivani, 1727). © Collection of the Museum of Sanitary Arts of Naples.

their interest was not artistic, but rather rooted in the requirements of the medical professions. This museum also hosted an extensive library, which unfortunately became dispersed during the years of the hospital's decline.

## The Museum's collection

Today, thanks to the patient work of private individuals who have amassed and then donated collections, the Museum of the Hospital of the Incurables has found a new lease of life. The museum's motto – 'The art of healing, healing through art' ('L'arte di guarire, guarire con l'arte') – underlines the dual nature of its collections, which include not only breath-taking pieces of art that can cure the soul, but also technical instruments used to practise the art of healing. It now offers its visitors a journey through the gradual evolution of the art of healing and the development of sanitary and pharmacological practices. The museum's permanent display is organised in chronological order and according to different medical fields. It aims to encourage the visitor to reflect on problems concerning health, illness and therapy (Fig. 9.15).

Currently the museum occupies three floors. Its exhibition spaces are named after some of the most important luminaries of the Neapolitan Medical School from the eighteenth century onwards, such as Domenico Cotugno (1736–1822),[15] Domenico Cirillo,[16] Ferdinando Palasciano (1815–1891),[17] Giovanni Ninni (1861–1922)[18] and Giuseppe Moscati.[19] The museum's collection is open to anyone, from curious visitors to specialists in the field. All are welcome to gather here to debate issues ranging from the ethics of treatment and research to the current relationship between society and health.

The exhibits that line the corridors illustrate how medicine has evolved over time and demonstrate how the science and medicine practised in this hospital have progressed. Currently the museum's collections number some 12,000 objects relating to various medical and pharmaceutical fields. They make it possible to reconstruct key stories in the history of medicine, including the origins of pharmacology, tracing its evolution from alchemy to modern pharmaceuticals, and the development of therapy, from the age of humoural medicine to that of the modern clinic (Figs 9.16 and 9.17). The objects also illustrate themes such as the use of surgery, vaccines, epidemiology and hygiene, as well as ethical issues such as the management of end of life care. The collections have been inventoried in the museum's database by volunteers, under the aegis of several regional institutions.

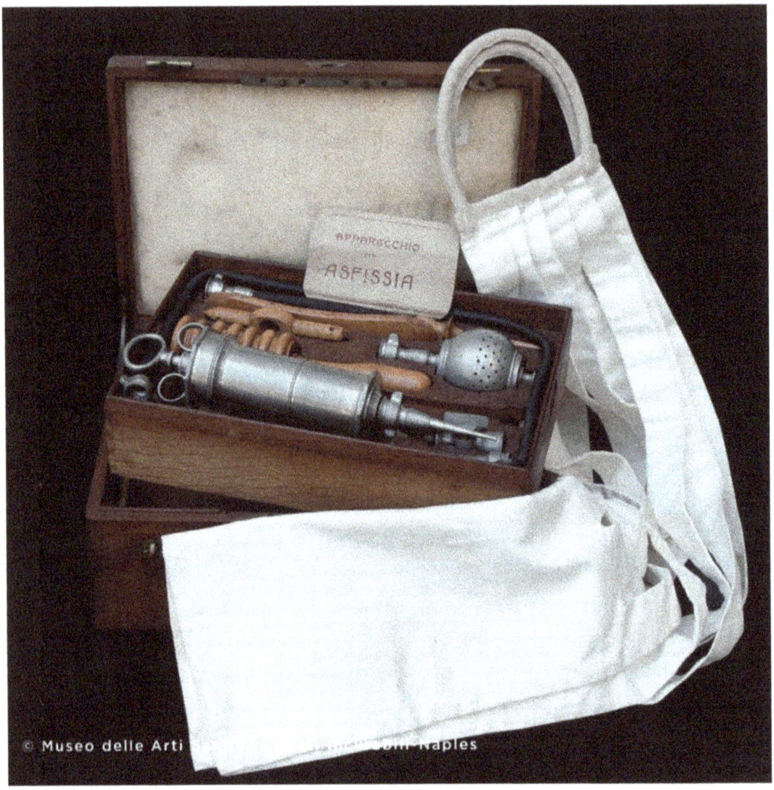


**Figure 9.15** Apparatus for asphyxiation, nineteenth century, located in the Museum of Sanitary Arts and History of Medicine in the Hospital of the Incurables. © Collection of the Museum of Sanitary Arts of Naples.

The museum presents a wide range of objects that reveal the most striking aspects of the history of medicine. Starting with the story of a barber who became a surgeon, the visitor is taken on a lively and detailed journey through the centuries. This portrays how anatomical knowledge has evolved, through exhibits such as eighteenth-century anatomical models, made of chalk and papier-mâché, and old drawings and prints (Fig. 9.18). Then, by observing portable pharmacies, old microscopes, ivory and tin enemas, the visitor can explore the close link between disease, health and social context. The journey also encompasses the beginnings of anaesthesia, already in use in the Hospital of the Incurables during the first half of the nineteenth century. All the objects on display bring us back to the history of the Hospital and its many innovations and contribution to medical history. In 1801, for example, the Hospital of the Incurables promoted a smallpox vaccination campaign.

**Figure 9.16** Portable pharmacy with landscape scenes, eighteenth century, located in the Museum of Sanitary Arts and History of Medicine in the Hospital of the Incurables. © Collection of the Museum of Sanitary Arts of Naples.

On the ground floor, the rooms reconstruct the history of medicine from the Schola Medica Salernitana to the Universitas Federiciana and the Collegio Medico Cerusico degli Incurabili. In addition, it hosts a gallery re-creating two nineteenth-century pharmacies, the Fra Nicola Pharmacy and the Irolla Pharmacy.

The exhibition also displays medical instruments such as amputation saws. Many of the artefacts are made from precious, sought-after materials and produced by craftsmen who worked in close contact with the most demanding clients: surgeons. These tools also illustrate the history of the Neapolitan industries that produced medical devices,

**Figure 9.17** Fra Nicola Pharmacy, nineteenth century, located in the Museum of Sanitary Arts and History of Medicine in the Hospital of the Incurables. © Collection of the Museum of Sanitary Arts of Naples.

such as the Gallo and Raimo cutleries; they made instruments in the first and second half of the nineteenth century respectively. The exhibition also displays the technological evolution of medical care, revealing how technology provided original solutions to specific problems. The idea of using light-transmitting instruments to explore the internal cavities of the human body, for example, benefited from systems of reflection such as a mirror with candles and reflective materials such as silver.

Other extraordinary objects include scalpels with blades made with special curvatures depending on their use. Ergonomic research lay behind the choice of material for the grips, which ranged from boxwood and ivory to turtle shell, turning these instruments into real works of art. Even an amputating scalpel or saw show the level of Neapolitan craftsmanship, although the collection also includes a large quantity of French-made instruments, especially Charrière cutlery. The originality of design also extends to pharmaceutical apparatus such as jars used to prepare and store medicaments. This display of centuries-old tools not only recounts the evolution of science and its impact on society, but also offers the opportunity to present current perspectives on bioengineering.[20]

On the ground floor visitors can view one of the museum's great jewels, 'Il presepe degli Incurabili' (Fig. 9.19). This unusual nativity

**Figure 9.18** Anatomical papier-mâché machines, eighteenth century, located in the Museum of Sanitary Arts and History of Medicine in the Hospital of the Incurables. © Collection of the Museum of Sanitary Arts of Naples.

scene also features figures representing incurable patients and charlatan healers. Made by eighteenth-century Neapolitan craftsmen, this nativity scene merges art and medicine. It features around one hundred figures suffering from afflictions of all kinds, which together represent an epidemiological overview of the eighteenth century. Here we find the representations of plague sufferers, as well as amputees, herniated patients, the blind and the obese.

We also discover the disturbing figure of 'La donna scandalosa' ('the scandalous woman'),[21] a woman sick with syphilis who is being devoured by rats and insects. The work has often been interpreted as

**Figure 9.19** 'Il presepe degli Incurabili', eighteenth century, located in the Museum of Sanitary Arts and History of Medicine in the Hospital of the Incurables. © Collection of the Museum of Sanitary Arts of Naples.

a warning about the perils of prostitution (Fig. 9.20). This image is a smaller representation of another sculpture preserved in the Chapel of the Bianchi della Giustizia, which is notable for the woman's expressive strength and somewhat macabre features (Fig. 9.21). Then we find the 'malformed pastors', for example the hunchback and the dwarf – figures who, according to Neapolitan legend, were good omens. We see also the figure of a tooth-puller, taken from the collection of Professor Fernando Gombos, which represents a charlatan pulling a tooth from a patient under the gaze of a little monkey. Moreover, the nativity scene features a small grocery store, in which monks prepare medicines for the sick, evoking healing.

The figures represented in this unusual nativity reflect scenes that a visitor would have witnessed in the old Neapolitan hospitals of San Gennaro, Real Santa Casa dell'Annunziata and Morificio di Aversa. They allow the visitor to discover, with a touch of irony, the variety of illnesses – some strange and unknown, others distressingly familiar and current – that afflicted people in the past. All pastors smile in the face of their misfortune, transmitting to the visitor a message of comfort, as well as the hope of promised healing.

On the first floor there is a series of rooms named after Ferdinando Palasciano (1815–1891), which are dedicated to the history of the Red Cross. They form an exhibition route that illustrates the history

**Figure 9.20** 'La donna scandolosa', wax model from the chapel of Santa Maria Succurre Miseris dei Bianchi della Giustizia in the Monumental Complex of the Hospital of the Incurables, seventeenth century. © Collection of the Museum of Sanitary Arts of Naples.

of military health through a collection of books, images and surgical instruments owned by Ferdinando Palasciano, a physician whose beliefs are often regarded as precursors to the founding ideals of the Red Cross. The motto of military medicine, 'the wounded in war are no longer the enemy', echoes in the testimonies kept in these rooms to convey a message of human solidarity that extends beyond national borders. The galleries also highlight the central role of medicine in military conflicts and the importance of military medicine in strengthening civilian medical practice. On the second floor the section of the 'Cavadenti' (literally, teeth-pullers) is based on the collection of Professor Gombos to illustrate the history of dentistry (Fig. 9.22). It displays dental forceps, as well as hand-held and pedal drill bits. The gallery culminates with the faithful reconstruction of a laboratory of a mid-nineteenth-century dental technician.

This floor of the museum also hosts a series of rooms dedicated to and named after the physician Giuseppe Moscati (1880–1927)

**Figure 9.21** Staircase to the chapel of Santa Maria Succurre Miseris dei Bianchi della Giustizia, located in the Monumental Complex of the Hospital of the Incurables, seventeenth century. © Collection of the Museum of Sanitary Arts of Naples.

(Fig. 9.23). Canonised in 1987, his life combined science and religion. In his medical practice, he embodied the ideal of the altruistic doctor for whom care is all that matters. The survival of his recipes and writings allows us to comprehend Moscati's attention not just to the physical health of his patients, but also to their mental and spiritual well-being. The reconstruction of his laboratory, which includes his anatomy room complete with tables, ornaments and desks, along with a rich collection of objects from his time, enables the study of this period of medical positivism represented by Antonio Cardarelli (1831–1927) and Moscati himself.

On the third floor it is possible, upon request, to visit collections of artefacts relating to anatomy, urology, anaesthesia, otorhinolaryngology (ENT) and ophthalmology. The museum's extensive anatomy collection is made up of hundreds of anatomical drawings, pieces of bone, numerous organs preserved in formaldehyde, a collection of malformed foetuses and anatomical wax models from the nineteenth century. The urology section includes books by the physician Michele Troja (1747–1827),

**Figure 9.22** Dental chair and pedal drill, located in the 'Il Cavadenti' section of the Museum of Sanitary Arts and History of Medicine in the Hospital of the Incurables, nineteenth century. © Collection of the Museum of Sanitary Arts of Naples.

nineteenth-century diagnostic and therapeutic instruments, cystoscopes, lithotrities, an operating table and several volumes and prints. The centrepiece of the otorhinolaryngology collection is the collection of surgical instruments from the first specialised ENT centre, founded at the end of the nineteenth century at the Ascalesi hospital of Naples. It is supplemented by around one hundred other instruments, books and prints. The ophthalmology section consists of numerous complete surgical boxes and diagnostic devices, dating from the early nineteenth century to the 1930s.

**Figure 9.23** Chemical laboratory, located in the section 'In the time of the Saint Professor Giuseppe Moscati' of the Museum of Sanitary Arts and History of Medicine in the Hospital of the Incurables, early twentieth century. © Collection of the Museum of Sanitary Arts of Naples.

## The museum's library

The museum's library contains an archive of documents and photographs that forms a research and documentation centre for the historical hospitals of southern Italy. The library collaborates with similar institutions, both in Italy and abroad. Its holdings are included in OPAC SBN (the catalogue of the Italian National Library Service). The core of the collection is the Gennaro Rispoli collection, but the library comprises in total about 8,000 volumes. One of the most important collections of historical and scientific documents relating to Naples, it is of incomparable cultural value at both regional and national level.

The library is organised thematically. A large section dedicated to the history of medicine contains monographs about the history of the Neapolitan Medical School. Another specific section is dedicated to publications on the history of Italian hospitals; it forms a research centre dedicated to surveying and preserving the memory of historical places of care. The museum is one of the promoters of ACOSI (the Cultural Association of Italian Historical Hospitals) and serves as its headquarters in the south of the country. The museum has also promoted publications on the medical history of these former hospitals and their future destination.

The library offers many services aimed at education, knowledge and scientific training. Its publications and cultural activities not only analyse the past, but also aim to find links to – and provide analogies with – the problems encountered in modern healthcare. The venue therefore represents an important meeting place for the medical faculties and professional orders of the regions of southern Italy.

Safeguarding the memory of the Neapolitan Medical School, which flourishes in the Hospital of the Incurables, as well as that of medicine itself, means both keeping alive and disclosing the memory of Italy's health history. The museum thus considers itself to be not only the guardian of this memory, and of the heritage contained in its collections, but also an active promoter of the knowledge that it has made.

## Notes

1 Rispoli and Valerio, *L'ospedale del Reame*, vol. 2, 19.
2 Rispoli and Valerio, *L'ospedale del Reame*, vol. 1, 33.
3 See also de Lièvre, 'Naples: History and botany in the nineteenth century', 46–62.
4 Rispoli and Valerio, *L'ospedale del Reame*, vol. 1, 96.
5 Caccioppoli and Rispoli, 'Nell'antica spezieria degli Incurabili', 2018, 192.
6 The pharmacy is currently closed for restoration.
7 Donatone, *La Farmacia degli Incurabili*, 33–49.
8 Fiorillo, *Gli incurabili*, 1136.
9 Rispoli and Oliviero, *Complesso monumentale di S. Maria del Popolo degli Incurabili*, 13–20.
10 On Cattaneo see Catapano, *Le reali case de' matti*.
11 Rispoli and Piedimonte, *La Collina Sacra*, 179–80.
12 The academy inherited a large part of the functions of the glorious Collegio Medico Cerusico, the first university medical school at a hospital. Here students were accompanied in their learning by ordinary doctors.
13 The *Annali Clinici dell'Ospedale degl'Incurabili* were officially established in 1835. Before that the hospital had already produced publications, such as the *Annali d'Ospedale* of 1820, founded by the surgeon and professor Francesco Petrunti. The publication of the *Annali* continued until 1939.
14 'Bello e lucente era chiuso a chiave nella raccolta degli strumenti dell'arte, che per semplice mostra trovasi esposta nella Santa Casa'. See Lepre, 'Allacciatura della arteria iliaca'.
15 Domenico Cotugno (1736–1822) was an Italian doctor from Apulia who studied and worked in Naples. He wrote a dissertation of 94 chapters titled *De aquaeductibus auris humanae anatomica dissertatio* (1871), with illustrations made in collaboration with Domenico Cirillo. In this work Cotugno described the structure of the ear, and analysed for the first time the aqueducts of the vestibule and cochlea and the ubiquitous presence of liquid in the labyrinth. Another important text is *De ischiade nervosa commentarius* (1764). This work consists of 57 illustrated chapters, which describe the sciatic nerve and examine sciatica from clinical and pathological perspectives. Cotugno's main writings are gathered in the collection *Opuscola Medica antehac seorism ab auctore in lucem edita, nunc primum in duo volumina collecta* (1826–7).
16 Domenico Cirillo (1739–1799) was a Neapolitan physician and botanist. He participated in the revolutionary movements of the Neapolitan Republic in 1799 and was beheaded when they ended in failure. Among his most celebrated writings is the *Accademici Discorsi* (1789), which addresses moral and social themes related to medicine.
17 Ferdinando Palasciano (1815–1891) was an Italian doctor and patriot, famous above all for anticipating the ideals of the Red Cross in military medicine. He was known in France in 1847 as 'surgeon of the military hospital of the Sacrament of Naples'.

18 Giovanni Ninni (1861–1922) was a doctor from Venosa who worked frequently in Naples in the Pellegrini hospital. His scientific output covers topics related to surgery and consists of 47 publications. Perhaps one of the most important is the *Compendio di medicina operatoria*, aimed at doctors and students.

19 Giuseppe Moscati (1880–1927) was a physician and saint, originally from Benevento, Campania. He was also a leading representative of Neapolitan medical positivism.

20 Rispoli and Oliviero, *Complesso monumentale di S. Maria del Popolo degli Incurabili*, 5–11.

21 Di Giacomo, *Luci e ombre napoletane*, 236.

## Bibliography

Caccioppoli, Carmen and Gennaro Rispoli. 'Nell'antica spezieria degli Incurabili: dall'Alchimia alla Chimica nel segno dei Secreta'. In *Atti del 68° Congresso Nazionale Rimini, 9–10 giugno 2018*.

Catapano, Vittorio D. *Le reali case de' matti nel Regno di Napoli*. Naples: Edizione Liguori, 1986.

Cotugno, Domenico. *Opuscola Medica antehac seorism ab auctore in lucem edita, nunc primum in duo volumina collecta*. Naples: Ex Officina Bibliographica et typographica, 1826–7.

Donatone, Guido. *La Farmacia degli Incurabili e la maiolica napoletana del Settecento*. Naples: Edizioni del Delfino, 1973.

Di Giacomo, Salvatore. *Luci e ombre napoletane*. Naples: Francesco Perrella editrice, 1914.

Fiorillo, Ciro. *Gli incurabili: l'ospedale, la farmacia, il museo*. Udine: Campanotto, 1991.

Lepre, A. 'Allacciatura della arteria iliaca esterna in seguito a ferita della crurale', *Annali Clinici dell'Ospedale degl'Incurabili* 27 (1881): 67–94.

Ninni, Giovanni. *Compendio di Medicina Operatoria per gli studenti di Medicina e Chirurgia*. Naples: Pasquale, 1898.

Rispoli, Gennaro and Sara Oliviero. *Complesso monumentale di S. Maria del Popolo degli Incurabili. Guida*. Naples: Rossi srl, 2014.

Rispoli, Gennaro and Antonio Emanuele Piedimonte. *La Collina Sacra. Passeggiate sull'Acropoli di Neapolis: l'altura di Caponapoli tra mito e storia, magia e religione, alchimia e scienza medica*. Naples: Museo delle arti Sanitarie, 2017.

Rispoli, Gennaro and Adriana Valerio. *L'ospedale del Reame. Gli Incurabili di Napoli. Volume 1: Storia e Arte a cura di Adriana Valerio*. Naples: il Torchio della Regina Editore, 2010.

Rispoli, Gennaro and Adriana Valerio. *L'ospedale del Reame. Gli Incurabili di Napoli. Volume 2: Documenti e riflessioni scientifiche a cura di Gennaro Rispoli*. Naples: il Torchio della Regina Editore, 2010.

# Index

Palmieri, Pasquale, 17n, 19n, 123n

www.ingramcontent.com/pod-product-compliance
Lightning Source LLC
LaVergne TN
LVHW052016300826
846742LV00020B/75